Ahmed Mahdi teaches at the Future University in Egypt. He holds a PhD from the University of Birmingham.

ENERGY AND US FOREIGN POLICY

The Quest for Resource Security after the Cold War

AHMED MAHDI

I.B. TAURIS

LONDON · NEW YORK

Published in 2012 by I.B.Tauris & Co Ltd
6 Salem Road, London W2 4BU
175 Fifth Avenue, New York NY 10010
www.ibtauris.com

Distributed in the United States and Canada
Exclusively by Palgrave Macmillan
175 Fifth Avenue, New York NY 10010

International Library of Security Studies vol.3

ISBN 978 1 84885 776 6

A full CIP record for this book is available from the British Library
A full CIP record for this book is available from the Library of Congress

Library of Congress catalog card: available

Typeset by Newgen Publishers, Chennai
Printed and bound by CPI Group (UK) Ltd, Croydon CR0 4YY

To my parents

CONTENTS

ACKNOWLEDGEMENTS

As this book is based on my PhD, which I finished at the University of Birmingham, UK, in 2010, I would like to thank my PhD supervisor, Professor William Scott Lucas, for his keen supervision and insightful feedback. He has been very patient and understanding with me, giving me advice on sources, and encouraging me to develop my own ideas and reflections. I also thank him for arranging for me to give presentations at seminars and conferences based on my thesis and for supporting my essays and publications. I would also like to thank my examiners, Dr. Trevor McCrisken of the University of Warwick, and Dr. Steve Hewitt of the University of Birmingham, for providing me with important insights on how to improve my work.

I would also like to thank a number of other scholars for their comments on my work, including Professor Robert Lieber of Georgetown University, author Jim Bovard, and Dr. Robert Crane, former advisor to Richard M. Nixon.

I would like to thank my editors at I.B.Tauris, Maria Marsh and Tomasz Hoskins, for patiently helping me with my manuscript. I would like to thank Allison McKechnie for her editing work.

Finally, I would like to thank my parents. I thank my father, who has been financially supporting me well beyond his retirement age. And I thank my mother, who has given me the psychological support which I needed during my stay in Britain.

INTRODUCTION

I think that in the modern world, if you don't understand the relationship between economics and politics, you cannot be a great statesman. You cannot do it with foreign policy and security knowledge alone.
Henry Kissinger, April 1986.[1]

The national policy of the United States should aim at securing for American nationals access to the world's oil resources.
Petroleum Industry War Council, December 1943.[2]

Security and economic considerations are inevitably linked and energy cannot be separated from either.
Richard Nixon, 1974.[3]

America imports 50% of its oil, more than 10 million barrels per day. And the figure is rising. This is dependence on foreign oil. And this dependence is a challenge to our economic security, because dependence can lead to price shocks and fuel shortages. And this dependence on foreign oil is a matter of national security. To put it bluntly, sometimes we rely upon energy sources from countries that don't particularly like us.
George W. Bush, 25 February 2002[4]

With its control of the world's most advanced military forces plus a nuclear arsenal, and with the US dollar as the global reserve currency, the United States emerged as the world's strongest superpower after World War II. This position was reinforced with the fall of the Soviet Union in December 1991,[5] making America, according to Zbigniew

Brzezinski, 'the first, only, and last truly global superpower',[6] as 'never before in history has a single power been so paramount'.[7]

This prompted authors to refer to the United States as an 'informal empire'. Susan Strange, for instance, argued that the reach of the United States' economic clout and global business corporations turned it into a 'non-territorial empire':

> It is that non-territorial empire that is really the flourishing economic base of US power ... The United States is still the largest and richest (and mostly open) market for goods and services under one political authority ... The world-wide reach of US controlled enterprises means that the capacity of the United States to exercise extraterritorial influence and authority is greater than that of any other government.[8]

Bill Emmott describes the United States as 'the mightiest colossus, in both absolute and relative terms, the world has ever seen', adding that it is indeed an empire, although in the informal sense in that it became an empire without having to occupy other countries like nineteenth-century Britain.[9] Similarly, Vassillis Fouskas and Bülent Gökay defined this American 'neo-imperialism' in terms of global military bases and economic domination without the actual need to invade countries as much as in traditional colonialism.[10]

This book is not proposing a new definition or interpretation of hegemony, imperialism or power. Instead, after an initial consideration of the role of economic factors in US global power, it focuses on the role of energy resources, in particular the role of oil resources and the procurement of foreign oil, within the context of the relation between economic, geopolitical and strategic factors in US foreign policy. A special focus is given to the George W. Bush administration (January 2001–January 2009). With a president and vice president (George W. Bush and Dick Cheney) coming from the energy industry, and with an American military invasion in two energy-rich regions (Central Asia and the Middle East) as a reaction to the 11 September 2001 attacks, there has been a great deal of talk about the imperial designs of the Bush foreign policy and the significance of energy resources within this imperial design.

This book starts by arguing that in order to fully understand the role of foreign energy supplies in US foreign policy, a discussion of 'American empire' is incomplete if it does not engage with the role of economic factors in US power. Many analysts have written about the role of economic interests as a determinant of US foreign policy and a necessity for US imperialism. This book sets its historical analysis within the tradition of William Appleman Williams, Gabriel Kolko and others on the informal 'Open Door' empire and the role of business corporations in US foreign policy and energy security. Based on this tradition, the book argues that American politicians, intellectuals and businessmen always sought the continued overseas expansion of the American marketplace as a solution to America's domestic economic problems, believing that the system of entrepreneurial capitalism could function successfully only if the marketplace constantly expanded, thus creating America's global (economic) empire without traditional colonialism.[11]

This global economic expansion was based on a marketplace conception, and on 'the struggle for economic empire', where the main strategy of American foreign policy became the Open Door policy, which was introduced with the advent of the Open Door Notes of 1899 and 1900, written by US Secretary of State John Hay. His Open Door Notes were influenced by politicians and corporation directors who advocated an American empire based on the 'corporation political economy', or 'the political economy of the large corporation', and were based on the economic definition of the world (which assumed that overseas economic expansion was the solution for America's social, political and economic problems). Thus, Hay wrote the Open Door Notes to define America's competition for empire,[12] turning it into America's version of the liberal policy of 'informal empire' or 'free trade imperialism' where the United States relied upon its economic power to control weaker or less developed nations, without the burdens and responsibilities of a formal annexation.[13] However, although the US empire was not a formal one, the Open Door policy had repercussions on US military policy abroad, as the Open Door Notes resulted in Washington sending troops to troubled areas around the globe, such as Morocco, Congo, Pearl Harbor, Hawaii, Venezuela, Brazil, Cuba and China in the nineteenth and twentieth centuries, to ensure the security of America's trade routes and economic interests[14] and maintain America's policy of 'sustained expansion'.[15] Thus, Williams argued that this 'marketplace image or conception of the world ... provided the dynamic causal force for a steady movement ... towards an imperial foreign policy', resulting in 'the formulation of a grand strategy for such imperial expansion of the free American marketplace'[16] that became 'the central feature of American foreign policy in the twentieth century'.[17] Williams added that 'the large corporations (and their leaders) dominated American history', leading to the creation of America's 'overseas economic empire',[18] since the interests of businesses and US foreign policy overlapped.

Gabriel Kolko, a student of Williams',[19] extended this analysis, arguing that politicians and entrepreneurs support a 'Hullian' view of the world, named after Secretary of State Cordell Hull (whose view of the world was based on Hay's Open Door policy). Hull saw that free trade was the solution to American and global problems, believing that wars and economic crises (such as the Great Depression and World War II) arose because of the lack of free trade and access to markets.[20] 'In this fashion', wrote Williams,

> [E]conomic expansion became both a means and an end for American policy-makers. Defined as a goal because of its vital importance to the continued success of the domestic economic system, it was also considered a means to build the empire of peace and prosperity which would secure the world for continued expansion in the years to come.[21]

Kolko paid special attention to America's increasing dependence on imports of raw materials to support its economy. America's ability to procure such raw materials as it needed, and at prices it could afford, was 'one of the keystones of its economic power' and maintaining these materials was 'vital to the future of its mastery of the international economy'.[22]

Paul Kennedy, too, cited William's thoughts on the Open Door in his famous book *The Rise and Fall of the Great Powers*, where he said that the Open Door policy was used to gain access to global markets as US producers feared that the domestic market might not be able to absorb the 'hyperproductivity of American factories and farms'.[23] Andrew Bacevich agreed, saying that the 'grand strategy' of the Open Door empire continued during the Cold War, and even after the Cold War as US presidents continued to push for global 'openness' to spread America's economic reach.[24]

Naomi Klein and Stephen Kinzer agreed on Williams' and Kolko's general arguments, but they focused on violent change as a tool for promoting US economic/business interests. Klein argued that US foreign policy aimed to promote a policy of maximum privatization through 'shock treatment' and 'disaster capitalism': using violent regime change and other similar 'disasters' to force change on target regions to impose US business interests. She cited Chile in 1973 and Iraq in 2003 as examples.[25] Similarly, Stephen Kinzer, following Williams' argument that the United States needs economic expansion to foreign markets to survive,[26] argued that 'most [US-backed] regime change operations fit within the larger category of resource wars'.[27]

Therefore, US foreign policy was based on expanding America's free trade empire, where the interests of politicians went hand in hand with business interests, in what Williams calls the 'political economy of the large corporation',[28] or what Kolko and David Painter called 'corporatism'; a 'public–private partnership' where people from large US corporations would temporarily leave their posts to work for the US Government, where they would continue to promote policies that would serve their previous industries (where there is usually no conflict of interest between government goals and business interests), then they would return to the US business sector once they left government office.[29] This was 'a reflection as well as the cause' of US foreign policy, which expressed the common ideologies and interests of 'the ruling class of American capitalism' which served as the 'fount' of foreign policy.[30]

This book extends all of the above analysis on US empire, the Open Door policy, public–private partnership and 'disaster capitalism' to US foreign policy on energy resources, specifically during the George W. Bush administration, under which oil considerations reached a level never seen before. To protect the interests of US business, the US Government intervened in foreign regions where there was no real or immediate threat to US national security, except that US business interests were threatened. I also argue that when the US Government intervened in such regions, it used (over-exaggerated) threats (whether terrorism, weapons of mass destruction or any other threat), in order to rally US public opinion behind the intervention. Kolko argued that, in order to protect access to raw materials,[31] the United States 'intervened continuously in numerous, diverse ways in regions where [internal political developments were] irrelevant to American security, save as they threaten US economic interests'.[32] Klein and Kinzer agreed that the United States cloaked its interventions abroad in the rhetoric of fighting a threat to US national security, even though the real aim was to promote US business interests abroad by overthrowing uncooperative foreign leaders.[33] I also argue that Iraq was a case study for this phenomenon due to its oil resources.

The significance of energy resources for the US empire

Oil lies at an intersection of politics and economics, due to the strategic importance of oil in the military and economic organization of society, the place of oil as the largest single component of world trade, the role of state-owned companies in the market, and the industry's periodic instability.[34] Oil historian Daniel Yergin argued that oil, as a commodity, is 'intimately intertwined with national strategies and global politics and power',[35] and 'has provided the point at which foreign policy, international economic considerations, national security and corporate interests would all converge'.[36] Therefore, an attempt to argue for the relation between oil resources and US hegemony, argued Simon Bromley, 'requires a high degree of cross-fertilization among the social sciences' of US foreign policy, security studies and energy studies.[37] As Susan Strange said, the political economy of the world energy system 'seems to be a classic case of the no man's land lying between the social sciences, an area unexplored and unoccupied by any of the major theoretical disciplines'.[38] There is a body of literature on the global energy system, but economic and international relations theories are ill-adapted to this field.[39] Economists who attempt to apply economic theory to the energy market are challenged by the fact that energy markets are highly susceptible to strong forces that are essentially political in nature,[40] while theorists in political science and international relations are unaccustomed to taking account of the powerful market forces associated with the global oil market, and political theorists who work on security matters still tend to think of strategy as something related mainly to military security and defence policy rather than to energy policy.[41]

Nevertheless, the limitations facing the studying of energy resources as a theoretical gap in social science is not the main argument of this book, which is more concerned with the historical question of how the procurement of foreign oil supplies was addressed in US foreign policy, especially under George W. Bush, with special focus on the invasions of Afghanistan and Iraq, and the negative impact of the Bush policy on the future of the American empire and its energy security.

There is no all-encompassing definition of 'energy security',[42] as, according to oil historian Daniel Yergin and security studies expert Michael T. Klare, energy security means different things to different people. For the United States, energy security is a 'geopolitical question', which may focus on geographic diversification of energy resources and protection from terrorist attacks. For Europe, energy security may focus on reducing dependence on foreign gas, especially from Russia.[43] But for all states, energy security means diversification; reducing dependence on one single source. If energy is imported, then 'energy security' would also have a foreign policy dimension.[44]

Throughout the cold war, and even after the cold war was over, the significance of energy control influenced US policy initiatives. The prominent role of energy resources in the American economic position affected America's geopolitical priorities, as many of the world's largest reserves of oil are located in areas that are unstable or rife with internal divisions and destabilizing factors. Three major wars in recent US history, the Gulf War (1991), the war in Afghanistan (2001), and the invasion of Iraq (2003) were in oil-producing regions and/or areas vital for energy distribution and transport routes.

Some historians have incorporated energy procurement in their narratives. William Appleman Williams included Middle East oil in his interpretation of American empire, with 'the Open-Door expansion in the Middle East'. According to Williams, 'oil was the major objective of American policy' in the Middle East, reflecting 'a successful example of corporate capitalism in action'.[45] Kolko also argued that oil, in particular, reflected America's economic aims in relation to the Hullian theory of US economic objectives and the Open Door policy.[46]

This book argues that America's (economic/informal) empire is based on controlling the global oil order. Bromley argued that, after World War II, the United States formed an 'informal empire of American hegemony', based on controlling the international oil order which was at the centre of American hegemony, displacing Great Britain as the dominant power over the Middle East and its oil reserves, with the help of the US major oil corporations who 'played a central role in this strategy'.[47] Similarly, Svante Karlsson's *Oil and the World Order: American Foreign Oil Policy* argued that 'the American-controlled world economic order was to a large extent built on the control of oil', as United States control over the international oil market was:

> intimately interconnected with the control of the world economic order, which was established after the Second World War. Oil is and has been the most important single commodity in the present world order. It became the very vehicle on which the unprecedented economic expansion of the postwar world has been riding. Thus, the control of the international oil market was one of the most important keys to the control of the world order.[48]

In addition to its importance for US power, energy is not just a goal in US foreign policy, it is also 'a powerful tool of US foreign policy'.[49] For instance, the United States used energy 'as a weapon, by sanctions that deny US investors and markets to countries that threaten America'.[50] Oil sanctions were used against the oil sectors of the 'rogue states' like Iraq, Iran, Sudan, Libya and others. Also, the US interest in controlling oil is not limited to America's own consumption, but also to denying it to others[51] in order to control the economic destinies of others. For instance, Noam Chomsky argued that the US has to control vital energy regions around the world, not necessarily for US consumption, but to control the supplies to other great powers in Europe and Asia.[52] Similarly, Ian Rutledge argued that the United States did not always aim to 'steal the oil', but to 'control' it without having to own it.[53] This book therefore argues that control over oil resources in US foreign policy is both an end and a means to other ends. An end because it is necessary for US economic performance and US business profit. A means to other ends, because control over energy resources paves the way for US imperial influence, whether by controlling the supplies to other great powers, or (as in the case of Iraq) using oil revenues to rebuild Iraq into a pro-American model country in the Middle East to reshape the region, as we will see.

The link between the oil industry and the US Government

This book also argues that the link between the US Government and the oil business is necessary to maintain control over the global oil system, maintain US empire, secure America's oil interests and promote foreign-policy decisions which advance the interests of the oil industry. Energy considerations and the need to control the global oil system affected the foreign oil policy of the United States during World War II and the cold war, as the US Government and the US oil industry cooperated in what Painter called a 'public–private partnership', or a 'symbiosis between public and private interests that safeguarded and advanced the private interests of the oil companies while furthering US efforts to control world oil reserves, combat economic nationalism, and contain the Soviet Union', in accordance with US ideological aims.[54] Based on his argument that the 'ruling class of American capitalism' is the 'fount' of US foreign policy,[55] Kolko argued that since World War II, the US State Department

> completely identified the national interest with that of American
> oil firms operating abroad, not merely because numerous former
> oil industry executives occupied key posts in the department,

but primarily due to traditional synthesis of private and public interest which had been the functional basis of American foreign economic policy for decades.[56]

Attempting to extend its influence over Middle East oil, the United States invoked the Open Door policy to abrogate agreements with Britain to limit US oil companies' access to Middle East oil in the 1930s and 1940s, allowing the US companies to extend to oilfields traditionally managed by Britain.[57] Also, both Painter and Kolko cited the Marshall Plan of 1947 as an obvious example where the interests of the US Government and the US oil companies intersected. More than 10% of the total aid extended under the European Recovery Program (ERP) was spent on oil, more than any other single commodity, and without the Marshall Plan, 'the American oil business in Europe would ... have been shot to pieces' said Walter Levy, head of the Marshall Plan's oil division, as the Marshall Plan purchased Middle East oil from American firms at the price of $2.65 a barrel, while Middle East oil cost only 50 cents a barrel.[58] (Levy had worked for Socony-Vacuum, the company later known as Mobil, but he resigned in July 1948.)[59]

Anthony Sampson's classic book on US oil corporations, *The Seven Sisters*,[60] extended this argument to the 1970s, arguing that besides invoking the Open Door policy to protect US oil business on several occasions during the first half of the twentieth century,[61] the US Government provided diplomatic protection to the US oil business after World War II,[62] where the US State Department would delegate diplomacy in the oil-producing countries, as far as possible, to the American oil corporations (due to the contradiction between Washington's support for Israel and the need to woo the oil-producing Arabs).[63] However, the oil companies started to get weaker in 1970 due to the Arab–Israeli conflict (and the power of the Israeli lobby in Washington), and the rising power of the oil-producing states.[64] After the 1973 oil boycott, it became clear that oil was too important to be left to the companies[65] as the Western governments tried to come up with serious oil policies following the oil crisis.[66] The US considered invading Saudi Arabia's oilfields in 1975, but rejected the idea because of the risks entailed.[67] However, attempts at energy conservation and regional diversification of foreign oil resources failed to reduce dependence on Middle East oil as desired. The best option, according to John Morrissey, was emphasizing military protection of Persian Gulf oil, as seen in the Carter Doctrine of 1980, and its applications in the 1991 Gulf War and the 2003 invasion of Iraq.[68]

Oil After the cold war

With the end of the cold war, competition over raw materials, especially oil and energy resources, rose to a new level as ideological conflicts ceased. Yergin argued that

> in the Cold War years, the battle for control of oil between international companies and developing countries was a major part of the great drama of decolonisation and emergent nationalism … With the end of the Cold War, a new world order [was] taking shape [where] economic competition, regional struggles and ethnic rivalries may replace ideology as the focus of international – and national – conflict.[69]

Along the same lines, Michael Klare's *Resource Wars* argued that conflict over resources would be a conspicuous feature of international security in the post-cold war world.[70] He believed that, among the US objectives after the cold war, 'none has so profoundly influenced American military policy as the determination to ensure US access to overseas supplies of vital resources',[71] and none of these natural resources was more likely to provoke conflict in the twenty-first century than oil.[72] George H.W. Bush used the 1991 Gulf War to protect the Gulf's oil and assert the US position in the post-cold war world. William J. Clinton's policy on foreign energy was not very different from his predecessor's, and he also used his wars in the Balkans to secure the region's energy routes and to secure the Albanian–Macedonian–Bulgarian Oil (AMBO) pipeline, an investment by BP and Halliburton.[73] According to Bacevich, both presidents were applying the Open Door policy. However, George W. Bush took this to a new level as he prioritized energy during his first days in office, and he had greater use for the military in oil-rich regions.

Oil, empire, the industry and the
George W. Bush administration

The importance of oil in maintaining the US empire, and the importance of the public–private partnership in the oil industry, rose to a new height during the George W. Bush administration. Bush brought oil to the fore in US foreign-policy making, raising the place of energy procurement in US foreign policy to a level never seen before. Bush introduced oil as a part of a strategic approach, for example through the National Energy Policy of May 2001 (on which his administration started working only a few days after he took office), and through the invasion of Iraq, which aimed to control its oil supplies, and the role that Iraq's oil revenues were meant to play in rebuilding Iraq, and thus the whole Middle East, into a pro-American region.

Furthermore, energy's central role in the Bush administration was linked to other foreign policy goals, such as promoting US hegemony and preponderance of power (which was stated as a goal in the Quadrennial Defence Review of 2001), and military advancement, which was also a major goal of the Bush administration as seen in promoting the National Missile Defence (NMD) system, with the War on Terror used as a justification to advance US energy interests. Because of the place of energy in that pursuit for power, certain cases became central to the Bush approach (most notably Iraq), acting as what I call 'points of intersection' where all of these goals were meant to be achieved.

Bush had more ties to the energy industry than any other American administration in history, as he used to be an oil executive himself (albeit not a very successful one), having established Arbusto Energy Inc., merged it with Spectrum Energy 7 Corp, and worked at Harken Energy. Furthermore, the Bush family had strong ties to the oil industry. His father, George H.W. Bush, established Zapata Offshore Drilling in the 1950s.[74] His Vice President, Dick Cheney, had previously been CEO of Halliburton, and he had a major role in forming the foreign and energy policies of the Bush administration as he headed several important committees which devised these policies. Government–oil industry ties were affecting Bush's foreign policy decisions. This prompted many authors, including Ian Rutledge, to argue that the George W. Bush administration was the most oil-dominated US administration in the history of the United States. This put 'oil capitalism' at the very heart of US power and high on the agenda of the Bush administration. This was especially true considering that Bush was elected at a time when the interests of the US oil companies and the oil consumers were rapidly moving even closer together, leading to a relentless drive for 'American Imperium' and 'Energy Security' in the Middle East, due to the increasing dependence on imports from the Persian Gulf.[75] (US dependence on Persian Gulf oil was decreasing in the 1990s, but began to rise again in the early years of the twenty-first century. Even as oil demand fell back slightly, to 19.6 mbpd (million barrels per day) in 2001, US oil imports continued to rise as they reached a record high of 10.9 mbpd, of which 2.8 mbpd were from the Persian Gulf. In 2001, 14.1% of total US oil consumption was supplied by the Persian Gulf, the highest level in America's history, and the share of imports supplied from the Gulf was at 25.3%.)[76]

Antonia Juhasz, too, focused on the Bush–energy industry relations,[77] where she linked the Bush administration to US empire and oil. She described the combination of imperialism, corporatism and oil interests as the 'Bush Agenda', or 'corporate globalization', which she defined as global economic policies designed to support key US multinational corporations by using military force, aiming for an American empire driven forward by US multinationals and military power,[78] whose main pillars are war, oil, imperialism and corporate globalization.[79] This policy was not new, since it has been a part of US policy for decades. But Bush was unique in his use of military power to advance economic interests, taking it to a new, radical level,[80] as he thought that he could build a new American empire by invading Iraq and having US bases in the oil-rich region. Juhasz's main failure, however, was that she did not mention the invasion of Afghanistan or the US plans for Central Asian oil as a part of Bush's global economic agenda.

This book argues that as the United States was in a phase of relative economic decline, and due to the fact that in 1998 it imported more than 50% of its oil[81] (and it is forecast that it will import 100%

of its oil by 2050),[82] Bush was aware of the increasing necessity to expand American economic reach and promote the Open Door policy for American business through military expansion.[83] Paul Kennedy famously argued that US power is declining, as its relative economic strength is declining,[84] and that the United States is now suffering from what he calls 'imperial overstretch', where the sum total of the United States' global interests and obligations is nowadays far larger than the country's power to defend them all simultaneously.[85] He also defined 'imperial overstretch' in the context of 'relative' economic change, compared to other nations.[86]

Bush saw that the inevitable relative economic decline of the United States made it even more necessary to transform the rest of the world in America's image.[87] This was why, after Iraq, Bush wanted to extend his agenda throughout the Middle East region, in order to expand the American empire,[88] as seen in his plans to democratize the whole region, and establish a Middle East Free Trade Agreement (MEFTA). Fouskas and Gökay argued that the relative decline in US economic power led to promoting the role of the US military to reshape the world (as seen in Central Asia and the Middle East after the 11 September attacks to control vital oil routes in Eurasia).[89] Chomsky and Chalmers Johnson argued that the invasion of Afghanistan was meant to establish US bases close to Central Asia's oil-rich region and help secure the region to build the Turkmenistan–Afghanistan–Pakistan (TAP) pipeline (a pipeline project to carry gas from Turkmenistan, through Afghanistan, to Pakistan), in addition to other important pipeline projects.[90] Similarly, Michael Klare,[91] Fouskas and Gökay,[92] and Lutz Kleveman[93] argued that the invasion of Afghanistan, and the establishment of US bases in the region for the first time, were beneficial for America's energy interests in the Caspian region. In fact, Ahmed Rashid,[94] Jean-Charles Brisard and Guillaume Dasquié argue that the Clinton and George W. Bush administrations initially negotiated with the Taliban over Osama bin Laden and over the TAP pipeline.[95]

Beyond the invasion of Afghanistan, the 11 September attacks had their influence on Bush's foreign policy agenda. When Bush first came to office, argued Klare, his foreign policy agenda sought two foreign-policy goals, or strands: military-power preponderance (as seen in Bush's

Citadel speech calling for 'agile' and 'readily deployable' US forces), and foreign-energy procurement. After 11 September, argued Klare, a third strand was added, which is the War on Terror.[96] However, I prefer to separate military preponderance (along with the NMD system) and global power projection into two strands, instead of merging them in one strand as Klare did, because I see that global power projection, as defined in the Quadrennial Defence Review (QDR) of September 2001, means 'preventing the rise of a rival power',[97] which may include economic as well as military factors. Thus, I present four strands of US foreign policy: military preponderance (along with the NMD system), energy procurement before 11 September, which were then joined by anti-terrorism and global power projection after 11 September. There were 'points of intersection' where the four strands met, such as the invasion of Afghanistan and Iraq.

Furthermore, after the 11 September attacks and the invasion of Afghanistan, Bush introduced the Bush Doctrine, which was America's first attempt at a grand strategy since the end of the cold war. The four strands of the Bush administration's foreign policy (energy procurement, military advancement, anti-terrorism and global power projection)[98] were reflected in the Bush Doctrine.

The Bush Doctrine was divided into several main elements or ideas. First, America was at war with global terrorism. Second, offence is the best defence; thus the United States should take the war to the enemy using pre-emptive strikes. Third, the doctrine gave the United States the freedom to act alone, so as not to be tied by multilateral commitments. Fourth, the United States was committed to maintaining its position as the world's sole superpower. Finally, the Doctrine stated that the United States should spread democracy in the Middle East as a means to fight terrorism.[99] The foreign-policy goals (or strands) of anti-terrorism, military preponderance and global power projection were clear in this definition of the Bush Doctrine. In an unprecedented manner, this book argues that the oil strand was also present in the Bush Doctrine (despite denials by the Bush administration) because, first, US global influence depends on control of global oil resources, and, second, the neo-conservative plans for democracy in the Middle East were unattainable without Iraq's oil revenues. These revenues were

needed to rebuild Iraq into a model democratic country for the Arabs
to follow (because the Iraqi economy was so dependent on oil revenues,
a strong democracy could not be built in Iraq without the oil revenues).
Furthermore, this book aims to provide an alternative framework for
thinking about US foreign policy and neo-imperialism in relation to
global oil resources, and to analyse the link between energy resources
and other US foreign-policy strands of military preponderance, global
power projection and anti-terrorism in the foreign policy approach of
the Bush administration and the Bush Doctrine.

The invasion of Iraq

This book argues that the invasion of Iraq was not for anti-terrorist
goals or to neutralize alleged weapons of mass destruction (as the Bush
administration claimed), as Iraq did not actually pose any immediate
or real threat to US security. Rather, the Bush administration used
these claims to cover up for its real aims: controlling Iraq's oil resources
(although not necessary for US consumption, but to control oil supplies
to other great powers), and build military bases in the region to protect oil interests. The invasion also aimed to rebuild Iraq, using its oil
resources and revenues, to turn it into a strong and stable pro-American
country to act as a model for the rest of the region, thus decreasing US
dependence on Saudi Arabia for oil and strategic assistance, and also to
secure Israel. Bromley argued that the war aimed to stabilize the 'arc of
instability' (the region which runs from the Middle East through Central
Asia to North-east Asia),[100] and use military power to provide for an
open, liberal international economic order and open oil industry.[101] Ian
Rutledge argued, within the context of the reasons mentioned above,[102]
that oil was the 'most important factor' for the invasion of Iraq, although
not to 'steal the oil', but to 'control' it without having to own it,[103] in
addition to the neo-conservative desire to remake the Middle East and
establish a 'new American Imperium in the Middle East: one in which
American-selected local rulers would invite American oil companies to
make super-profits for American investors under the protective shield
of the American military, while at the same time satisfying the vora-
cious demands of the motorized American oil consumer.'[104] Rutledge's

weakness was that he did not analyse Bush's National Energy Policy (NEP), the plan devised by oil industry executives and Bush administration officials in May 2001 to advise Bush on domestic and foreign energy policy, to the same extent that Michael Klare did.[105] Also, Rutledge did not link the different strands of Bush's foreign policy as deeply as Klare did. However, Rutledge's analysis of Washington's plans for Iraq's oil focused on 'Production Sharing Agreement' (PSA) contracts,[106] which Klare missed out (the significance of which will be discussed in detail).

Noam Chomsky, Chalmers Johnson and Stephen Pelletière argued that the war aimed to strengthen the system of US control over the region's oil resources, and to strengthen what Johnson calls an 'empire of bases' near vital oil routes and resources.[107] Similarly, Roger Burbach and Jim Tarbell argued that Bush's wars aimed to advance the interests of a 'petro–military complex' and to advance the Open Door policy in the twenty-first century.[108] Klein argued that Iraq fitted the 'shock doctrine' formula, as Washington was concerned over Saddam's oil contracts with Russian, French and Chinese oil companies.[109] Linking 'disaster capitalism' and the war on Iraq to American neo-imperialism, she claimed that the Washington 'hawks' and neo-conservatives were part of a 'disaster capitalism complex' (all members of which were allied on a US imperial/military/neo-conservative project) 'committed to an imperial role for the United States in the world and for Israel in the Middle East'. She added that 'nowhere has the merger of these political and profit-making goals been clearer than [in] Iraq'.[110] She cited an interview that right-wing economist Milton Friedman, the father of the 'shock doctrine',[111] gave to the German magazine *Focus* in April 2003, where he said that the war would 'undoubtedly stimulate the [US] economy', that 'US President Bush only wanted war because anything else would have threatened the freedom and prosperity of the US', that Bush 'didn't even have to consult the UN at all' regarding the war because 'the end justifies the means', and that the cost of the war would be 'only marginal'.[112] Both Klein and Kinzer briefly discussed the role of oil in invading Iraq, but they did not tackle it in the detail which this book intends.

Ron Suskind's account of Paul O'Neill's time in the Bush administration was an important insider's account of Bush's first days in office and his initial focus on Iraq, as National Security Advisor Condoleezza Rice noted that 'Iraq might be the key to reshaping the entire [Middle East]'.[113] Maps of Iraq's oilfields which, according to O'Neill, were used by Secretary of Defence Donald Rumsfeld[114] were the same maps used by Vice President Dick Cheney as he laid out the National Energy Policy (NEP) of May 2001 (as revealed by court order, as will be discussed), showing that energy procurement had a military dimension. O'Neill's accounts were supported by Greg Palast's BBC *Newsnight* interviews with oil executives.[115]

Apart from invading Afghanistan and Iraq to establish a military presence in oil-rich regions, this book argues that the invasion of Iraq also meant to protect the petrodollar system from the euro threat. US influence over the global oil order is supported by two factors on which the US position in the world economy depends. The first factor is the interdependence between the oil industry and US hegemony in ensuring US energy security, US global power based on control of oil resources, and the interests of the corporations[116] (reflecting Williams' and Kolko's business–government relations). Second, the dominant role of the US dollar in international financial and monetary relations is linked to the fact that, since the end of World War II, the US dollar has been the primary currency used in the international oil trade. (This was further facilitated by the agreement between the US and Saudi Arabia in 1974–1975, which stipulated that OPEC would price oil exclusively in US dollars; the so-called 'petrodollar agreement'.)[117] This helped create a global demand for the US dollar,[118] as every nation sought to maximize its dollar holdings (by buying dollar-based assets from the 'issuer' of the dollar; the United States), thus preserving the dollar's high liquidity value and America's strong economy,[119] and allowing the US to create credit[120] to finance its huge deficit while other countries accept the dollar as the global medium of exchange.[121] Thus, the US dollar as a dominant global currency gets its power from the oil system.

Vassilis Fouskas and Bülent Gökay linked US neo-imperialism to control over global oil resources, the role of the US dollar and petrodollar, military preponderance and the threat from the euro as an alternative currency in the international oil trade (the so-called 'petroeuro') to the dollar's status as the global reserve currency.[122] William Clark wrote along the same lines.[123] They argued that America's dominant position as the sole superpower rests upon two pillars: first, its over-whelming military superiority and, second, its influence over the global economic system through the unique role of the dollar as the world's reserve currency,[124] where America's entire monetary suprem-acy depends on dollar 'recycling';[125] the operation by which oil-export-ing countries receive payments for petroleum exports in US dollars, then invest these petrodollars in US and European banks.[126]

This book argues that Saddam Hussein's decision in November 2000 to switch the payment for Iraqi oil exports from US dollars to euros was one of the reasons for the US invasion of Iraq, which aimed, among other things, to return Iraq's oil exports to the dol-lar system.[127] However, while Clark and Fouskas and Gökay briefly mentioned the neo-conservative dream to use Iraq as a model for the rest of the region to weaken OPEC, secure Israel, fight terrorism and impose a pro-US agenda, they gave only little attention to these neo-conservative goals.

Finally, I believe that the Open Door policy and the thoughts of William Appleman Williams were still relevant under the George W. Bush administration. This was seen in the increased American military involvement in the world after the 11 September attacks to make up for America's relative economic decline, especially with the wars in Afghanistan and Iraq, and as Washington (especially under the Bush administration) replaced access to Asian (and Chinese) markets with access to Persian Gulf oil through building military bases in and around Afghanistan and Iraq.[128] Some authors did link the Open Door to Bush's oil policies. For instance, former CIA analyst Stephen Pelletière argued that Bush's War on Terror was a 'variant' of the Open Door policy to expand the US empire,[129] but he did not go into the details of the origins of the Open Door, and he only mentioned it briefly. Similarly, Burbach and Tarbell argued that Bush's wars aimed to advance the interests of a 'petro–military complex' and to advance the Open Door policy in the twenty-first century,[130] but their analysis did not focus on energy and oil resources as this book does.

* * *

Chapter 1 focuses on the role of oil and energy resources in the New World Order of George H.W. Bush, and in the 'Engagement and Enlargement' policy of William J. Clinton, where I argue that both administrations proceeded with the Open Door policy, that oil did increase in significance after the end of the cold war, but that, in both policies, oil did not rise to the level it did with the George W. Bush agenda. This was also the period where the US paid more attention to Caspian oil as an alternative to diversify away from Middle East oil.

Chapter 2 focuses on the George W. Bush administration before 11 September, where I argue that oil and Iraq were among Bush's highest priorities from his first days in office. This was because of his administration's particularly close links with the oil industry, America's increased dependence on oil imports and pressure from the military hawks and neo-conservatives for more focus on Iraq. I also argue that the two foreign-policy aims of energy procurement and military advancement were 'melded' under Bush, as seen in official documents.

Chapter 3 focuses on the effect of 11 September 2001 and the addition of the War on Terror as a primary goal of the Bush administration, in addition to global power projection which was mentioned in the QDR of 30 September 2001. I argue that the invasion of Afghanistan was a point of intersection of all foreign-policy strands of oil procurement, military advancement, anti-terrorism and global power projection, although the Bush administration did not have plans to remake Central Asia, as it did with the Middle East. I also argue that the Bush administration used 11 September as an 'opportunity' to promote an invasion of Iraq, which would also serve as a point of intersection of all strands, in addition to the neo-conservative goal of reshaping the Middle East into a pro-American mould to secure Israel, fight terrorism and open markets for US business.

Chapter 4 focuses on the invasion of Iraq, where I argue that the Bush administration aimed to secure its grip on the Iraqi economy (and its oil) to open exceptional opportunities for US business. However, the plan for Iraq did not succeed, because the Bush administration failed to plan properly for post-invasion Iraq, and this led to the failure to reshape the Middle East through reshaping Iraq.

Finally, I argue that, due to failure in Iraq, and due to the financial strain of the war, the US got weaker in terms of global clout and in terms of energy security. Furthermore, the diversification of US energy supplies away from the Middle East will fail, as all other regions have less oil reserves than the Middle East, and are not necessarily very stable or friendly towards the US. I also argue that the US is facing strong competition over energy resources from rising powers like China and Russia, and that the Bush administration still followed an ad hoc, case-by-case approach in foreign policy towards the Middle East and other oil-producing regions, as Iraq failed to provide the US with the strategic and energy security that it aimed for. (The book's concluding chapter includes a coda on the Barack Hussein Obama administration, to further explain the consequences and legacy of Bush's policies.)

CHAPTER 1

FOREIGN ENERGY RESOURCES IN THE POST-COLD WAR DECADE

The Democracy Conundrum, Regional Diversification, and an Ad Hoc Policy on Energy Resources

I put it this way. They got a president of the United States that came out of the oil and gas industry, that knows it and knows it well.

George H. W. Bush, on the eve of his inauguration as President.[1]

Prosperity at home depends on stability in key regions with which we trade or from which we import critical commodities, such as oil and natural gas.

William J. Clinton, December 1999.[2]

During his State of the Union speech in February 1989, President George H.W. Bush expressed optimism that 'we meet at a time of extraordinary hope ... And it's a time of great change in the world'.[3] Despite this optimism, however, there were authors who thought that the new world was rife with instability. Charles Krauthammer challenged the conventional wisdom which said that the threat of war would decrease after the fall of the Soviet Union,[4] arguing that Saddam Hussein's invasion of Kuwait and plans to develop weapons of mass destruction using his oil wealth were proof that the threat of war was not declining, and that weapons of mass destruction were increasingly a global threat. Moreover, said Krauthammer, Iraq was not the only state seeking weapons of mass destruction; so were Syria, North Korea, Libya and Iran. He argued that the post-cold war era could be called the era of weapons of mass destruction.[5] Michael Cox agreed that the end of the cold war made the world more complicated, due to the great shifts in the balance of global power, politically as well as economically, which made it more difficult for policy makers to come up with solutions to global problems.[6] Bush himself would say during his State of the Union speech in January 1992: 'As we seek to build a new world order in the aftermath of the Cold War ... the enemy we face is ... instability itself.'[7] Moreover, William J. Clinton added during his presidential campaign that the factors which could make the new era a time of peace were the same factors which could make it a time of uncertainty and increased danger.[8]

On the other hand, the end of the cold war, despite ushering in a more complex world, also ushered in the Age of Globalization, where doors were open for the free movement of US capital, trade, principles and culture. The two American administrations which presided during the post-cold war decade (the George H.W. Bush and William J. Clinton administrations) were similar in that they both pursued a policy of global economic openness. Thus, an American policy of global economic openness replaced the policy of Containment (although, as seen in the writings of William Appleman Williams, Bush and Clinton did not invent openness; they just revived a strategy that predated the cold war by several decades).[9]

Even though George H.W. Bush failed to come up with a foreign policy 'vision' and failed to uphold the New World Order, he did adhere to the patterns identified by William Appleman Williams. Similarly, Clinton's doctrine of Engagement and Enlargement followed Bush's footsteps on emphasizing global openness and free trade and investment as prerequisites for prosperity at home, as Bush pushed for NAFTA and APEC, and Clinton continued on this path, again building on Williams' views. Openness was also a matter of national security, as Larry Summers, Clinton's Treasury Secretary, argued that trade promoted peace among nations, that the American market was saturated, and that the American economy could not grow without the international market.[10]

This shows that both Bush and Clinton followed on with a global Open Door imperialism, based on corporatism, or the political economy of the large corporation, or a corporate globalization policy. In both administrations, the US military had a great role in supporting the American policy of openness in regions critical to the strategy of openness, such as the Persian Gulf and Asia-Pacific; as Secretary of Defence William Cohen pointed out, 'economists and soldiers share the same interest in stability'.[11]

While Bush and Clinton tried to come up with global doctrines – Bush's New World Order and Clinton's Engagement and Enlargement – which promoted American global leadership, a policy of democratization and promotion of market openness, both doctrines failed to promote democracy in the Middle East, due to the Democracy Conundrum, which is the fear that attempts at democratization or political reform would threaten the stability of the pro-American, oil-producing Gulf regimes.[12] Another name for this phenomenon is the 'Arab despotic exception', where America opposes democracy in the Middle East due to fears over oil supplies, and the fear that anti-American parties might win elections.[13] Thus, Bush and Clinton's strategy of openness did not apply to the Middle East, where Washington's aim was to maintain the status quo because it feared that attempts to change the region would threaten the flow of oil.[14] Also, both had a policy of regional diversification of energy resources away from the Middle East. For example, NAFTA, which gave much attention to North American energy cooperation,[15] started in the Bush era, US energy companies started doing business in the Caspian region in 1991,[16] and Clinton followed through with this diversification policy.

Also, both initially supported an anti-democratic regime which was vital for America's oil supplies, which later turned out to be anti-American, as Bush initially supported Saddam Hussein in order to promote the interests of American oil corporations, before the invasion of Kuwait, and urged Iran to stop its anti-American and anti-Iraq behaviour in the region. Similarly, Clinton supported the Taliban, in order to promote the TAP gas pipeline which American corporations were building in the region,[17] before the 1998 bombings in Kenya and Tanzania. They both maintained oil sanctions on other rogue states. Thus, both administrations had an ad hoc policy on foreign energy resources. They also both followed what David Painter called the 'public–private partnership' in the field of energy, where the US Government promotes and protects the aims of the oil industry, in order to secure America's national and energy security.

With the end of the cold war, and as George H.W. Bush succeeded Ronald Reagan as President of the United States, Yergin argued that 'the confrontations of the future would surely become global competitions for money and markets'[18]:

> The competition among nations in the years ahead, some predicted, would no longer be ideological but instead primarily economic ... If that would be the case, oil as a fuel would certainly remain a vital commodity in the economies of both the industrialized and the developing nations of the world. As a bargaining chip among the producers and consumers of oil, it would also remain of paramount importance in the politics of world power.[19]

Similarly, Michael Mandelbaum wrote that 'as security issues lose some of their previous significance, economic questions will assume a new international importance ... Promoting capitalism, then, is a plausible goal for American foreign policy in the post-Cold War world.'[20]

A leading author who agreed that conflicts after the cold war were mainly over resources, especially oil, was Michael Klare. He wrote in *Resource Wars* (2002) that resources were always critical to US foreign policy, but during the cold war their emphasis in US foreign policy diminished as attention to Europe and Asia increased. After the end of the cold war, the issue of resources reassumed their central role in US military planning.[21] Among the US objectives after the cold war, 'none has so profoundly influenced American military policy as the determination to ensure US access to overseas supplies of vital resources', especially Persian Gulf oil, on which US dependence is growing, and the US presence in the region is dictated by its need to control its oil supplies.[22]

Due to the importance of energy, a major concern for the George H.W. Bush administration when it took office in 1989 was Persian Gulf security through cooperation with Iraq. Before the Iraqi invasion of Kuwait, the Bush administration remained committed to a policy of engagement with Iraq, as it thought that continued engagement would lead to the moderation of Iraqi behaviour.[23] As the Iraqis emerged with the upper hand from the Iraq–Iran war, US policy makers saw Baghdad as a potential ally against Iran. Furthermore, in a show of corporatism, a secret policy review of Iraq was performed shortly after Bush took office, promoting US business interests, saying that Iraq's 'vast oil reserves [promised] a lucrative market for US goods', and adding that US oil imports from Iraq skyrocketed after Iraq began offering American oil companies 'large incentives'.[24] This led to National Security Directive 26, 'US Policy Toward the Persian Gulf', dated 2 October 1989. The first paragraph of the directive emphasized the importance of Persian Gulf oil:

> Access to Persian Gulf oil and the security of key friendly states in the area are vital to US national security. The United States remains committed to defend its vital interests in the region, if necessary and appropriate through the use of US military force, against the Soviet Union or any other regional power with interests inimical to our own.[25]

The directive also encouraged normal relations with Saddam Hussein and Iraq, linking Baghdad to the wider issues of the Middle East, such as Arab–Israeli peace, Lebanon, Iran, WMDs and human rights:

> Normal relations between the United States and Iraq would serve our longer-term interests and promote stability in both the Gulf and the Middle East. The United States Government should propose economic and political incentives for Iraq to moderate its behaviour and to increase our influence with Iraq. At the same time, the Iraqi leadership must understand that any illegal use of chemical and/or biological weapons will lead to

economic and political sanctions ... Human rights considerations should continue to be an important element in our policy toward Iraq. In addition, Iraq should be urged to cease its meddling in external affairs, such as in Lebanon, and be encouraged to play a constructive role in negotiating a settlement with Iran and cooperate in the Middle East peace process.[26]

On Iran, NSD 26 recognized the opportunities presented by the death of Ayatollah Khomeini in June 1989, and held out the possibility of normalized relations with Iran if the latter stopped its anti-American behaviour and signed a peace treaty with Iraq:

The United States should continue to be prepared for a normal relationship with Iran on the basis of strict reciprocity. A process of normalization must begin with Iranian action to cease its support for international terrorism and help obtain the release of all American hostages, which will not be a matter for bargaining or blackmail. Other criteria Iran must meet before full normalization of US–Iranian relations include halting its subversive activities and improving relations with its neighbours, making a good faith effort toward a peace treaty with Iraq, and improving its human rights practices.[27]

However, the document was vague on what US policy would be if Tehran did not change its behaviour. Also, Washington failed to obtain Baghdad's cooperation on the Arab–Israeli peace process, as Iraqi foreign minister Tariq Aziz refused James Baker's offer of an aid package of $1.2 billion in return for Iraq's assistance in Arab–Israeli negotiations.[28]

Nevertheless, despite the lack of progress on these issues, American business interests in Iraq, especially in the field of energy and reconstruction, were encouraged:

We should pursue, and seek to facilitate, opportunities for US firms to participate in the reconstruction of the Iraqi economy, particularly in the energy area, where they do not conflict with our non-proliferation and other significant objectives.[29]

Oil was a significant factor in the Bush administration's relationship with Baghdad. From only 80,000 barrels a day in 1987, US imports of Iraqi oil jumped to 675,000 barrels in 1990, and the figures kept rising rapidly from month to month. By July 1990, it had leapt to 1.1 mbpd – more than a quarter of Iraq's total oil exports.[30] Bush, an old oilman himself (having founded Zapata Offshore Drilling in the 1950s, with James Baker as his lawyer),[31] understood the needs and importance of oil considerations, and Baker admitted in his memoirs, *The Politics of Diplomacy*, that policy toward Iraq was 'not immune from domestic economic considerations'.[32] ('When you look at NSD 26, you find out that it was the Administration's sole desire and policy to aid and abet Saddam Hussein,' said Congressman Sam Gejdenson, a Connecticut Democrat and one of Bush's harshest critics.)[33]

Naturally, American businessmen supported doing business with Iraq. The US–Iraq Business Forum, an American trade association which represented 60 US companies (including Bechtel, Lockheed, Texaco, Exxon and Mobil) was established by Marshall Wiley, a lifelong US foreign-service officer. Arguing that the United States should not prosecute Saddam Hussein for his human rights violations, since Iraq was important as a trading partner and as a guardian of US business interests and energy resources in the Gulf region, Wiley said in 1989 that:

> Iraq plays an important role now in bringing about stability in the Gulf region as an offset to Iran. This balance of power is important to us because of our interest further down the Gulf in the countries that are friendly to us and where we have substantial energy reserves that must be protected. So Iraq plays an important political and geopolitical role in addition to being an important trading partner for us, and we can't push all these considerations to the side and shut our eyes to them because of some perceived human rights violations.[34]

Moreover, military cooperation with Iraq was encouraged in NSD 26, which said that 'as a means of developing access to and influence with the Iraqi defence establishment, the United States should consider sales of non-lethal forms of military assistance, e.g., training courses and medical exchanges'.[35]

As late as April 1990, four months before the Iraqi invasion of Kuwait, Assistant Secretary of State for Near Eastern and South Asian Affairs, John Kelly, explained that the US policy on Iraq was based on an attempt to gradually develop a mutually beneficial relationship that would strengthen the positive trends in Iraq's foreign and domestic policies. He also said that the principal American interest in the region was energy, and that the importance of Iraq and the Persian Gulf to the United States was likely to increase in the coming decade, due to the increased dependence of the United States on oil from the region.[36] However, Saddam's invasion of Kuwait on 2 August 1990, caused the first post-cold war crisis and exposed the weakness of depending on Iraq for securing the Persian Gulf and its energy supplies.

The crisis was partly explained in economic terms. If successful in holding on to Kuwait, Saddam would directly control Iraq and Kuwait's oil resources, which together add up to 20% of OPEC production and 20% of world oil reserves. He would also be in a position to threaten his oil-producing neighbours, because if he captured Saudi Arabia as well, the sum of world oil under his control would rise to about 45%, and he would wield unparalleled influence over the world oil market and have the economic freedom to take even larger steps, raising the price of oil to unprecedented levels. Secretary of Defence Dick Cheney, Deputy Secretary of State Lawrence Eagleburger and National Security Advisor Brent Scowcroft argued to Bush that Saddam's invasion was a far more important issue than Kuwait, as it was about oil. Even if Saddam was not going to invade Saudi Arabia, there was still a risk of oil being denied to the US and its allies.[37]

In response to the invasion, Bush released NSD 45, titled 'US Policy in Response to the Iraqi Invasion of Kuwait', on 20 August 1990. The first paragraph of NSD 45 was very similar to that of NSD 26 of October 1989 as it emphasized the importance of securing Persian Gulf oil:

> US interests in the Persian Gulf are vital to the national security. These interests include access to oil and the security and stability of key friendly states in the region. The United States will defend its vital interests in the area, through the use of US military force

if necessary and appropriate, against any power with interests inimical to our own.[38]

NSD 45 then listed the 'four principles' that would 'guide US policy during this crisis': the immediate, complete, and unconditional withdrawal of all Iraqi forces from Kuwait; the restoration of Kuwait's legitimate government to replace the puppet regime installed by Iraq; a commitment to the security and stability of the Persian Gulf; and the protection of the lives of American citizens abroad.[39]

NSD 45 also allocated a whole paragraph to the procedures that the United States would take to protect its energy supplies, calling on oil-producing countries to raise oil production:

> The United States now imports nearly half the oil it consumes and, as a result of the current crisis, could face a major threat to its economy. Much of the world is even more dependent on imported oil and more vulnerable to Iraqi threats. To minimize any impact that oil flow reductions from Iraq and Kuwait will have on the world's economies, it will be our policy to ask oil-producing nations to do what they can to increase production to offset these losses.[40]

Thus, Saudi Arabia raised its production of oil in order to offset the rise in prices that resulted from the crisis. Due to the sanctions on Iraq, which pulled Iraqi oil out of the global market (thus decreasing the global supply of oil), plus Saddam's threats to destroy the Saudi oil supply system, oil prices jumped to $40/barrel, approximately double what they were before the invasion.[41] This sharp rise in oil prices led to an increase in prices at the pump and for home heating.[42] However, the Saudis quickly raised production to push the prices back down to around $20/barrel.[43]

Bush continued to emphasize the importance of the oil factor in this crisis, as he said on 8 August 1990 that 'our nation now imports nearly half the oil it consumes and could face a major threat to its economic independence ... the sovereign independence of Saudi Arabia is of vital interest to the United States'.[44] He also said in a speech on 15 August 1990, to gather US public support for the upcoming Gulf War, that 'our jobs, our way of life, our own freedom and the freedom of friendly countries around the world would all suffer if control of the world's greatest oil reserves fell in the hands of Saddam Hussein'.[45] Secretary of Defence Dick Cheney highlighted the threat of oil on his first major statement on the crisis, on 11 September 1990, before the Senate Armed Services Committee, saying that once Saddam acquired Kuwait and deployed his massive army, he would be 'in a position to be able to dictate the future of worldwide energy policy, and that [would give] him a stranglehold on our economy'.[46]

On the other hand, some analysts believed that oil was not the number one reason for American involvement in the 1991 Gulf War, although they did not deny oil's importance. Apart from oil, the United States had to use the military option against Saddam in order to maintain and expand American influence on the global arena. Bush himself acknowledged that preserving the global system of US influence was a main factor in the war, as he said in the January 1991 State of the Union address: 'What is at stake is more than one small country; it is a big idea: a new world order, where diverse nations are drawn together in common cause to achieve the universal aspirations of mankind – peace and security, freedom, and the rule of law.'[47]

According to Graham E. Fuller and Ian O. Lesser, oil was not the main reason for the war, as US global leadership was more important:

> Seizure of one Gulf state by another à la Saddam is, of course, unacceptable, but that has less to do with energy security than with preventing a grab for power by hostile states in a strategic region. 'Oil' may have been part of the early rationale for US intervention after Iraq invaded Kuwait ... but the real stakes were larger: US leadership and global order.[48]

Moreover, Fuller and Lesser emphasized that preventing the rise of a regional hegemonic power and preserving US hegemony in the Gulf were more vital factors:

> Some US strategists suggest that America has a broader interest in preventing the rise of any regional hegemonic power anywhere, especially one capable of threatening global stability through the

use of force. Implicit in this formulation is retention of the US role as the primary security arbiter in the Gulf and maintenance of the US presence there as a symbol of a global American security commitment. In some respects it would be more accurate to describe this objective as preserving US hegemony in the Gulf.[49]

Thomas McCormick considered both the oil factor and the global US leadership factor, as he acknowledged that 'short-term concern over the American and global economies' had prompted the United States to intervene to liberate Kuwait, since 'Iraq's conquest of Kuwait's oil production, and its threat to Saudi Arabia's, left [Iraq] in a position to profoundly alter the structure of world oil prices'. Iraq was in a position to raise oil prices, and cause a new oil shock that would deepen the US and global recession.[50] (He also mentioned Bush's political need to wage a quick and victorious war in order to improve his re-election chances in 1992 and overcome the so-called 'wimp' factor.)[51] Having said that, McCormick acknowledged that longer-term goals regarding the preservation of US hegemony were more important,[52] as US hegemony was required to protect 'global capitalism':

> While ... short-term political and economic factors were undoubtedly important in promoting American military action, they pale in significance alongside the [Bush] administration's long-term desire to perpetuate the credibility of American hegemony and to continue with its ongoing hegemonic project. Notwithstanding the end of the Cold War, the American government believed that the structural imperatives of global capitalism required a hegemonic centre to make and enforce the international rules of liberal capitalism.[53]

Similarly, Zbigniew Brzezinski thought that Saddam's invasion of Kuwait was a challenge to America's global power, which was 'more important' than just oil, saying that:

> Saddam's action was not only a challenge to the traditional US position in the Persian Gulf (especially to America's oil interests

in Saudi Arabia and the United Arab Emirates) but – perhaps even more important – to America's new dominance in the world and Bush's new global status. Whatever the legitimacy of Iraq's historic claims to Kuwait, the invasion was an act of defiance.[54]

Thus, according to William Hyland, the Iraqi invasion of Kuwait was also an important opportunity for the United States, as 'the United States was essentially free to intervene abroad without fear of confrontation with Moscow. This was a new and important dimension in foreign policy.'[55] Therefore, the economic and energy considerations were indeed an important factor in deciding to wage war on Iraq. However, the more important factor was emphasizing a post-cold war global system based on US leadership and interests.

On 15 January 1991, NSD 54, titled 'Responding to Iraqi Aggression in the Gulf' was adopted to implement the US policy on the crisis within the context of the New World Order and coalition building. NSD 54 set itself out as an extension of NSD 26 and NSD 45, again stressing the importance of Persian Gulf oil:

Access to Persian Gulf oil and the security of key friendly states in the area are vital to US national security. Consistent with NSD 26 of October 2, 1989, and NSD 45 of August 20, 1990, and as a matter of long-standing policy, the United States remains committed to defending its vital interests in the region, if necessary through the use of military force, against any power with interests inimical to our own.[56]

In order to protect the US interest, Bush authorized US military action, with the statement: 'I hereby authorize military actions designed to bring about Iraq's withdrawal from Kuwait.'[57] The directive also said that the United States would change the Iraqi regime if Saddam used weapons of mass destruction or destroyed Kuwait's oilfields:

Should Iraq resort to using chemical, biological, or nuclear weapons, be found supporting terrorist acts against US or coalition partners anywhere in the world, or destroy Kuwait's oilfields, it shall become an explicit objective of the United States to replace the current leadership of Iraq.[58]

Saddam did burn 700 of Kuwait's oilfields, causing $20 billion in damage.[59] However, the US military did not follow through with the regime change threat in NSD 54, refraining from entering Baghdad after the liberation of Kuwait.[60] In his memoirs, General Norman H. Schwarzkopf, Centcom Commander-in-Chief during Operation Desert Storm, explained the reasons for not entering Baghdad. UN resolutions, he said, only called for liberating Kuwait from the Iraqi invasion and this gave the United States the authority to 'take whatever actions were necessary to accomplish that mission, including attacks into Iraq'. The Americans, therefore, had no authority to invade Iraq for the purpose of capturing the entire country or its capital, and breaching UN resolutions would have caused the collapse of the coalition that the US had built on the basis of UN legitimacy. This would have cost the United States a lot in terms of increased financial burden and loss of American troops. Citing the defeat in Vietnam as an example, he said that lack of international legitimacy would have caused a defeat to the US military if they had invaded Iraq.[61]

Furthermore, Bush and Scowcroft admitted in their joint book *A World Transformed* that they preferred to leave Saddam in power, as removing him would have led to the division of Iraq, or, as they called it, to a 'Lebanonization of Iraq', thus destabilizing the region. Colin Powell made a similar assessment in *My American Journey*: 'It would not contribute to the stability we want in the Middle East to have Iraq fragmented into separate Sunni, Shia and Kurd political entities.' Powell added that 'our practical intention was to leave Baghdad enough power to survive as a threat to an Iran that remained bitterly hostile to the United States'. What the Bush team wanted was Iraqi and Persian Gulf stability, even if it had to be enforced by keeping Saddam in power. Thus Washington allowed Saddam to crush the Shiite and Kurdish uprisings following the war.[62] According to Richard Haass, it was 'unlikely that anyone else would have the same cult of personality' to hold the country together.[63] Also, the Bush administration feared that a US military invasion of Iraq would not help Arab–Israeli peace.[64]

Even though it stopped short of an invasion of Iraq, the Gulf War showed the mutual need for cooperation between the Arab oil exporters and the Western importers, as the importers needed oil, and the exporters needed military protection.[65] The war also provided Washington with enhanced influence in the Middle East. It enabled the United States to deploy its forces in the Gulf Arab states and to strengthen its ties and/or sign bilateral defence agreements with the region's major powers: Egypt, Israel, Kuwait, Qatar, Turkey and Saudi Arabia. Following the end of the 1991 Gulf War, 25,000 US troops remained in the region, 5,000 of whom were deployed in Saudi Arabia.[66] Colin Powell, Chairman of the Joint Chiefs of Staff, said in February 1991: 'I cannot believe the lift that this crisis and our response to it have given our country. This is the way the world's only remaining superpower is supposed to behave.'[67]

Furthermore, Bush used the liberation of Kuwait to promote the idea of a New World Order, saying 'what is at stake is more than one small country; it is a big idea: a new world order, where diverse nations are drawn together in common cause to achieve the universal aspirations of mankind – peace and security, freedom, and the rule of law'.[68] Shortly after the war, Bush said before the US Congress on 6 March 1991:

> We can see a new world coming into view, a world in which there is the very real prospect of a new world order ... a world in which freedom and respect for human rights find a home among all nations.[69]

However, he stressed that the United States should be the nation to dictate the form of the New World Order, saying in his 1991 State of the Union speech that 'American leadership is indispensable' and that 'we must step forward and accept our responsibility to lead the world' because 'we're the only nation on earth that could assemble the forces of peace [and] freedom'.[70]

In tandem with this vision, his 6 March 1991 speech outlined the challenges that faced the Persian Gulf following the war. First, he cited the need to create shared security arrangements in the region, in which the United States would maintain a strong military presence. Second, he wanted to prevent a nuclear arms race from developing in the region, meaning that the Iraqi nuclear arsenal had to be dismantled.[71] Third, Bush said that the Gulf War presented new opportunities for a comprehensive Arab–Israeli peace agreement. Finally, the United States would foster economic development in the Middle East using the region's oil resources.[72]

These concerns were also outlined in the National Security Strategy of August 1991, in the context of the Gulf War and the promise of the New World Order, including oil supplies, Arab–Israeli peace, fighting terrorism and weapons of mass destruction.[73]

In the State of the Union speech of January 1992, Bush recognized that the United States should not withdraw from international issues just because the Soviet Union has disintegrated, as the Soviet threat was replaced by the threat of 'instability itself':

> As the war in the Gulf made clear, the easing of the Soviet threat does not mean an end to all hazards. As we seek to build a new world order in the aftermath of the Cold War, we will likely discover that the enemy we face … is instability itself. And, in the face of multiple and varied threats to stability, we will increasingly find our military strength a source of reassurance and a foundation for security, regionally and globally.[74]

New World Order rhetoric and promises were included, in the form of a promise for 'a stable and secure world, where political and economic freedom, human rights and democratic institutions flourish', since 'our interests are best served in a world in which democracy and its ideals are widespread and secure'.[75]

The Gulf War also made the Bush administration realize the importance of Arab–Israeli peace, since Saddam used the Palestinian cause as a tool to rally supporters around him, enabling him to win the Palestinians to his side against the United States. This showed that the Arab–Israeli conflict was linked to any cause of instability in the region, which could be hazardous to American interests, especially access to oil. Thus, Washington saw that solving the Arab–Israeli problem would help secure vital US interests, easing the challenges to Israel's survival and to vital energy supplies.[76]

The NSS 1991 also renewed the US pledge to protect energy resources using military means and ties to the oil industry:

> Security of oil supplies is enhanced by a supportive foreign policy and appropriate military capabilities. We will work to improve understanding among key participants in the oil industry of the basic fundamentals of the oil market. We will also maintain our capability to respond to requests to protect vital oil facilities, on land or at sea, while working to resolve the underlying political, social and economic tensions that could threaten the free flow of oil.[77]

Special attention was paid to the protection of the energy resources in the Persian Gulf region, since it 'contains two-thirds of the world's known oil reserves', and thus the region 'is of fundamental concern to us'. It was also recognized that events in the region would affect US interests, as 'political and military turbulence in the region has a direct impact on our economy, largely through higher oil prices and potential supply disruptions'.[78]

The 1991 Gulf War did not fundamentally change the US policy on the Middle East (as it was actually an application of the Carter Doctrine). The US policy in the Middle East remained committed to Arab–Israeli peace and the protection of oil resources, both of which became easier to advance after the end of the Cold War, the collapse of the Soviet Union and the defeat of Saddam Hussein.[79] However, there was no real follow-up on the Arab–Israeli front after the Madrid Conference which Bush organized in October 1991, as the conference did not result in an agreement between the Arabs and the Israelis on steps towards a peace deal.

In his frequent references to a New World Order, Bush seemed to be on the verge of setting a new doctrine for US policy. However, Bush did not establish a new doctrine, and a rare moment of opportunity passed without seizing it.[80] Bush's ambition of realizing a New World

Order failed to get off the ground, due to the lack of a clear enemy in the wake of the Cold War,[81] and the inward-looking American public opinion which was increasingly unwilling to interfere in other people's affairs.[82] Furthermore, Bush's vision for global coalition-building based on a New World Order was mainly focused on the events of the Persian Gulf rather than on the international system as a whole. The New World Order reflected the conduct of American foreign policy in the Gulf but did not necessarily reflect the general changes in international politics,[83] and Bush's skilful coalition-building was not achievable under different conditions in the future where the world's energy resources were not at an immediate or direct risk.[84]

Furthermore, Bush showed little interest in promoting the moral values enshrined in the New World Order such as spreading human rights, democracy, free market economies and other similar values. This was especially true in the Middle East, as Washington feared producing potentially unfriendly regimes (the symptom known as the Democracy Conundrum; the contradiction between American democracy and American support of non-democratic regimes, where the United States cannot avoid calling for democratic values abroad and yet it cannot allow its non-democratic allies to be undermined).[85] Thus, the New World Order failed to materialize, due to concerns about the security and price of Persian Gulf oil supplies.[86] So the concern for oil stability, and thus the stability of the Gulf's non-democratic regimes, was a reason for the failure of the New World Order to promote human rights and democratization in the Middle East.

Bush generally followed a policy of global economic openness, based on maintaining US preponderance in order to protect the Open Door empire. This was evident in his initial support of Saddam, especially in the field of energy where the political economy of the large corporation was seen in the use of corporate interests to promote America's energy security. The 1991 Gulf War was a war to protect energy supplies, and to maintain American dominance in a system of global openness. Echoing William Appleman Williams and Gabriel Kolko, Noam Chomsky said that the Bush administration needed the 1991 Gulf War to deal with domestic economic problems in the US, by 'selling protection' and acting 'like mercenary states' in order to protect the petrodollar recycling process necessary to 'prop up' the American economy which was suffering recession,[87] and to protect America's 'imperial settlement' based on control of oil resources, especially in the Middle East.[88] This was the situation which President Bill Clinton inherited, where Clinton followed in the footsteps of Open Door imperialism.

Clinton did raise the importance of economic issues in US national security,[89] and the slogan of his 1992 campaign was 'It's the economy, stupid!' During a speech in December 1991 at Georgetown University, Clinton called for the integration of US foreign policy and domestic economic concerns, to enable productive US economic competition and a stable international system:[90]

> We face two foreign policy challenges today. First, we must define a new national security policy that builds on freedom's victory in the Cold War ... And second, we must forge a new economic policy to serve ordinary Americans by launching a new era of global growth. We must tear down the wall in our thinking between domestic and foreign policy.[91]

Furthermore, 'in a throw back to the Open Door policy, Clinton declared that it was time "to make trade a priority element of American security"'.[92] As he took office in January 1993, Clinton saw that with the fall of the Soviet Union, the United States was able to expand global democracy while prioritizing economic affairs in US national security.[93] Clinton thus introduced the Engagement and Enlargement policy, which called for America to remain 'engaged' in world affairs while 'enlarging' and spreading market democracy worldwide, leading to the promotion of market capitalism and a policy of global economic openness and American trade that would sustain the US economy. The Clinton administration did set out clear objectives in the Engagement and Enlargement policy, linking democratization to security and economic objectives,[94] where the US would 'support the consolidation of market democracy where it is taking new root'. However, he added limitations, saying that 'I do not mean to announce some crusade to force our way of life and doing things on others'.[95] National Security

Advisor Anthony Lake agreed that it would not be a democratic crusade, or a one-size-fits-all policy, saying that 'this is not a democratic crusade; it is a pragmatic commitment to see freedom take hold where that will help us most'.[96]

Therefore, Engagement and Enlargement did not apply to the Middle East. As the administration made it clear that the democratization drive would apply only where it would benefit US interests, what took priority in the Middle East was not economic and political reform, but the search for an Arab–Israeli settlement[97] which would stabilize the region and its oil resources. In the words of former Assistant Secretary of State for Near Eastern Affairs Martin Indyk, 'the engine for change in the region was not democratization, but peace'.[98] Indyk added that the United States did not want to call for democratization in the Middle East lest it would destabilize the oil-producing regimes:

> The United States could not afford the destabilizing impact that pressure for reform would generate in deeply traditional and repressed societies. Pushing hard for political change might not only disrupt the effort to promote peace but could also work against vital US interests: stability in the oil-rich Persian Gulf and in strategically critical Egypt.[99]

Thus, instead of democratization, Arab–Israeli peace was seen as a realistic option that would ensure unimpeded access to the oilfields of the Persian Gulf,[100] again reflecting the Democracy Conundrum. And since Washington saw that the threat to this peace came from Iraq and Iran, Clinton tried to link the policy on both countries to a new approach on the Persian Gulf. On 18 May 1993, Indyk spoke at the Washington Institute of Near East Policy and introduced the policy of Dual Containment on Iraq and Iran, where instead of supporting one nation against the other, the Clinton administration would seek to rein in both countries, which were both viewed as 're-arming ... fervently anti-Western ... potential threats to Persian Gulf oil supplies'.[101] Lake said that the Clinton administration saw that Iraq and Iran could disrupt regional and global stability, thus the US needed to contain them,[102] and according to Indyk, 'our strategy for the whole region was based on pursuing a comprehensive Middle East peace at the same time as containing Iraq and Iran'.[103]

However, the Clinton administration would not yet support regime change in either country[104] because, first, the US was still committed to the territorial integrity of Iraq,[105] and, second, regime change in Iraq or Iran would require an armed conflict that had little or no backing from the US Congress, the American people or the White House itself. Dual Containment would not provide a long-term solution, but it would keep Saddam 'in his box' for the rest of the decade, even though the policy failed to obtain support from America's allies.[106] (Furthermore, Dual Containment also aimed to protect Saudi oil from competition from Iranian and Iraqi oil and keep supplies at a level that would maintain moderate oil prices.)[107]

Beyond Dual Containment, US relations with Saudi Arabia, the world's largest oil producer, deteriorated under Clinton. The tension started with problems over Saudi payments for weapons imported from the US.[108] But the main point of dispute was that the United States no longer coordinated oil policy with Saudi Arabia, partly because of the attacks on US bases in November 1995 (the Saudi Arabian National Guard, SANG) and June 1996 (Khobar),[109] and partly because of Saddam's propaganda war against the United States, which prompted the Saudis to withdraw some basing rights and over-flight privileges.[110] US–Saudi relations were strained further after the transfer of power from the ailing King Fahd to his half-brother, Crown Prince Abdullah, who was not as pro-American.[111] For example, under King Fahd, Saudi Arabia sold oil at a discounted rate to the United States in order to maintain a strategic relationship, a concession that was stopped under Abdullah.[112] Furthermore, when George H.W. Bush left the White House in 1993, the bulk of American imports of oil were from the Gulf. Under Clinton, however, the United States decreased its dependence on Persian Gulf oil and moved to getting the bulk of its oil from Venezuela, Mexico and Canada (although US dependence on Persian Gulf oil would rise again early in the twenty-first century.)[113]

The preference for stabilization over democratization in the Middle East shows that the Middle East did not fit into the policies of engagement and democratic enlargement, as the United States would only pursue Engagement and Enlargement where possible and where desirable. But this Middle East policy started to crack, as it became clear in the mid 1990s that Dual Containment was coming under strain because it lacked international support as French, Russian, Chinese and other firms were still dealing with the Iraqi regime.[114] Thus, by November 1998, the Clinton administration replaced its policy of Dual Containment towards Iraq and Iran with, in the case of Iraq, a policy of 'containment combined with regime change' and, in the case of Iran, a policy of 'containment until the Iranians are ready for engagement'. According to Indyk, the reason for this change in policy was that it had become more difficult to contain Iraq, while, in Iran, the election of the moderate Mohammad Khatami as president gave new hope for a change in Iranian policy.[115]

The defining moment in the Clinton administration's policy on regime change in Iraq came with the authorization of the Iraqi Liberation Act on 31 October 1998. Clinton officially stated:

> Today I am signing into law H.R. 4655, the 'Iraq Liberation Act of 1998.' This Act makes clear that it is the sense of the Congress that the United States should support those elements of the Iraqi opposition that advocate a very different future for Iraq than the bitter reality of internal repression and external aggression that the current regime in Baghdad now offers ... The United States wants Iraq to rejoin the family of nations as a freedom-loving and law-abiding member. This is in our interest and that of our allies within the region.[116]

Regarding Iran, 'Dual Containment' never meant that the policy was the same for Iran and Iraq. Iran posed a different challenge for the Clinton administration. In contrast to domestic pressure to work on regime change in Baghdad, the policy on Iran did not aim for regime change. Instead, the Dual Containment policy towards Iran focused more on economic sanctions,[117] especially in the energy field. This was evident in US policy on pipelines in the region. In January 1994, Iran, Turkmenistan and Turkey signed an agreement to construct a pipeline from Turkmenistan to Turkey passing through Iran, but US opposition delayed the project by blocking its finance.[118] However, in July 1997, amidst growing scepticism in Washington that Pakistan and the Taliban would actually help unite Afghanistan (as the Clinton administration hoped),[119] the United States announced that the pipeline would not violate the American embargo on Iran, but, in an apparent U-turn, subsequent American opposition dashed hopes for early implementation of the pipeline project.[120]

Despite the US sanctions imposed on Iran in 1980, Iran had continued to export oil to the United States, with sales peaking at $1.6 billion worth in 1987, until the 'Tanker War' in that year (an American military operation which involved the re-flagging of Kuwaiti oil tankers to protect them from Iranian attacks) ended the import of Iranian oil.[121] With the measures taken by the George H.W. Bush administration to scale back the sanctions on Iran, the United States became the largest single buyer of Iranian oil, through the overseas subsidiaries of American companies, and became Iran's third-largest trading partner.[122] In 1995, however, Clinton issued an executive order to ban all trade with Iran, including oil deals involving the Iranian energy sector, citing Hezbollah's attacks on Israel and on Israeli and Jewish interests in Argentina.[123] Executive Order 12957 prohibited 'the entry into or performance by a United States person, or the approval by a United States person of the entry into or performance by an entity owned or controlled by a United States person' of a contract that would include the supervision, management or financing of 'the development of petroleum resources located in Iran' or 'a guarantee of another person's performance under such a contract'.[124] Among the deals suspended was a billion-dollar agreement between Iran and Conoco.[125]

Clinton's order was complemented by the presentation of the Iran and Libya Sanctions Act of 1996 (H.R. 3107), which aimed to 'punish both states for supporting terrorism and aiming to acquire weapons of mass destruction',[126] including sanctions on the Iranian petroleum sector:

a) **Policy With Respect to Iran**: The Congress declares that it is the policy of the United States to deny Iran the ability to support acts of international terrorism and to fund the development and acquisition of weapons of mass destruction and the means to deliver them by limiting the development of Iran's ability to explore for, extract, refine, or transport by pipeline petroleum resources of Iran.

(b) **Policy With Respect to Libya**: The Congress further declares that it is the policy of the United States to seek full compliance by Libya with its obligations under Resolutions 731, 748, and 883 of the Security Council of the United Nations, including ending all support for acts of international terrorism and efforts to develop or acquire weapons of mass destruction.[127]

The Iran and Libya Sanctions Act (ILSA) imposed severe penalties on non-US firms that had invested more than $40 million in Iran's oil industry.[128] Nevertheless, the Act had a loophole, since it did not ban foreign subsidiaries of US corporations from working with Iran, as long as the subsidiaries did not employ US citizens and were not a front for the parent company. For example, the US Department of Justice, a federal grand jury in Texas, and the Securities and Exchange Commission subsequently launched formal investigations to determine if Halliburton's use of subsidiaries in Iran was in conflict with the Act. Halliburton responded in early 2005, saying that it would not seek new contracts in Iran, but would maintain the 'existing contracts and commitments which the subsidiaries have previously undertaken' there.[129] In February 2007, Hallilburton issued a statement which said that it 'intend[ed] to wind up its work in Iran and not enter into any other future contracts', and that it would 'exit upon the completion of existing commitments'. On 9 April 2007, Halliburton issued a brief press release which said that it was 'no longer working in Iran'.[130]

The significance of Iran is that it is not just situated on the Persian Gulf; it is also situated on the Caspian Sea, which was a key area for Clinton's policy for regional diversification of energy resources away from the Middle East. The Middle East, despite its importance, was not the only region which provided Washington with foreign oil. Washington sought to geographically diversify its foreign sources of oil, in order to decrease risks of political or economic upheaval which may lead to a reduction of oil supplies or even a complete cutoff. The regions which attracted Washington's attention for diversification away from the Middle East included Central Asia, Africa and the Western Hemisphere.

Regional diversification of US foreign energy sources

In a world of growing energy demand, our nation cannot afford to rely on any single region for our energy supplies.

William J. Clinton, August 1997.[131]

I cannot think of a time when we have had a region emerge as suddenly to become as strategically significant as the Caspian.

Richard Cheney, CEO, Halliburton, 1998.[132]

The long-term stability of the Gulf region is elusive, due to the Iranian threat, militant Islamist attacks on US interests, instability caused by declining living standards, disputes within the ruling families of the Gulf States and border disputes among these states.[133] Therefore, the United States sought to diversify its foreign energy sources, in order to decrease the risks of supply disruptions. The United States saw the oil-rich Caspian region as an alternative to the Gulf.[134] In describing the current geopolitics of the Caspian and Central Asian region, it became customary to refer to the 'Great Game' – the nineteenth-century struggle between Great Britain and Tsarist Russia for control over Central Asia.[135] In Central Asia today, 'big powers such as Russia, China, the United States, Iran, Pakistan, Afghanistan, Turkey, the Central Asian states themselves and the most powerful players of all, the oil companies, compete in … the New Great Game'[136] over the energy resources of the region. However, the Caspian region suffered from the same problems as the Persian Gulf, and US involvement in the Caspian region would exacerbate problems with Moscow, as the United States and Russia compete over the region's oil resources, pipeline routes and proxies in the 'New Great Game'.[137]

The policy of regional diversification was not clearly defined in a formal policy document, but Energy Secretary Bill Richardson spoke in 1998 about 'America's energy security which depends on diversifying our sources of oil and gas worldwide', adding that regional diversification was 'also about preventing strategic inroads by those who don't share our values'.[138] Thus, diversification of energy resources became a matter of national security, especially considering that America's oil imports increased by more than 50% during the Clinton era[139] (so that the United States imported more than 50% of its oil in 1998).[140]

US policy makers started to pay more attention to the Caspian region after oil contracts were signed between US oil firms and Kazakhstan and Azerbaijan in 1993–1994,[141] in a show of the public–private partnership between the US Government and the US oil corporations. According to Fiona Hill:

> It was not until major oil contracts were signed between US oil companies and the governments of Kazakhstan and Azerbaijan in 1993–1994 that the region really began to register on the radar screens of the American public. The commercial interests of US oil companies in exploiting new energy reserves gave US policymakers a specific interest to protect in the Caucasus and Central Asia. The US has come to see Caspian resources as one of the few prospects for diversifying world energy supply away from the Middle East.[142]

In an April 1997 report to Congress, the State Department explained the strategic importance of the Caspian region, and US policy began to 'promote rapid deployment of Caspian energy resources' to 'reinforce Western energy security'.[143] Clinton was a vigorous advocate of American companies which sought drilling rights in the Caspian basin, and he explicitly designated the area's energy resources a matter of national security. He said to President Heydar Aliyev of Azerbaijan in the White House in August 1997: 'In a world of growing energy demand, our nation cannot afford to rely on any single region for our energy supplies.' By helping Azerbaijan to develop its untapped oil reserves, 'we not only help Azerbaijan to prosper, we also help diversify our energy supplies and strengthen our nation's security'.

Clinton reiterated this view in conversations with other presidents from the region, including Nursultan Nazarbayev of Kazakhstan and Saparmurad Niyazov of Turkmenistan.[144] Citing the US–Russian competition over the region's energy resources and routes, Sheila Heslin of the US National Security Council told a Senate committee on 17 September 1997 that the goal of American policy in the Caspian is 'in essence to break Russia's monopoly of control over the transportation of oil from the region'.[145]

Clinton also ordered the establishment of a Caspian task force headed by Deputy Secretary of State Strobe Talbott, and composed of officials from the State, Energy and Commerce Departments, the National Security Council, and the CIA. On 21 July 1997, Talbott announced that the United States wanted economic and political reform to fight terrorism and to make the region a valuable trade route and a stable energy supplier:[146]

> If economic and political reform in the countries of the Caucasus and Central Asia does not succeed – if internal and cross-border conflicts simmer and flare – the region could become a breeding ground of terrorism, a hotbed of religious and political extremism, and a battleground for outright war. It would matter profoundly to the United States if that were to happen in an area that sits on as much as 200 billion barrels of oil. That is yet another reason why conflict resolution must be job one for US policy in the region: It is both the prerequisite for and an accompaniment to energy development.[147]

America's support to the states of the region would 'have four dimensions: the promotion of democracy, the creation of free market economies, the sponsorship of peace and cooperation within and among the countries of the region, and their integration with the larger international community'.[148]

American interest in the region was seen in visits by First Lady Hillary Clinton to the region, and joint military exercises (codenamed CENTRAZBAT) in September 1997 between the United States and Central Asian states.[149] In May 1998, during the 'Crossroads of the

World' conference in Istanbul, US Energy Secretary Federico Peña announced the Caspian Sea Initiative

> [bringing] together, for the first time, the heads of the US Government's three independent trade and investment agencies – EXIM [Export-Import Bank], OPIC [Overseas Private Investment Corporation] and TDA [Trade and Development Agency] – to coordinate the development and support of project opportunities in the Caspian.[150]

Peña said that the Initiative emphasized the role of US private investment in promoting Caspian energy resources and transport routes:

> Through these agencies, the United States will dedicate and coordinate the diplomatic and financial resources needed to promote the rapid and effective development of Caspian energy resources and multiple export routes. This focus will help to mobilize and leverage private sector capital from the United States and other countries, as well as government participation from the countries in the region ... This high-level policy attention will help ensure that American commercial interests and diplomatic interests work together in the Caspian region.[151]

In March 1999, Ambassador Richard L. Morningstar, Special Advisor to the President and Secretary of State for Caspian Basin Energy Diplomacy, testified before the Senate Subcommittee on International Economic Policy, Exports and Trade Promotion that, 'in general', the American 'Caspian energy policy [addressed] four strategic objectives in the Caspian region': strengthening the independence, sovereignty and prosperity of the new Caspian states and encouraging political and economic reform; mitigating regional conflicts by building economic linkages between the new states of region; bolstering the energy security of the United States and its allies and the energy independence of the Caspian region by ensuring the free flow of oil and gas to the world marketplace, and enhancing commercial opportunities for US companies, where the US Government's role was to facilitate discussions between US firms and Central Asian governments.[152]

With this linkage of geopolitics, competition with Moscow, and the need to secure Caspian energy supplies and routes, the US Government became more involved in the deals which the US oil companies signed with the region's countries, such as the 'Contract of the Century'; a Production Sharing Agreement (PSA) signed in September 1994 between Azerbaijan and a consortium of eight oil companies, including three US oil multinationals: Amoco, Pennzoil and Unocal.[153]

Another major project in the region was the Baku–Tbilisi–Ceyhan (BTC) pipeline to carry Caspian oil and gas, passing through the Azeri capital, Baku, then the Georgian capital Tbilisi, to the Turkish port of Ceyhan, avoiding Russian- and Iranian-controlled territory. Energy Secretary Bill Richardson said in 1998 that the BTC pipeline was vital for American national security: 'This is about America's energy security, which depends on diversifying our sources of oil and gas ... It is also about preventing strategic inroads by those who don't share our values.'[154] He added in 1999 that 'this is not just another oil and gas deal ... It is a strategic framework that advances America's national security interests'.[155]

The BTC pipeline was not the only pipeline planned by US oil corporations and supported by the US Government. Argentina's Bridas and USA's Unocal planned to build two pipelines from Turkmenistan, through Afghanistan, to Pakistan (the Turkmenistan–Afghanistan–Pakistan Pipeline, or TAP Line). Pakistan's Prime Minister Benazir Bhutto supported this idea. On 21 October 1995, Turkmen President Saparmurad Niyazov signed an agreement with Unocal in New York, where Henry Kissinger was present, acting as an advisor to Unocal, and the US Government supported the deal.[156] The pipeline project had an impact on Washington's policy towards the Taliban, who were in control of Kabul.

According to Rashid, the strategy over pipelines was 'the driving force behind Washington's interest in the Taliban', and US policy on the Taliban went through several phases 'attempting quick-fix solutions rather than a strategic policy'. Initially, from 1994 to 1998, the United States supported the Taliban, believing that they had anti-Iranian and pro-US business inclinations which would help the Unocal pipeline. However, the US still 'had no strategic plan towards accessing Central Asian energy and thought that pipelines could be built without resolutions to regional civil wars',[157] especially the civil war in Afghanistan.

Furthermore, Zalmay Khalilzad, as a senior strategist at the RAND Corporation and Unocal official, expressed his support for the Taliban in October 1996:

> Based on recent conversations with Afghans, including various Taliban factions, and Pakistanis, I am confident that they would welcome American reengagement. The Taliban does not practice the anti-US style of fundamentalism practiced by Iran – it is closer to the Saudi model.[158]

Similarly, Robin Raphel, Assistant Secretary of State for Southeast Asian Affairs, told the UN that the Taliban provided 'a measure of peace and security'.[159] She was a supporter of the TAP pipeline, saying at a press conference in Pakistan in April 1996 that 'we have an American company which is interested in building a pipeline from Turkmenistan through to Pakistan [via Afghanistan] ... The pipeline project will be very good for Turkmenistan, for Pakistan, and for Afghanistan as it will not only offer job opportunities but also energy for Afghanistan.'[160]

Moreover, John J. Maresca, Vice President of International Relations, Unocal Corporation, testified to the Committee on International Relations of the House of Representatives in February 1998 that Afghanistan was the most practical and economic path for the pipeline between Turkmenistan and Pakistan (the TAP pipeline), but since the country was not stable, the United States had to work on stabilizing it. His testimony focused on three issues: first, the need for multiple pipeline routes for Central Asian oil and gas resources; second, the need for US support for international and regional efforts to achieve balanced and lasting political settlements to the conflicts in the region, including Afghanistan; third, the need for structured assistance to encourage economic reforms and the development of appropriate investment climates in the region.[161]

Thus, said Steve Coll, author of *Ghost Wars*, 'American policy had been captured by the language of corporate deal making' as the State

Department took up Unocal's agenda as its own, and 'American tolerance of the Taliban was ... linked to the financial goals of an oil corporation'.[162]

Both the Taliban and the Northern Alliance (the loose group of Afghan warlords who were fighting the Taliban) were interested in the TAP pipeline project. The Taliban were offered $250 million/year in transit fees, and both groups were told that the pipeline would give them money, jobs and gas, but it would not be feasible as long as the civil war went on. Also, as long as the Taliban were not recognized by the West as a legitimate government, Unocal was unable to secure loans from international financial institutions. But the Northern Alliance was not ready to make peace with the Taliban just for the pipeline.[163]

However, after the attacks on the American embassies in Kenya and Tanzania, and the American bombing of Afghanistan (and Sudan) in August 1998, the US–Taliban negotiations did stop, but for only six months,[164] as the United States still needed a strong government in Afghanistan (which could only be provided by the Taliban) to preserve stability for US energy investment. Thus, even after the August 1998 bombings, negotiations between the United States and the Taliban continued.[165] Senior American officials were dispatched to meet senior Taliban officials in Islamabad in February 1999. But the Americans did pressurize the UN every now and then to impose sanctions on the Taliban, to try to force the Taliban into giving up Osama bin Laden.[166] In July 1999, Clinton cut off commercial ties with the Taliban and froze all of their US assets to press them to hand in Osama bin Laden,[167] but secret meetings with Taliban officials continued nevertheless.[168]

To facilitate the passage of oil and gas pipelines from the Caspian region to the West, and to secure America's oil interests in Eurasia, it was essential to ensure US influence in the Balkan region. During Operation Allied force, the US-led NATO war on Serbia in 1999, critics of NATO argued that the United States was in fact seeking to secure a passage for oil and gas from the Caspian region. In 1997, BP and Halliburton proposed the Albanian–Macedonian–Bulgarian Oil (AMBO) pipeline that would transport oil from Burgas in Bulgaria, through Skopje in Macedonia to Vlorë, a port in Albania on the Adriatic Sea, where supertankers would load the oil from Vlorë and ship it to the

West. After the NATO bombing campaign ended, the US built Camp Bondsteel in southern Kosovo, the largest American foreign military base constructed since Vietnam. Camp Bondsteel was built by Brown & Root, a division of Halliburton, and is located close to vital oil pipelines and energy corridors, such as the AMBO pipeline.[169] In a demonstration of what may be dubbed the 'military–petroleum complex,' Brown & Root performed the initial feasibility study of the AMBO pipeline,[170] where the link between AMBO and Camp Bondsteel was a part of what Chalmers Johnson described as an American 'empire of military bases', where the United States is building military bases in the oil-rich region of southern Eurasia, including the Middle East and Central Asia, in order to control oil supplies going to other countries.[171] Of course, the war on Serbia had other reasons which had to do with preserving Europe's security and stability, and maintaining NATO's prestige and leadership, but Operation Allied Force was also necessary to preserve US primacy in Europe, and to preserve openness and economic integration.[172]

Despite all of these efforts, however, the Clinton administration failed to establish an alternative source of energy in the Caspian region. A major reason for this failure was the over-optimistic estimations of Caspian energy resources, leading to unrealistic expectations that dependence on Middle East oil could decrease significantly. An Energy Information Administration report in 1998 described Central Asia as a 'new Middle East'[173] with over-optimistic estimates forecasting reserves as large as Kuwait's or even the Persian Gulf's.[174]

Another reason for the failure to provide an alternative to the Middle East in the Caspian was Washington's lack of a coherent strategy towards the region, which led to the rise of anti-American Islamic militancy. As Ahmed Rashid noted, Washington was never serious about developing a new strategic framework or vision for Central Asia after the end of the Cold War. The United States dealt with issues as they came up in a haphazard, piecemeal fashion, pursuing constantly changing single-issue agendas that were driven more by American politics than the goal of ending the civil war in Afghanistan.[175] Rashid added that:

> Had the United States been serious about its strategic vision
> for Central Asia, policymakers should not only have talked

about conflict resolution; they should have insisted that it be the number one priority. In particular, the United States should have lent political muscle to the UN attempts to end the civil war in Afghanistan, which was posing the key threat to Central Asia.[176]

Fiona Hill added in February 2001 that 'United States policy towards the Caucasus and Central Asia over the last ten years has been marked by a distinct lack of direction. The US stumbled into the region.' Like Rashid, she believed that decisions on the region were 'ad hoc', saying that 'all of these policy decisions [on pipelines in the region] were ad hoc responses to increasing US engagement in Caspian oil development. They were not part of a grander strategy for the Caucasus and Central Asian states.'[177]

Like the Middle East, Central Asia would not fit in Clinton's Engagement and Enlargement, where Washington continued to support non-democratic regimes as long as they provided a stable supply of oil and gas for Western economies. In this sense, there was no difference between US policy in the Middle East and that in the Caspian.

The Caspian was not the only region which Washington saw as an alternative oil source to the Middle East. There were other regions, including the Western Hemisphere and Africa. With the Free Trade Agreement (FTA) of 1989 and the North American Free Trade Agreement (NAFTA) of 1994, the United States ensured the integration of Canada in a continental energy market.[178] Things were not as successful with Mexico, however, as the United States failed to pressurize Mexico to open its energy sector for US investment during NAFTA negotiations. The failure with Mexico was followed by the failure with other potential hemispheric partners such as Venezuela, and by 2001, the hemispheric solution had not been successful in Mexico or in Hugo Chavez's Venezuela.[179]

As for Africa, America's other option for regional diversification, there was no unified agreement that would link countries in the same way that NAFTA had tied the US to Canada and Venezuela,[180] although the US Government did turn its attention to the continent and its energy resources.[181]

'We have substantially changed the way the US government is structured to deal with Africa' said the Assistant Secretary of State for African Affairs in 1999. 'There was a time not long ago when Africa was the exclusive domain of one understaffed bureau at Foggy Bottom', but now, he said,

> virtually every government agency is building the capacity to implement new programs that support our policy of comprehensive engagement with Africa ... We have important strategic interests in Africa ... Africa is the source of over 16% of our nation's imported oil ... Within the next decade, oil imports from Africa are expected to surpass those from the Persian Gulf.[182]

In April 1999, Secretary of Energy Bill Richardson visited Africa to discuss Western energy investments,[183] and during Clinton's second visit to Africa in August 2000 there was a major focus on energy concerns. In Nigeria, Clinton appealed to the Nigerian leaders to increase their production of oil in order to alleviate the worldwide shortage of petroleum.[184]

Conclusion: The US Open Door empire continues to depend on oil, but democratization fails and an ad hoc policy continues

It may not be surprising that the Iraqi invasion of Kuwait, 'the first post-Cold War crisis turned out to be a geopolitical oil crisis'.[185] The US reaction to the crisis was meant to secure access to the region's oil, and to maintain US prestige in order to protect the global system of Open Door imperialism.

George H.W. Bush and William J. Clinton worked on sustaining America's Open Door empire based on the political economy of the large corporation, and the control of energy resources using a public–private partnership between the US Government and the US oil corporations, the role of which was vital in ensuring US energy security and maintaining the international oil system on which the US empire is based.

The Middle East and the Caspian region did not fit in the democratizing efforts of the New World Order, nor in Clinton's Engagement and Enlargement, as the Democracy Conundrum hindered a real effort to reshape the Middle East, for fear of any instability which might threaten the pro-US regimes and the Middle East's energy resources. Instead, Bush and Clinton preferred to support undemocratic regimes like the Gulf States, the Caspian states and the Taliban, while limiting themselves to a case-by-case approach to deal with specific issues. These specific, case-by-case issues included securing energy resources, the Arab–Israeli conflict (which did not progress after Oslo), containing Iraq and Iran (where Dual Containment was failing due to the lack of European support), and the rise of al Qaeda, as seen in the bombings in New York (1993), Saudi Arabia (1995 and 1996), Kenya and Tanzania (1998) and Yemen (2000). However, Bush and Clinton failed to come up with a comprehensive policy to link the securing of energy resources to democratization and Arab–Israeli peace.

As the United States attempted to diversify America's foreign energy resources away from the troublesome Middle East, it faced problems in the alternative regions of the Caspian, the Western Hemisphere and Africa. However, the US did not have a real comprehensive policy on these regions, and would thus fail to make them suitable enough as alternatives to the Middle East. Clinton bequeathed these problems to George W. Bush, a new president who had foreign energy procurement high on his agenda since his early days in office. This would be evident in the formation of the Energy Task Force only a few days after he took office, and the subsequent National Energy Policy (NEP) of May 2001.

CHAPTER 2

THE UNILATERAL OILMAN

Energy Procurement in
George W. Bush's Foreign Policy
(January–September 2001)

In his memoirs, Bill Clinton wrote about a visit from President-elect George W. Bush to the White House on 19 December 2000. According to Clinton, Bush's foreign policy team 'believed that the biggest security issues were the need for national missile defence and [action against] Iraq'. In reaction, Clinton wrote:

> I told him [Bush] that based on the last eight years, I thought his biggest security problems, in order, would be Osama bin Laden; the absence of peace in the Middle East; the standoff between nuclear powers India and Pakistan and the ties of Pakistanis to the Taliban and al Qaeda; North Korea; and then Iraq. I said that my biggest disappointment was not getting bin Laden, that we still might achieve an agreement in the Middle East, and that we had almost reached a deal with North Korea to end its missile program, but that he would probably have to go there to close the deal.
>
> He listened to what I had to say without much comment, then changed the subject to how I did the job.[1]

Beyond Iraq and the NMD program, it is more accurate to say that the Bush administration had two 'strategic priorities', or two foreign policy strands, upon taking office in 2001: the modernization of US military capabilities, and the procurement of additional oil from foreign sources.[2] Anti-terrorism was added as a third major strand after the 11 September 2001 attacks, and there was also a fourth strand after 11 September: the global power projection envisaged by the QDR of September 2001. Bush's focus on Iraq aimed to serve these foreign-policy goals.

The United States had not had an energy policy since the days of Jimmy Carter, during the years of soaring oil prices. There were two main reasons why the Bush administration wanted a return to an energy policy: the increased US dependence on foreign oil (as US oil imports exceeded the psychological 50% threshold), and the increased clout of the US energy industry in the Bush administration.[3] No other candidate in federal office received as much money from the oil industry as Bush did in 2000[4] (making Bush, according to Craig Unger, 'the candidate of Big Oil' and 'the ultimate insider'[5]). From the energy and natural resources sector as a whole, Bush received almost $3 million, of which $1.9 million came from the oil and gas sector, which was 13 times as much as the money received from the oil and gas industry by the Gore campaign.[6] (From 1998 to 2003, Halliburton's contributions to the Republican Party totalled $1,146,248, and $55,650 went to the Democratic Party.[7] The coal industry contributed $700,000 to the Bush campaign.[8])

Bush's links to the energy lobby were evident early in his administration, as nine days before he was inaugurated energy lobbyists gathered at the American Petroleum Institute (API) offices on K Street in Washington and drew up a wishlist: looser rules for drilling on federal lands, more drilling for oil and gas in Alaska and the Gulf of Mexico, lower royalty payments for tapping offshore wells, among other suggestions. These and many other recommendations made it into the Bush energy plan of May 2001.[9] Also, key leaders from the API were appointed to pivotal positions in Bush's administration. 'Not since the rise of the railroads more than a century ago has an industry placed so many foot soldiers at the top of the new administration' wrote *Newsweek*.[10] Industry leaders, who donated $22.5 million to the Republicans in the 2000 elections, enjoyed constant contact with the Energy Task Force,[11] a committee headed by Vice President Dick Cheney to come up with an energy plan for the Bush administration. Controversially, no one enjoyed better access than Enron CEO Ken Lay, who was a good friend of Bush.[12] Moreover, as soon as he was in office as Vice President, Cheney began a series of meetings with the chief executives of major oil companies who had supported the Republicans in 2000, showing the clout of the US energy corporations in the working of the Energy Task Force.[13] Furthermore, Energy Secretary Spencer Abraham hosted 109 representatives of energy firms between January and May 2001, including ChevronTexaco, ExxonMobil and Enron, who were lavish contributors to the Bush campaign in 2000.[14] Cheney defended these meetings, saying that 'just because somebody makes a campaign contribution doesn't mean that they should be denied the opportunity to express their view to government officials'.[15] Nevertheless, the influence of business interests on the Bush administration's energy plans, devised by a task force headed by Cheney, would be evident.

These ties to the oil industry date back to the earlier careers of Bush and Cheney before they took office. Bush, who aspired to be an oilman like his father, established his first energy company, Arbusto (Spanish for 'Bush') in Texas in 1977. However, the company was struggling, as it drilled one dry hole after another. In 1984, in need of more financing, Bush sold Arbusto to another oil company, Spectrum 7, where he was given the job of CEO with a salary of $75,000 and 16.3% of the company's stocks. To many, the Spectrum–Arbusto transaction was seen as a bail-out of the younger Bush by his father's friends. But Spectrum 7, too, was suffering as global oil prices were as low as $10 in the mid-1980s, and oil companies were suffering from layoffs and losses. In 1986, Harken, an oil-and-gas company based in Texas and partially owned by George Soros and Harvard University, bought Bush's failing company for $2.25 million in stock. Bush got roughly $600,000 out of the deal, a seat on the board, and a consultancy paying between $50,000 and $120,000 annually. Harken's motivation to bail Bush out was its eagerness to develop ties with the son of the US vice president. But Harken, too, was suffering from dry wells and financial problems. However, governments of the Gulf States, eager to foster their ties with the son of the US president, started to invest in Harken and grant it drilling rights.[16]

Similarly, Dick Cheney had a background in the oil sector. As Secretary of Defence in 1990, he argued to the elder President Bush that Saddam's invasion was a far more important issue than Kuwait, as it was about oil.[17] On leaving office in 1993, Cheney became a senior fellow at the American Enterprise Institute (AEI), a conservative Washington-based think tank.[18] In 1994, thanks to the global relations which he developed as Secretary of Defence, he was appointed by Kazakh President Nursultan Nazarbayev to the 12-member Kazakhstan Oil Advisory Board, where Cheney helped broker the deal which set up the Caspian Pipeline Consortium (CPC), a 1,580 km pipeline to transport oil from Kazakhstan's Tengiz oilfield (operated by Chevron) to the Russian Black Sea port of Novorossiysk, and Halliburton built the refineries on the Tengiz oilfield.[19] Cheney also became one of the seven-member Honorary Council of Advisors for the Azerbaijan–US Chamber of Commerce, which focused on promoting the interests of US oil corporations in Azerbaijan.[20] He then took the job of CEO at Halliburton from October 1995 to August 2000.[21] Halliburton appointed him CEO because, during his days as Secretary of Defence, he developed good relations with the Saudis and considerable leverage in Washington which Halliburton could use.[22] According to the *Washington Post*: 'soon he was on first name basis with oil ministers all over the world, building on the ties he had developed in the Middle East during the Pentagon days.' According to David Lesar, who became Halliburton's CEO after Cheney left, '[Cheney] never pitched a particular contract or even closed a piece of business. He [just] opened the door.' Between 1994 and 2001, Halliburton's revenues increased by 127%.[23]

As he took his vice-presidential post in 2001, Cheney asserted that there would be no conflict of interest on his part because, as he said, 'since I left Halliburton to become George Bush's vice president, I've severed all my ties with the company, gotten rid of all my financial assets'. Indeed, two days before the inauguration, Dick Cheney and his wife gave away an estimated $8 million in stock options from Halliburton and six other companies. (Although he neglected to say that he was still due approximately $500,000 in deferred compensation from Halliburton and could potentially profit from his $433,333 shares of unexercised Halliburton stock options. A Congressional Research Service report said that unexercised options in a private corporation, as well as deferred salary received from a private corporation, were 'retained ties' or 'linkages' and should be reported as 'financial interest'.)[24] Nevertheless, Halliburton and its subsidiaries would benefit greatly during the George W. Bush administration from non-bidding contracts following the invasion of Iraq, as Cheney's role was vital in providing for the no-bid contracts awarded to KBR, a subsidiary of Halliburton, in Iraq.[25] Cheney's role was also vital in devising the administration's global energy policy, as he headed the Energy Task Force responsible for devising this policy, resulting in a global energy policy which reflected his worldview,[26] his belief in the vital role of energy in US national security, and his belief that the US Government should ease the restrictions on the work of the energy sector.[27]

Bush's National Security Advisor, Condoleezza Rice, also had ties to the oil industry. Rice finished her PhD in Soviet studies from the University of Denver, and received a fellowship at Stanford University in 1981. Brent Scowcroft (former National Security Advisor under Gerald Ford) was attending a conference at Stanford in 1985 when he first met Rice, and was impressed by her knowledge and personality. When Scowcroft was appointed George H.W. Bush's National Security Advisor in 1989, he appointed Rice as his Soviet expert. This put her in direct contact with Bush, who was heavily dependent on her knowledge of Soviet affairs. However, she was on the wrong side of policy when she called for supporting the Soviet leader Mikhail Gorbachev in his competition against Boris Yeltsin, and Yeltsin won. In March 1991 she left Washington and returned to pursue her academic career at Stanford, fearing that she would lose her tenure if she did not return. There she met George Shultz (former Secretary of State under Ronald Reagan), who was a member of Stanford's board of advisors, a board member at Bechtel and a board member at Chevron. He, too, was impressed by Rice's knowledge and personality. He introduced her to Chevron, which was working on oil contracts in the former Soviet Union, especially the Tengiz oilfield in Kazakhstan. Chevron used Rice's knowledge of the former Soviet Union to help manage the deal. He was impressed by Rice's work, and appointed her a board member. In accordance with the long-standing practice at Chevron to name oil tankers after board members, Chevron named a 129,000 ton supertanker the *SS Condoleezza Rice*. In addition to Chevron, she also joined the boards of the San Francisco insurance company Transamerica in 1991 and Hewlett-Packard in 1992. In 1994 she joined J.P. Morgan in a paid advisory position and became a board member of the Charles Schwab Corporation in 1999. When Rice became National Security Advisor under George W. Bush, she had annual Chevron board fees of $60,000 and over $250,000 in stock, in addition to a Stanford faculty salary of about $125,000. Rice resigned from Chevron's board six days before becoming National Security Advisor, and in the spring of 2001, Chevron renamed the *SS Condoleezza Rice* the *Altair Voyager*, in the face of criticism of Bush's ties to the oil industry.[28]

As President (and Vice President), the elder Bush perfected 'access capitalism' and the so-called 'revolving door' in Washington, where individuals move between the government sector and the business sector, especially the oil and defence industries.[29] The younger Bush would take business–government relations to a new level, as never before had the highest level of an administration so nakedly represented the oil industry,[30] leading to an extraordinary confluence of power in the public and private sector.[31]

The unilateral oilman: A general introduction to Bush's foreign policy agenda

Beyond the foreign-policy goal of energy procurement, Bush had an existing set of goals even before the 11 September attacks. The Bush administration believed that the advancement of its military capabilities, namely through the introduction and development of the NMD system, would give it protection against rogue states that possessed weapons of mass destruction, and, presumably, preserve the American freedom of military action in a conflict with any such state by backing up an American first strike against the state concerned. Iraq was the rogue state which attracted most attention during the early days of the Bush administration. Iraq could be regarded as a point of intersection between the two strands of energy procurement and military advancement. That was evident in Bush's desire to install the NMD system to face threats from rogue states, including Iraq, and evident in Secretary of Defence Donald Rumsfeld's recommendation to pursue a military action against Iraq that would demonstrate the American intention to face the threats against US interests, and, at the same time, secure Iraq's oil resources for foreign companies which looked forward to invest in it. Because of this intersection of foreign policy goals, Iraq became Bush's number one concern in the Middle East, instead of the Arab–Israeli peace process.

Despite having the goals of energy procurement and military advancement, the Bush team did not produce any clear policy documents or statements of 'grand strategy' or 'national security strategy' (such as George H.W. Bush's New World Order or Clinton's Engagement and Enlargement) before 11 September 2001. According to Zbigniew Brzezinski, 'there had been little reason to expect ... grand historical swings from the new president' when he first took office in January 2001,[32] as he 'initially focused on unfinished business: missile defence, military transformation, big power politics'.[33] The Bush administration thus lacked a grand foreign policy vision, limiting itself to promoting the NMD, devising a National Energy Policy (with domestic and foreign elements for energy procurement), and focusing on the Iraqi question and big power politics.

During the 2000 presidential campaign, Bush emphasized the need for America to conduct what he called a 'humble' foreign policy, promoting what Halper and Clarke described as an 'interest-driven foreign policy', as he withdrew from commitments such as peacekeeping and nation-building.[34] As Bush ran for president, he argued that the focus of American efforts should shift away from Clinton-era preoccupations with nation-building, international social work, and the incoherent use of force, and toward cultivating great-power relations and rebuilding the nation's military,[35] as he said that he 'would be very careful about using [US] troops as nation builders'[36] and that 'we can't allow ourselves to get overextended'.[37] (He was therefore eager to extract the US military from its nation-building efforts in the Balkans.)[38] In the words of Jacob Weisberg, Bush was a 'Unipolar Realist' who dismissed peacekeeping and nation-building not driven by the American interest, had little use for the 'smiles and scowls of diplomacy' and sought to abrogate the Anti-Ballistic Missile (ABM) Treaty of 1972, in favour of deploying the NMD[39] (as he thought that the ABM Treaty was a relic of the past).[40] The United States has previously resorted to unilateralism at many points in history, but Bush took it to a new level, not just as an ad hoc policy, but a new and wider strategic orientation,[41] in what Charles Krauthammer called a 'new unilateralism'. Krauthammer praised the Bush administration for its willingness to 'assert American freedom of action and the primacy of American national interests ... rather than contain American power within a vast web of constraining international agreements'.[42]

Nevertheless, Bush acknowledged that 'America must be involved in the world'. He set out his foreign policy priorities in a speech on 19 November 1999, focusing on relations with rising powers such as Russia, China and India, and discussing the expansion of NATO and weapons proliferation.[43] He stopped short of declaring a grand vision of his own, preferring instead to focus on these unfinished issues. He also rejected the notion of an American empire, and called for a 'humble' foreign policy: 'Let us reject the blinders of isolationism, just as we refuse the crown of empire. Let us not dominate others with our power ... And let us have an American foreign policy that reflects American character. The modesty of true strength. The humility of real greatness.'[44]

Contrary to popular belief, and despite the fact that many of Bush's initial unipolar policies coincided with the beliefs of the neo-conservatives, his administration was not simply 'neo-conservatives came to office'.[45] 'When campaigning, Bush did not articulate a foreign policy that reflected the neo-conservative view of America's role in the world', according to Halper and Clarke.[46] Bush himself was not a neo-conservative, as David Frum said that Bush 'was not at all an ideological man'.[47] Vice President Dick Cheney, Secretary of Defence Donald Rumsfeld, Republican Senators, various interest groups and Bush's personal beliefs all had an influence on Bush's position on Kyoto, the ICC and the ABM treaty.[48] Indeed, there were different visions among the members of the Bush administration. Vice President Dick Cheney was focused on the promotion of a national energy strategy (as evident in his meetings with energy executives days after Bush's inauguration), and on the Iraqi problem, as seen during the first NSC meeting on 30 January 2001. Cheney was 'uncharacteristically excited' during the meeting as CIA Director George Tenet showed photos of what he said were Iraqi WMD factories.[49] This was also evident in Cheney's request to outgoing Secretary of Defence William Cohen to arrange a security briefing for the president-elect, with a focus on Iraq.[50] Secretary of State Colin Powell's priorities were multilateral agreements and using diplomacy 'to repair rips in the status quo rather than chart a bold course' for US foreign policy.[51] Secretary of Defence Donald Rumsfeld was promoting military preponderance through the NMD system, in order to deter against threats from rogue states and to take action against what he perceived as the Iraqi threat. National Security Advisor Condoleezza Rice was interested mainly in China and Russia, as she wrote that 'for America and our allies, the most daunting task is to find the right balance in our policy toward Russia and China. Both are equally important to the future of international peace.' She referred to China as a 'strategic competitor' instead of a 'strategic partner'.[52] She did note, however, that 'the challenges they pose are very different', as China is 'a rising power' while, in the case of Russia, America's security is more threatened by Russia's weakness and incoherence than by its strength.[53] Also, along with Rumsfeld, she focused on the importance of the NMD system.[54]

Foreign energy, energy crisis and national security

The perceived relationship between energy sufficiency and US national security had emerged as a significant issue during the 2000 presidential campaign,[55] due to soaring petrol prices in the summer of 2000 and the prospect of a heating-oil shortage in the winter. During the presidential campaign, Dick Cheney accused Bill Clinton and Al Gore of leaving the United States without an energy policy. Senate Majority Leader Trent Lott also emphasized the energy theme, promising that Bush and Cheney would establish a clear approach to energy issues.[56] In part, the Bush administration's claim that America was suffering from an 'energy crisis' was due to the power cuts in California, promoting Bush and Cheney to say that they were 'deeply concerned' about a broader energy shortage: 'It's becoming very clear in our country that demand is outstripping supply.'[57]

On 14 March 2001, during a visit to the Youth Entertainment Academy in New Jersey, Bush repeatedly declared that the United States was suffering from an 'energy crisis', adding that 'the reality is, the nation has got a problem when it comes to energy. We need more sources of energy.'[58] A few days later, on 19 March, Bush once again brought up the subject of an energy crisis after meeting the National Energy Policy Development Group (the group responsible for devising a national energy plan, headed by Cheney), saying that 'demand for energy in the United States is increasing, much more so than production is, and, as a result, we're finding in certain parts of the country that we're short on energy', and adding that 'one thing is for certain, there are no short-term fixes; that the solution for our energy shortage requires long-term thinking and a plan that we'll implement that will take time to bring to fruition'.[59] Furthermore, Energy Secretary Spencer Abraham warned the US Chamber of Commerce in March 2001 that the United States was facing the most serious energy shortage since the 1970s, citing the energy crisis as a threat to US national security. Without a solution, he said, the energy crisis would threaten prosperity and national security and change the way Americans live.[60]

In April, the report of an independent task force co-sponsored by the James A. Baker III Institute and The Council on Foreign Relations (CFR), titled *Strategic Energy Policy: Challenges for the 21st Century* was submitted to Vice President Cheney.[61] It echoed Bush's apocalyptic language, citing the energy crisis as a threat to US national security and its effects on US foreign policy:

> As the 21st century opens, the energy sector is in critical condition. A crisis could erupt at any time from any number of factors and would inevitably affect every country in today's globalized world. While the origins of a crisis are hard to pinpoint, it is clear that energy disruptions could have a potentially enormous impact on the US and the world economy and would affect US national security and foreign policy in dramatic ways.[62]

The *Strategic Energy Policy* report made it clear that the most serious threat came from the oil sector, and that America and the world were facing the most serious oil crisis since the 1970s, saying that:

> the world is currently precariously close to utilizing all of its available global oil production capacity, raising the chances of an oil-supply crisis with more substantial consequences than seen in three decades. These limits mean that America can no longer assume that oil-producing states will provide more oil. Nor is it strategically and politically desirable to remedy our present tenuous situation by simply increasing our dependence on a few foreign sources.[63]

The CFR report also emphasized the mission of Cheney's Energy Task Force to work out 'how best to cope with high energy prices and how best to cope with reliance on foreign oil'.[64] Eventually, Cheney's efforts with the Energy Task Force led to the *National Energy Policy: Report of the National Energy Policy Development Group*, released in May 2001, with the warning that dwindling supplies of oil and gas, an antiquated power grid and burdensome regulation threatened to drag the United States into the worst energy-supply crisis since the 1970s. Taking the interests of the energy industry into consideration, the National Energy Policy (NEP) also 'promised something for everyone in the world of energy business', according to Rutledge.[65]

Bush's public introduction of the NEP came on 17 May 2001. After a visit to a high-tech energy generation plant he gave a speech at the River Centre Convention Centre in St Paul, Minnesota, promoting his energy plan and stressing the link between foreign oil procurement and national security:

> My administration has developed a sane national energy plan to help meet our energy needs this year and every year. If we fail to act on this plan, energy prices will continue to rise ... If we fail to act, Americans will face more and more widespread blackouts. If we fail to act, our country will become more reliant on foreign crude oil, putting our national energy security into the hands of foreign nations, some of whom do not share our interests.[66]

He added that, to protect national energy security from foreign threats and energy blackmail, the National Energy Policy aimed to:

> Expand and diversify our nation's energy supplies. Diversity is important not only for energy security, but also for national security. Over-dependence on any one source of energy, especially a foreign source, leaves us vulnerable to price shocks, supply interruptions, and in the worst case, blackmail.[67]

Bush also supported drilling for oil in the Arctic National Wildlife Refuge (ANWR) as a means of reducing America's dependence on foreign oil, saying that 'America today imports 52% of all our oil ... we should produce more of it at home', adding that 'ANWR can produce 600,000 barrels of oil a day for the next 40 years. What difference does 600,000 barrels a day make? Well, that happens to be exactly the amount we import from Saddam Hussein's Iraq'[68] (under the UN's Oil-For-Food programme). Similarly, the NEP stated that 'ANWR production could equal 46 years of current oil imports from Iraq',[69] and recommended 'that the President direct the Secretary of the Interior to work with Congress to authorize exploration and, if resources are discovered, development' of the ANWR energy reserves.[70]

Justifying the need for a national energy strategy and the need for action on the energy procurement front, the NEP started with a grim picture of America's energy situation:

> America in the year 2001 faces the most serious energy shortage since the oil embargos of the 1970s. The effects are already being felt nationwide. Many families face energy bills two or three times higher than they were a year ago. Millions of Americans find themselves dealing with ... blackouts ... Drivers across America are paying higher and higher gasoline prices. Californians have felt these problems more acutely.[71]

The NEP acknowledged the importance of oil, since it is 'the largest source of primary energy, serving almost 40% of US energy needs',[72] forecasting that, in 2020, oil would 'account for roughly the same share of US energy consumption as it does today'.[73] It also warned that 'estimates indicate that over the next 20 years, US oil consumption will increase by 33%, natural gas consumption by well over 50%, and demand for electricity will rise by 45%'.[74] Citing that 'today, oil accounts for 89% of net US energy imports',[75] and that 'US oil consumption will continue to exceed production',[76] the NEP warned that the impact of rising demand and declining domestic production would lead to increased need for energy imports, with negative repercussions on America's foreign policy:

> We produce 39% less oil today than we did in 1970, leaving us ever more reliant on foreign suppliers. On our present course, America 20 years from now will import nearly two of every three barrels of oil – a condition of increased dependency on foreign powers that do not always have America's interests at heart.[77]

The first seven chapters of the NEP focused on boosting domestic energy output, particularly by removing the legal barriers on greater exploitation of domestic oil, gas and coal, and by increasing dependence on nuclear energy. In the eighth and final chapter, however, the NEP shifted emphasis from conservation and energy efficiency to the need for foreign oil. Chapter Eight, titled 'Strengthening Global Alliance: Enhancing National Energy Security and International Relationships', started by saying that 'US national energy security depends on sufficient energy supplies to support US and global economic growth'.[78] The NEP pointed out the grim fact that 'US oil consumption will continue to exceed production':

> Over the next 20 years, US oil consumption will grow by over 6 mbpd [million barrels per day].[79] If US oil production follows the same historical pattern of the last 10 years, it will decline by 1.5 mbpd. To meet US oil demand, oil and [oil] product imports would have to grow by a continued 7.5 mbpd. In 2020, US oil production would supply less than 30% of US oil needs.[80]

Between 2000 and 2020, US imports of foreign oil would rise by 68%, from 11 mbpd to 18.5 mbpd[81] (and it is forecast that the United States will import 100% of its oil by 2050.)[82] Therefore, the first recommendation in Chapter Eight was that 'the NEPD Group recommends that the President make energy security a priority in our trade and foreign policy'.[83] The fact that the NEP made 35 foreign policy recommendations (a third of its total recommendations) regarding oil imports, calling for stronger ties between the United States and oil rich countries,[84] and the overcoming of obstacles to US investments in these countries,[85] showed the extent of the attention which the NEP paid to foreign energy sources and its impact on US foreign policy.

Beyond national energy security, however, there were the interests of the powerful lobby of economic interests and oil companies, to which Bush, Cheney, Rice and others in the administration were closely linked.[86] The NEP advocated the effort of foreign energy procurement, in cooperation with major oil firms:

> American energy firms remain world leaders, and their investments in energy-producing countries enhance efficiencies and market linkages while increasing environmental protections. Expanded trade and investment between oil importing and exporting nations can increase shared interests while enhancing

global energy and economic security. Promoting such investments will be a core element of our engagement with major foreign oil producers.[87]

It thus recommended opening international markets to energy firms
(in a language similar to William Appleman Williams's references
to an Open Door policy, Milton Friedman's call for a large role for
American corporations, or Kolko's writings on corporatism):

> The NEPD group recommends that the President direct the
> Secretaries of State, Commerce and Energy to continue support-
> ing American energy firms competing in markets abroad, and
> use our membership in multilateral organizations ... to level
> the playing field for US companies overseas, and to reduce barri-
> ers to trade and investment.[88]

Thus, the Energy Task Force and the NEP were 'creatures of Cheney's
worldview' which sought to redefine the relation between regulation
and the marketplace.[89] Cheney was a true believer that national
security and economic health required a boost in energy production,
and this, in turn, required a rollback of stifling rules and greater
government cooperation with the energy industry.[90] He also resisted
legal action to release the documents of the energy executives and
National Energy Policy Development Group (NEPDG) meetings,
citing his belief that such revelations would limit the powers of the
executive and of the President of the United States, preventing the
privacy and flexibility necessary to do the job.[91]

Experts disagree, however, on whether the United States was
indeed suffering from an energy crisis in early 2001. Some government experts doubted that a crisis existed. *The Economist* magazine
said that Bush 'has insisted, on absolutely no evidence whatsoever,
from his first days in office, that America was mired in a serious
energy supply crisis'.[92] Furthermore, the NEP suggested drilling on
ANWR to solve California's power cuts, even though California did
not burn oil in its power plants.[93] Moreover, Paul Krugman wrote
on 20 May 2001 that Cheney had 'fabricated an energy crisis'.[94]

Also, Judy Pasternak wrote in the *Los Angeles Times* in August 2001 that the basic assumptions in the NEP were 'tailored to the [energy] industry's measures'. She stated that a briefing paper prepared for a 19 March task force meeting with Bush said that 'on the whole, US energy markets are working well, allocating resources and preventing shortages'; a situation far from the NEP invocation of the worst energy crisis since the 1970s. Pasternak added that one staffer recalled seeing a memo that discussed 'utilizing' California's blackouts and the summer 2000 high prices of gasoline to press for more drilling for gas and oil.[95]

One of the obvious regions affected by Bush's energy plan for America's foreign policy was the Middle East, especially Saudi Arabia; a friend in the region, which also happened to be the on-off number-one supplier of America's foreign oil. During a meeting with the NEPDG on 19 March 2001, Bush commended the Saudi efforts to maintain oil prices within a reasonable range, saying that

> the Saudi [oil] minister made it clear that he and his friends would not allow the price of oil, crude oil to exceed $28 a barrel. That's very comforting to the American consumer, and I appreciate that gesture. I thought that was a very strong statement of understanding, that high prices of crude oil will affect our economy.[96]

Also, Baker's *Strategic Energy Policy* report praised Saudi Arabia's role in acting as a 'swing producer' when Saddam Hussein cut oil supplies, especially given Saddam's attempts to use oil as a weapon and become a 'swing producer', and the domestic pressure on the Gulf Cooperation Council states not to increase oil production:

> Over the past year, Iraq has effectively become a swing producer, turning its taps on and off when it has felt such action was in its strategic interest to do so. Saudi Arabia has proven willing to provide replacement supplies to the market when Iraqi exports have been reduced. This role has been extremely important in

avoiding greater market volatility and in countering Iraq's effort to take advantage of the oil market's structure.[97]

However, troubles in US–Saudi relations started just before the 11 September attacks, as the United States and Saudi Arabia clashed over the Arab–Israeli conflict and global oil prices.[98] In late August 2001, Crown Prince Abdullah bin Abdul Aziz (then the de facto ruler of Saudi Arabia due to King Fahd's ailing health) dispatched Prince Bandar bin Sultan, the Saudi ambassador to the United States,[99] to give the Americans a harsh message: 'If the United States continued to permit Israel to wage war on Palestine, Saudi Arabia would have to heed Arab public opinion.'[100] He added that 'it is time for the United States and Saudi Arabia to look at their separate interests',[101] and that the relations between the two nations 'were at a crossroads, and Saudi Arabia will now look after its own interests'.[102]

Thus, Bush discovered during his first months in office that he could be challenged by his regional allies, not only because of disagreements with Saudi Arabia but also due to the Gulf States' refusal to submit to America's energy demands. Furthermore, Baker's *Strategic Energy Policy* report warned that there were dangers in taking the cooperation of Gulf Arab oil producers for granted,[103] saying that 'Saudi Arabia's role in this needs to be preserved, and should not be taken for granted. There is domestic pressure on the GCC leaders to reject cooperation to cool oil markets during times of a shortfall in Iraqi oil production.'[104] The *Strategic Energy Policy* report cited the 'possibility that Saddam Hussein may remove Iraqi oil from the market for an extended period of time and that Saudi Arabia will not or cannot replace all of the barrels', saying that this 'is a contingency that continues to hang over the market'.[105] Thus, the NEPDG advised the President to 'support initiatives by Saudi Arabia, Kuwait, Algeria, Qatar, the UAE and other suppliers to open up areas of their energy sectors to foreign investors'.[106]

At the same time that the NEPDG recognized the central importance of the Middle East for global energy supplies, it devoted attention to exploring ways in which US dependence on the region could be reduced. A report issued by EMAP Business International Ltd on

1 July 2001, highlighted the extent of US dependence on oil imports from the Middle East:

> The United States now imports about 10 million barrels a day (b/d) of oil, compared with just 4.3 million b/d in 1985. Imports account for 51.6% of total oil consumption, compared with 34.8% in 1973. The Middle East supplies 24% of US oil imports, with Saudi Arabia and Iraq the leading suppliers, at 1.6 million barrels per day and 600,000 barrels per day respectively in 2000.[107]

Furthermore, the NEP noted that Middle East oil resources will always be vital to the United States:

> By 2020, Gulf oil producers are projected to supply between 54% and 67% of the world's oil. Thus, the global economy will almost certainly continue to depend on the supply of oil from Organization of Petroleum Exporting Countries (OPEC) members, particularly in the Gulf. The region will remain vital to US interests. Saudi Arabia, the world's largest oil exporter, has been a linchpin of supply and reliability to world oil markets. Saudi Arabia has pursued a policy of investing in spare oil production capacity, diversifying export routes to both of its coasts, and providing effective assurances that it will use its capacity to mitigate the impact of oil supply disruptions in any region.[108]

Due to the importance of Middle East oil, the NEP called on Bush to urge the Gulf States to open their energy sectors for foreign investment: 'the NEPD Group recommends that the President support initiatives by Saudi Arabia, Kuwait, Algeria, Qatar, the UAE and other suppliers to open up areas of their energy sectors to foreign investment.'[109] Similarly, the *Strategic Energy Policy* report recommended that 'the Department of State, together with the National Security Council, the Department of Energy and the Department of Commerce should develop a strategic plan to encourage the reopening of foreign investment in ... important states of the Middle East Gulf'. Furthermore,

said the *Strategic Energy Policy* report, the United Sates should 'initiate efforts to spur the reopening of countries that have nationalized
and monopolized their upstream sectors',[110] because 'if political factors
were to block the development of new oilfields in the Gulf, the ramifications for world oil markets could be quite severe'.[111]

What made matters worse for the West was that, according to the
International Energy Agency (IEA) the Persian Gulf producers would
have to spend $523 billion on new equipment and technology between
2001 and 2030 to increase their output as required.[112] They were not
willing to spend such a huge sum,[113] and even if they were, it was
unlikely that they would be able to assemble such a huge sum of money
without foreign investment in their state-controlled oil industry, which
would contradict their policies of full control over their nationalized
energy sectors.[114] Furthermore, the Gulf States did not have the desire
to increase oil production, since an increase in supply would decrease
the price of oil, and this was simply not in their self-interest.[115] Saudi
Arabia was reluctant to increase production, and the Kuwaiti constitution prohibited foreign ownership of petroleum reserves while the government limited the participation of foreign firms in other activities.
Qatar and the UAE were more open to foreign investment but legal
and technical obstacles stood in the way of increased production.[116]

However, the country which really attracted the attention of Bush
and the administration hawks during the first few months in office
was Iraq, due to the fact that it was a rogue state, and due to the
importance of its energy resources. Even before taking office, Cheney
asked the outgoing Secretary of Defence William Cohen to arrange
a security briefing for the president-elect, with a focus on Iraq.[117]
During the first NSC meeting, on 30 January 2001, Bush's policy
on rogue states came down to one: Iraq, as Bush opened the meeting
with a main focus on the Middle East. Bush wanted to disengage from
the Arab–Israeli conflict and let both sides work it out themselves.
'We're going to tilt back toward Israel and we're going to be consistent.
Clinton overreached, and it all fell apart', he said. Secretary of State
Colin Powell said that this was hasty since it would unleash Israeli
Prime Minister Ariel Sharon and the Israeli army, with dire consequences for the Palestinians. Bush replied: 'Maybe it's the best way to

get things back in balance ... Sometimes a show of strength by one side can really clarify things.'[118]

The Arab–Israeli peace process was not the only commitment from which Bush withdrew. During the first eight months in office, Bush withdrew from foreign policy commitments that he inherited from Clinton.[119] He withdrew the US from its commitments to the Arab–Israeli conflict, the Kyoto agreement, the International Criminal Court, the Korean Sunshine Policy, and peacemaking in the Colombian civil war.[120]

Bush, supported by others like Secretary of Defence Donald Rumsfeld, saw that it would be more feasible to start with Iraq.[121] As Secretary of the Treasury Paul O'Neill noted, a 'major shift' was under way in US foreign policy. After more than 30 years of intense engagement, from Nixon to Clinton, America was now washing its hands of the Arab–Israeli conflict, and focusing on Iraq. Rice showed a report titled 'How Iraq is destabilizing the region', in which she noted that 'Iraq might be the key to reshaping the entire region'. To support Rice's argument, CIA Director George Tenet produced a picture of what he said was a chemical or biological weapons factory.[122]

After the 30 January meeting, Bush assigned members of his administration to different approaches to the Iraqi issue. Powell would work on sanctions, and Rumsfeld would examine military options which included rebuilding the military coalition of the 1991 Gulf War, examining the use of US ground forces in the north and south of Iraq and studying how the armed forces could support groups inside the country that could help challenge Saddam. Tenet would work on improving intelligence and covert operations. Treasury Secretary Paul O'Neill would investigate how to squeeze Saddam's regime financially.[123]

In the NSC Principals meeting on 1 February 2001, Rumsfeld said that a regime change in Iraq would set an example to the rest of the region. He said 'Imagine what the region would look like without Saddam and with a regime that is aligned with US interests. It would change everything in the region and beyond it. It would demonstrate what US policy is all about.' The hanging question was not 'why should Iraq be targeted', but 'how'. Condoleezza Rice, Donald Rumsfeld and JCS Chairman Henry Shelton discussed rebuilding

the 1991 Gulf War coalition, though an invasion was never specifically mentioned. Rumsfeld added: 'It is not my specific objective to
get rid of Saddam ... I am after the weapons of mass destruction.
Regime change is not my prime concern.' Rumsfeld believed that uni-
versally available technologies enabled small or medium-sized states
to perform dangerous asymmetrical attacks on US forces around the
globe. His military ideology thus included the need to 'dissuade' others from creating asymmetrical threats and to 'demonstrate' America's
unilateral resolve, as he wrote in January 2001. He recommended
increasing the US ability to deter the use of WMD and long-range
missiles against US interests, i.e., the promotion of the NMD system. O'Neill thought that Rumsfeld's ambitions on Iraq were a part of
Rumsfeld's broader ideology: the need to dissuade others from creating asymmetrical threats in order to demonstrate America's unilateral
resolve.[124] Rumsfeld was also working on a new strategy to enforce
the No-Fly-Zones as a means of weakening Saddam's regime.[125] Thus,
the Pentagon started working on developing military options for Iraq
months before the 11 September attacks.[126]

Furthermore, insiders interviewed by Greg Palast for BBC *Newsnight*
added that Bush's secret planning for Iraq's oil started 'within weeks'
after the Bush administration took office, long before the September
11 attacks.[127] Moreover, linked to Rumsfeld's plans for Iraq were the
Secretary of Defence's specific intentions towards Iraq and its oil:

> Documents were being prepared by Defence Intelligence Agency,
> Rumsfeld's intelligence arm, mapping Iraq's oilfields and explo-
> ration areas and listing companies that might be interested in
> leveraging the precious asset. One document, headed 'Foreign
> Suitors for Iraqi Oilfields Contracts' lists companies from thirty
> countries – including France, Germany, Russia, and the United
> Kingdom – their specialities, bidding histories, and in some cases
> their particular areas of interest. An attached document maps
> Iraq with markings for 'supergiant oilfields', 'other oilfields' and
> 'earmarked for production sharing' while demarking the largely
> undeveloped southwest of the country into nine 'blocks' to des-
> ignate areas for future exploration.[128]

The link between the energy strand and other national security strands was evident in a top-secret document dated 3 February 2001, where a high NSC official directed the NSC staff to cooperate with the NEPDG to assess the military implications of the Bush administration's energy plan. Jane Mayer of *The New Yorker* said that this document showed the 'melding' of two seemingly unrelated top priorities: 'the review of operational policies towards rogue states', such as Iraq, and 'actions regarding the capture of new and existing oil and gas fields'.[129] Mark Medish, who served as a senior NSC director during the Clinton administration, told Mayer in response to this document that 'if this little group was discussing geostrategic plans for oil, it puts the issue in the context of the captains of the oil industry sitting down with Cheney and laying grand, global plans'.[130]

Another sign of the link between energy and other foreign policy strands was that Cheney headed the committee that oversaw energy, as well as installing himself as overseer of defence and foreign policy portfolios and sitting in on the weekly lunches held by Rice, Rumsfeld and Powell even before the 11 September attacks.[131] Moreover, in May 2001, Bush asked Cheney to create a new office within the Federal Emergency Management Agency (FEMA) to prevent terrorist acts and/or WMD attacks within the United States, and to oversee these defence preparations himself. The task force was just getting underway when the 11 September attacks occurred.[132] Furthermore, Cheney's interest in the Iraqi problem was seen during the first NSC meeting on 30 January 2001,[133] and in his quest to outgoing Secretary of Defence William Cohen to arrange a security briefing for the president-elect, with a focus on Iraq.[134] Thus by mid-2001, Cheney was in a position to oversee discussions of policy on Iraq, discussions of policy on energy, and discussions of policy on counter-terrorism, indicating that the strands of military preponderance, energy procurement and anti-terrorism (and possibly the fourth strand of overall global power projection, as seen in the QDR of September 2001) were related even before 11 September. (The Cheney vice presidency was unique in American history, as all foreign policy issues went through Cheney. Even though he was not the only source of foreign-policy strategy, he was nevertheless the 'vortex' of foreign policy and the 'tie-breaker' in foreign policy making disputes.)[135]

Also, the fact that the same maps and lists were used by Rumsfeld and by the NEPDG shows the linkage between the strands of military preponderance and foreign energy procurement.[136] Furthermore, Rumsfeld's interest in Iraq's oilfields was linked to his support for pre-emptive strikes and to his policy to use military force to 'dissuade' and 'demonstrate':

> The desire to 'dissuade' countries from engaging in 'asymmetrical challenges' to the United States – as Rumsfeld said in his January [2001] articulation of the demonstrative value of a pre-emptive attack – matched with plans of how the world's second largest oil reserve might be divided among the world's contractors made for an irresistible combination, O'Neill said.[137]

The Bush administration's concern over Persian Gulf oil supplies resulted in an increased focus on Iraq, as the CFR's *Strategic Energy Policy* report recommended that, in order to reduce Saddam's oil threat, the United States should:

> Review policies towards Iraq with the aim of lowering anti-Americanism in the Middle East and elsewhere, and set the ground to eventually ease Iraqi oilfields investment restrictions. Iraq remains a destabilizing influence to US allies in the Middle East, as well as to regional and global order, and to the flow of oil to international markets from the Middle East. Saddam Hussein has also demonstrated a willingness to threaten to use the oil weapon and to use his own export program to manipulate oil markets. This would display his personal power, enhance his image as a 'Pan Arab' leader supporting the Palestinians against Israel, and pressure others for a lifting of economic sanctions against his regime.[138]

Furthermore, said the *Strategic Energy Policy* report, the US should review its policy on Iraq, taking into consideration military, economic, energy and diplomatic factors, especially as, the report implied, sanctions were not always effective:

> The United States should conduct an immediate policy review toward Iraq, including military, energy, economic and political/diplomatic assessments … Sanctions that are not effective should be phased out and replaced with highly focused and enforced sanctions that target the regime's ability to maintain and acquire weapons of mass destruction.[139]

The *Strategic Energy Policy* report also recommended the resumption of investment in the Iraqi oil sector when the time was right (especially considering that sanctions on oil were actually strengthening Saddam's grip on his country, and that he still had sources of revenue despite the sanctions), while considering the risks that lifting the sanctions would entitle, like encouraging Saddam to use his oil to challenge the US interests, and offending the Gulf States that did not want to face competition from an Iraqi oil-production increase. However, said the report, action on Iraq's oil was necessary, especially given the global importance of Iraq's oil reserves:

> Once an arms-control program is in place, the United States could consider reducing restrictions on oil investments inside Iraq. Like it or not, Iraqi reserves represent a major asset that can quickly add capacity to world oil markets and inject a more competitive tenor to oil trade. However, such a policy will be quite costly as this trade-off will encourage Saddam Hussein to boast his 'victory' against the United States, fuel his ambitions, and potentially strengthen his regime. Once so encouraged and if his access to oil revenues were to be increased by adjustments in oil sanctions, Saddam Hussein could be a greater security threat to US allies in the region if weapons of mass destruction (WMD) sanctions, weapons regimes, and the coalition against him are not strengthened. Still, the maintenance of the continued oil sanctions is becoming increasingly difficult to implement. Moreover, Saddam Hussein has many means of gaining

revenues, and the sanctions regime helps perpetuate his lock on the country's economy. Another problem with easing restrictions on the Iraqi oil industry to allow greater investment is that GCC allies of the United States will not like to see Iraq gain larger market share in international oil markets ... These issues will have to be discussed in bilateral exchanges.[140]

Despite the *Strategic Energy Policy* report's focus on Iraq, the National Energy Policy did not recommend any policies specifically for Iraq. 'It is unbelievable how little attention was paid to Iraq', said Amy Myers Jaffe, senior energy advisor at the Baker Institute and one of the authors of the *Strategic Energy Policy* report, adding that 'the problem of Iraq is the major underlying problem in the oil market'.[141]

Nevertheless, between January and September 2001, the Bush administration was still undecided on how to deal with Saddam. Bush had three options on the table: enforcing the No-Fly-Zones to protect the Kurds in the north and the Shiites in the south, regime change through an armed coup by Iraqi opposition groups, and enforcing UN economic sanctions. He therefore empowered three working groups to study each of these three options.[142]

One of the first things that George W. Bush did when he took office was to launch air strikes against Iraq in February in areas outside the No-Fly-Zones (NFZ). Speaking during his first presidential visit abroad, to Mexico, Bush stressed that the raids were just a 'routine' enforcement of the NFZs, and that they were a 'part of a strategy' towards Iraq. But while Washington talked down the significance of the raid, analysts said that it looked like a warning to the government in Baghdad.[143]

Parallel to enforcing the NFZs, the administration worked on the new package of 'smart sanctions' which aimed to increase the amount of legal trade with Baghdad while also cutting smuggling routes. The new sanctions would ease restrictions on civilian goods but retain bans on military hardware, reviewing a list of 'dual use' supplies that could be used for both military and civilian purposes,[144] and dropping embargoes on all non-military imports to Iraq,[145] in the hope of restoring international solidarity against Iraq's acquisition of military equipment or materials for weapons of mass destruction.[146] On 22 May

2001, the United States and Britain formally presented the new Iraq sanctions proposal to the UN Security Council.[147] Iraq's neighbours, Jordan, Turkey and Syria, would be allowed to import 150,000 barrels of oil a day under the plan.[148] China and France agreed that Iraq would not be allowed to import a core list of goods which could potentially help Iraq in building WMDs.[149] France, which was sympathetic to Iraq, tried to persuade Baghdad to agree to the plan.[150] Iraq, which produced nearly 5% of world exports at the time, halted its oil exports on 4 June 2001, in protest against the US–British proposal, triggering a wave of panic buying. In early June, Brent crude futures for July delivery rose by 43 cents to reach $29.5/barrel on the International Petroleum Exchange in London.[151] On 3 July, however, Washington and London dropped their proposal after Russia threatened to veto the idea at the UN,[152] ending the immediate possibility of smart sanctions as proposed by Powell. On 6 July Baghdad announced that it would resume oil exports.[153]

A third approach was advocated by the administration hawks, like Rumsfeld and his deputy Paul Wolfowitz, who preferred to depend on increased support to the Iraqi opposition groups[154] in order to change the regime in Baghdad.[155] (Wolfowitz was the greatest supporter of removing Saddam.[156] He believed that it was possible to send in the military to overrun and seize Iraq's southern oilfields (about 1,000 wells, two-thirds of Iraq's oil production) which were 60 miles from Kuwait's borders. He thought that if this happened, the Iraqi people would be encouraged to remove Saddam. Powell, on the other hand, thought that this was 'lunacy'.)[157] Another hawk was Richard Perle, a leading neo-conservative and a self-described advocate of 'regime change' in Iraq, who said at the AEI in April 2001 that there was 'an obvious alternative to sanctions and that is to support the internal opposition to Saddam' and that 'German intelligence has estimated that by 2003, Saddam Hussein will have two or three nuclear weapons'. Perle embodied the hawks' rejection of sanctions, saying that 'the sanctions policy has failed and will be even less effective in the future', and adding that 'we probably can't police the current sanctions, much less the so-called smart ones'.[158] Both Wolfowitz and Perle were 'democratic imperialists', calling for the active deployment of overwhelming

American military, economic and political strength to remake the world in its image, not just by toppling tyrants but also by creating democracies in their wake.[159] Cheney, too, favoured military action, as evident in his request to outgoing Secretary of Defence William Cohen to arrange a security briefing for the president-elect, with a focus on Iraq, before Bush took office.[160]

However, the strategy was still in development stage, and on 25 April 2001, in a Deputies meeting which included Wolfowitz, Deputy Secretary of State Richard Armitage, and Cheney's national security advisor Lewis Libby, a draft for a military coup was set down for further revisions.[161] On 1 June 2001, Rice chaired a meeting of the Principals Committee, where four options were on the table: continuing the current containment strategy, continuing containment while actively supporting Saddam's opponents, setting up a safe haven for insurgents in southern Iraq, and planning a US invasion. On 1 August the Deputies presented the Principals with a secret document, 'A Liberation Strategy', which envisioned heavy reliance on the Iraqi opposition to pressure Saddam's regime. However, a formal policy recommendation for attacking Iraq was never forwarded to Bush, no policy was set, and administration officials continued to pursue their separate agendas.[162] Bush himself was reluctant to define a course of action. A White House official stated that 'faced with a dilemma, [Bush] has this favourite phrase he uses all the time: Protect my flexibility'.[163] To preserve this flexibility, the NSC authorized the three working groups to consider covert action to topple Saddam, to consider enforcing the NFZs and to evaluate smart sanctions.[164]

Nevertheless, even though Bush considered Saddam a threat to be kept in check,[165] and even though he told Clinton that Iraq, along with the missile-defence system, were the two largest foreign concerns that America faced, it seemed that the Bush administration did not yet consider the Iraqi WMDs an immediate threat. Colin Powell said in a press conference in Cairo in February 2001 that Saddam 'has not developed any significant capability with respect to weapons of mass destruction. He is unable to project conventional power against his neighbours.'[166] He also said 'We kept him contained, kept him in his box'.[167] When asked by *Time* magazine in late August 2001 'Do you see Saddam as a threat?' Powell replied 'I do not lose a lot of sleep about him late at night'.[168] Moreover, Condoleezza Rice, despite her NSC presentation of January 2001, said in a press conference in July 2001 'We are able to keep arms from him [Saddam]. His military forces have not been rebuilt.'[169] Tenet, too, said that during the spring and summer of 2001, the topic of Iraq 'faded into the background' of his priorities as he paid more attention to 'plenty of other issues'.[170]

This shows that there was no coherent or unified policy on Iraq among the members of the Bush administration before the September 11 attacks. Thus, the attention given to Iraq did not result in a clear or precise decision on what was to be done. Rather, its immediate significance was the marginalizing of the Arab–Israeli peace process. In a sharp break with Clinton, who was personally involved in negotiations between Israelis and Palestinians, Bush declined to even send an envoy to the last-ditch Israeli–Palestinian peace talks at Taba, Egypt, in late January 2001. The White House eventually eliminated the post which Dennis Ross had held for eight years, and three months into the administration the NSC still did not have a senior director for Middle East affairs.[171] In late February, Colin Powell went on his first trip abroad since taking office: a six-nation tour of the Middle East, including Egypt, Saudi Arabia and Syria. Iraq was the main issue of the trip, specifically the sanctions on Iraq and the American raid on Iraq in February 2001.[172] The Bush administration did not give up on containing Saddam, but they were still undecided on how exactly to get rid of him.

Iraq was not the only rogue state with oil reserves against which the Bush administration had plans. Libya and Iran, too, were rogue states whose significant oil reserves played a role in the Bush foreign policy. US policy on rogue states was intertwined with the NMD system. The Bush administration had motives, other than defence, for building the missile shield. The NMD system would back up an American first strike against a state armed with ballistic missiles and/or WMDs. This would 'preserve freedom of action' in a regional conflict involving North Korea or any other potential state armed with WMDs.[173] This was linked to the Nuclear Posture Review (NPR), which was 'unveiled at the end of 2001 in the highly charged post-September 11 political climate, [which represented] an abrupt departure from the policies of prior post-Cold War administrations, Republican and Democratic alike'. This 'integrated, significantly expanded planning of doctrine for using nuclear weapons against a wide range of potential adversaries ... reverse[d] an almost two-decades-long trend of relegating nuclear weapons to the category of weapons of last resort',[174] and was explicitly directed at China, Iran, Iraq, Libya, North Korea, Russia and Syria. Indeed, Laurence Kaplan wrote in March 2001 that 'missile defence [was not] really meant to protect America', and that it was 'a tool for global dominance'.[175]

Beyond the specific issue of missile defence, the United States did not significantly shift its position towards rogue states under the Bush administration, as Bush's policy on rogue states was essentially the same as his predecessor Clinton, apart from Iraq to which Bush gave much more attention than Clinton.[176] (Indeed, seven rogue states were mentioned in the 2000 campaign: Iran, Iran, Sudan, Syria, Libya, North Korea and Cuba.)[177] Focused mainly on Iraq, the Bush administration wanted to soften its line on Iran and Libya, (as seen, for example, in Bush's initial desire to extend the Iran–Libya Sanctions Act for only two years instead of five).[178]

Vice President Cheney expressed his opposition to the sanctions on Iran, a position complementing that of the US oil companies during his tenure as CEO of Halliburton.[179] There were also signs that the Bush administration was ready to initiate dialogue if Iran took steps to change its behaviour.[180] Despite these initial signs, however, it was evident from the first days in office that Iran policy had sunk quickly to the bottom of the administration's lists of priorities.[181] There was a policy review on Iran, but it was never concluded, and by 11 September 2001 Iran policy was officially 'still being studied'.[182]

On Libya, Bush was very much following the Clinton policy.[183] He was anxious not to be seen as soft on rogue states in his first term in office, but he was also under two conflicting sets of pressure from the relatives of the American victims of the Pam Am flight which was bombed over Lockerbie in 1988 on the one hand, and oil companies that were anxious to secure lucrative contracts in Libya on the other hand. The Washington-based lobby group USA-Engage suggested that the sanctions cost American companies billions of dollars in lost exports every year.[184] Pro-trade groups such as USA-Engage and other business groups depended on Cheney as their ally on the issue of the Iran and Libya Sanctions Act, for his work against unilateral US sanctions when he was CEO of Halliburton.[185]

The NEP called for a general review of US sanctions on oil-producing states, but no specific example was offered. Neither Iraq, Iran or Libya were mentioned by name in the 21-page chapter on 'National Energy Security and International Relationships'. Instead, there was only a vague reference to reviewing overall sanctions policy,[186] with the statement that:

> Sanctions should be periodically reviewed to ensure their continued effectiveness and to minimize their costs on US citizens and interests ... The NEPD Group recommends that the President direct the Secretaries of State, Treasury and Commerce to initiate a comprehensive review of sanctions. Energy security should be one of the factors considered in such a review.[187]

The Bush administration did review sanctions on Iran, but indicated that it would make no decisions until after the Iranian elections scheduled for June 2001. The pro-Israel lobby in Congress pressed for reauthorization of the Iran–Libya Sanctions Act, which in theory had prohibited foreign companies from making new investments in Iran of more than $40 million had never been enforced in practice.[188] The call for renewal caused a political battle between the pro-Israeli groups, who were in favour of renewal, and the US oil industry which argued that the law had been bad for business and was a political failure.[189] In July 2001 the US Senate extended the sanctions for a further five years, but the Bush administration argued for only two years, as this would give the US some flexibility to change its policy towards Libya and Iran when necessary. The extension of the law was also opposed by many European countries that had companies involved in the energy sector in Iran and Libya.[190] The new law was criticized by the European Commission, as the EU's External Relations

Commissioner Chris Patten said that such measures threatened the open international trading system, and that the EU would take measures against the US through the World Trade Organization if any action was taken against European companies operating in Iran and Libya. Nevertheless, Bush signed the five-year extension into law in August 2001.[191]

Due to the problems of relying on Middle East oil, regional diversification of oil sources was also pursued. The NEP said that, despite the Persian Gulf's importance, the United States had to continue to diversify its resources away from the Middle East, through global engagement in other regions:

> Middle East oil production will remain central to world oil security. The Gulf will be a primary focus of US international energy policy, but our engagement will be global, spotlighting existing and emerging regions that will have a major impact on the global energy balance.[192]

In an effort to achieve diversification, one of the solutions that Bush proposed was a 'hemispheric policy' to the energy problem, involving cooperation with countries of the Western Hemisphere. During the first presidential candidates' debate in October 2000, Bush called for 'a hemispheric energy policy where Canada and Mexico and the United States come together'. He restated this during the California energy crisis in January 2001: 'The quickest way to have impact on the energy situation is for us to work with Mexico, and to a certain extent Canada, to build a policy for the hemisphere.' Furthermore, prior to his visit to Mexico in February 2001, he announced his hopes that energy cooperation between the United States and Mexico could help reduce the impact of debacles like the California burnout and liberate the United States from its addiction to Persian Gulf oil.[193] Similarly, the NEP recommended steps towards hemispheric energy cooperation:

> The NEPD Group recommends that the President direct the Secretaries of State, Commerce and Energy to engage in a

dialogue through the North American Energy Working Group to develop closer energy integration among Canada, Mexico and the United States to identify areas of cooperation, fully consistent with the countries' respective sovereignties.[194]

Bush, at least initially, paid more attention to Latin America than to Washington's traditional European allies. Indeed, Bush's first foreign trip after taking office was a one-day visit to Mexico in February 2001, where Bush held talks with the Mexican president, Vicente Fox, discussing illegal drug-trafficking and immigration, and reaffirmed their common desire to boost the Mexican economy, bilateral trade, and Mexican energy exports to the United States. Bush and Fox spoke of a 'shared prosperity' between the US and Mexico, but stopped short of specific commitments.[195]

Regarding Venezuela, the NEP acknowledged that 'the United States, with Venezuela, [was] a coordinator of the Hemispheric Energy Initiative Process', and that Venezuela was 'the third largest oil supplier to the United States. Its energy industry is increasingly integrated into the US marketplace. Venezuela's downstream investments in the United States make it a leading refiner and gasoline marketer here.' It also acknowledged the increasing US and international investment in Venezuelan energy resources, and said that Venezuela was 'also moving to liberalize its natural gas sector'.[196] Thus, the NEPDG recommended 'that the President direct the Secretaries of Trade and Commerce to conclude negotiations with Venezuela on a Bilateral Investment Treaty'.[197] It also gave similar recommendations regarding Brazil and its energy cooperation with the United States.[198]

To a lesser extent, the United States also had hopes for Africa, with the NEP noting that 'Sub-Saharan Africa holds 7% of world oil reserves and comprises 11% of world oil production. Along with Latin America, West Africa is expected to be one of the fastest growing sources of oil and gas for the American market.'[199] Bush did not have a policy on Africa, as he indicated in during his 2000 presidential campaign that Africa was not a priority,[200] remarking during the campaign that 'while Africa may be important, it does not fit into the national strategic interest, as far as I can see them'.[201] Nevertheless, there was indeed increased military interest in the continent, as Deputy Commander in Chief of Eurcom (European Command, whose area of responsibility included West Africa) visited the small island-state São Tomé e Principe in July 2001 to look for possible US military-base locations, since the island was close to the major West African oil-producing countries and yet had largely escaped the violence and conflict that have plagued mainland Africa.[202]

One problem with the regional diversification of energy resources, however, was that it would not work, as regions outside the Middle East did not have enough supplies to break the reliance on Middle East oil. Of all the oil-producing regions in the world, the Persian Gulf alone had enough untapped reserves to satisfy American and global demand for oil, and none of the other regions would ever produce enough to reverse, or even slow down, dependence on Persian Gulf producers.[203]

To fully understand the challenges facing regional diversification, one has to take a look at the data available to the authors of the NEP in 2001, the year the plan was devised. According to the data provided by the US Department of Energy in 2001, total world oil production would have to grow by 60% between 1999 and 2020 to meet anticipated world consumption of 119 mbpd. But because of flat or declining production in many other regions of the world, output in the Gulf would have to climb by 85% to satisfy the enormous rise in demand. This meant that combined Persian Gulf production would have to increase from 24 mbpd in 1999 to 44.5 mbpd in 2020.[204]

The Bush administration concluded that since the Gulf States were unable or unwilling to raise production, and since US interests in the region were facing dangers from Iraq and Iran, and terrorist attacks on US installations, then the United States would have to be the dominant power in the region, overseeing the politics, security and oil output of the area. The build-up of American power and influence in the region had begun with Presidents Roosevelt, Truman and Eisenhower. It took a new level under Jimmy Carter, who said in December 1979 following the Soviet invasion of Afghanistan: 'The Soviet effort to dominate Afghanistan has brought Soviet military forces within 300 miles of the Indian Ocean and close to the Strait of Hormuz, a waterway through which most of the world's oil must flow.' He then announced the Carter Doctrine of January 1980, where he said:

> The Soviet Union is now attempting to consolidate a strategic position that poses a grave threat to the free movement of Middle East oil ... Let our position be absolutely clear: an attempt by any outside force to gain control of the Persian Gulf region will be regarded as an assault on the vital interests of the United States of America, and such an assault will be repelled by any means necessary, including military force.[205]

Since then, American foreign policy was a continuation of the Carter Doctrine to secure control over Middle East oil by any means necessary.[206]

The Carter Doctrine was applied in the 1991 Gulf War and the subsequent deployment of 25,000 US troops in the region after the war. But if the George W. Bush administration was to secure the oil supplies of the region, and also follow its military preponderance agenda, then this build-up had to be taken to an entirely new level[207] never seen before (which would happen in 2003, through the military invasion of a major Arab oil-producing state: Iraq). The Bush administration, therefore, aimed to solve this energy dilemma through:

> the establishment of a new American Imperium in the Middle East: one in which American-selected local rulers would invite American oil companies to make super-profits for American investors under the protective shield of the American military, while at the same time satisfying the voracious demands of the motorized American oil consumer.[208]

However, given the fact that there was no justification for a further military build-up or increase of US influence in the Middle East, the United States had to pay more attention to the so-called 'new Middle East'; i.e. the Caspian region.[209]

According to Klare, Bush 'embraced [Clinton's] strategic endeavours in the Caspian Sea region'.[210] Similarly, CSIS analyst Doug Blum wrote in March 2001 that there was 'no indication of any major change in America's Caspian policy under the Bush administration'. Thus, wrote Blum:

> it appears likely that the Caspian region will remain a distinctly secondary concern for American policymakers, one whose signifi-cance is determined largely by the role of the strategically impor-tant regional actors (Russia and Iran) [i.e., keeping these two actors in check] as well as by prevailing perceptions of national security [i.e. energy and perhaps militant-Islamic threats].[211]

It seemed, wrote Blum in March 2001, that the key goals of US policy remained intact:

1. Energy diversification;
2. Increasing economic opportunities for US firms;
3. Containing Russian and Iranian influence;
4. Promoting the independence, democracy and development of the Newly Independent States (NIS).

Bush also 'remained committed to the BTC pipeline' which would 'solidify relations with Turkey, sideline Iran, prevent Russia's monopoly on transportation, and promote the independence of the NIS'. Thus, the Bush administration continued to pressure US oil companies to participate by contributing finances and/or projected oil volumes, and also continued to exert pressure against routes which might pass through Iran or Russia.[212]

This fitted with the recommendations of the NEP. Describing the Caspian Sea region as 'a rapidly growing new area of supply',[213] the NEP recommended 'that the President direct the Secretaries of State, Commerce and Energy to support the BTC oil pipeline as it demon-strates its commercial viability'.[214] Moreover, the NEP recommended strengthening ties with countries in the Caspian region:

> The NEPD Group recommends that the President direct the Secretaries of Commerce, Energy and State to deepen their

commercial dialogue with … Caspian states to provide a strong, transparent, and stable business climate for energy and related infrastructure projects.[215]

To this end, US State Department spokesman Richard Boucher said on 29 August 2001 that the Bush administration was supporting the BTC pipeline and wider relations with Central Asian countries as recommended by the NEP:

> Georgia and Azerbaijan are negotiating the terms for the intergovernmental agreement that allows the construction of a gas pipeline through Georgia to Turkey [i.e. the BTC pipeline] … We do support the development of the pipeline on commercially viable basis, as part of a broader energy policy which fits with the national energy policy [NEP] that we have, that we released this spring.[216]

On the Taliban, Bush's initial strategy was to be more openly friendly with the Taliban than Clinton, wooing them to establish the pipeline from Turkmenistan to the Indian Ocean, to put pressure on Iran and to open negotiations on handing in Osama bin Laden. Bush invited a Taliban representative, Sayed Rahmatullah Hashemi, to the United States in March 2001.[217] He resumed talks on peace with the Northern Alliance, the establishment of the TAP pipeline, and the handing in of bin Laden.

In balancing oil diplomacy and anti-terrorism necessities, Bush used a two-track policy on the Taliban where, on one hand, he used diplomacy and negotiations, and on the other hand he used pressure and coercive measures. In March 2001, for instance, the United States supported UN sanctions against Afghanistan because of the Taliban's refusal to extradite bin Laden.[218] Additionally, in the summer of 2001, the UN (with Washington's consent) imposed two sets of sanctions over Afghanistan for providing sanctuary to terrorists and refusing to surrender bin Laden.[219] On the other hand, the Taliban were promised billions of dollars in commissions if they were to form a national unity government with the anti-Taliban Northern Alliance, extradite

Osama bin Laden and protect the TAP pipeline,[220] as Washington still saw the Taliban as a source of stability in Afghanistan that could help the TAP pipeline.[221] Also, US ambassador to the UN Nancy Soderberg said on 12 February 2001 that the United States would 'find a way to have a continuing dialogue on humanitarian issues with the Taliban'.[222]

However, by the summer of 2001, the limited results of negotiations with the Taliban were frustrating the Americans. In July 2001, at a UN-sponsored meeting in Berlin, the Taliban were invited to attend the meeting, but they declined to send a delegate,[223] saying that the West was too biased towards the Northern Alliance.[224] The Bush administration did explore policy options on Afghanistan, including the option of supporting regime change.[225] A draft presidential directive circulated in June 2001 directed Rumsfeld to 'develop contingency plans' to attack both al Qaeda and Taliban targets in Afghanistan. However, Rumsfeld did not order his subordinates to begin preparing any new plans against either al Qaeda or the Taliban before the 11 September attacks.[226] In July 2001, the deputies committee recommended a comprehensive plan not just to roll back al Qaeda but to eliminate it, by going on the offensive to destabilize the Taliban. On 4 September the principals approved and recommended a plan that would give the CIA $125–$200 million a year to arm the Northern Alliance.[227]

By 10 September the deputies formally agreed on a three-phase strategy. First, an envoy would give the Taliban a last chance. If that failed, continued diplomatic pressure would combine with a covert action program encouraging anti-Taliban groups in Afghanistan to attack Taliban and al Qaeda targets. In phase three, if the Taliban still did not change their policies, the deputies agreed that the United States would try covert action to topple the Taliban's leadership from within.[228]

As in Iraq, there was no clear-cut decision on what to do about Afghanistan, as energy interests, which required the Taliban's cooperation on the TAP pipeline, conflicted with the need to persuade the Taliban to hand in bin Laden, whether through negotiations or coercion, and the need to defeat al Qaeda. There was also a conflict between negotiations, economic sanctions and military plans (including Rumsfeld's contingency plans, plans on an offensive to destabilize Afghanistan and plans to arm the Northern Alliance). Washington therefore decided on a two-track policy where negotiations and the threat of sanctions (and perhaps military action) went hand in hand.

Conclusion: Bush and foreign energy before 11 September

The Bush administration did not enter office with a comprehensive foreign policy agenda or a 'grand strategy' on foreign policy. It focused on specific cases like the challenge from rogue states and foreign energy procurement. It was also concerned with the unilateral drive to free America from its foreign commitments such as the ABM treaty, the Kyoto agreement on climate change, the Middle East peace process, and peace in Colombia, as it thought that freedom from these international constraints would give America the freedom of action to move unilaterally in the foreign arena. Anti-terrorism was also a concern, but not a high-level priority.

During the months between taking office and the 11 September attacks, there were specific cases which symbolized the quest for American global power in the Bush administration. For instance, the NMD system was seen as a symbol of military power, and foreign energy procurement was seen as a symbol of economic power, while focusing on the Iraqi question and reshaping the Middle East were seen as a symbol of Rumsfeld's 'dissuade and demonstrate' vision of global geopolitical power, or global power preponderance. This was especially true after the embarrassing EP-3 incident with China, where the Bush administration felt the need to assert military power through the pursuit of NMD, and geopolitical power through the Iraq question. But the Bush administration could not, on the practical level, tie these factors together, until 11 September 2001.

There was convergence in Iraq as a point of intersection, or as a point of 'melding', the issues of oil, rogue states and the NMD. Since taking office, Bush had been seeking ways to undermine Saddam Hussein.[229] The fact that Bush gave everyone assignments on 31 January 2001 showed that no clear policy towards Iraq had yet been drawn up.

No real policy was made on Iraq due to divisions among the administration's members.[230] Beyond Iraq, there was no clear, unified policy on how to implement the American ambition for unipolar dominance. Instead, before 11 September, Bush was unsuccessful in moving on key issues. The Americans were unable to establish these symbols of global unipolar dominance, as there was no clear policy on rogue states (especially on Iraq, where there were three working groups, none of which came up with a triumphant approach) and no clear policy on the Middle East. In addition, they were unclear on how to implement the recommendations of the NEP (an energy policy which clearly adopted the Open Door approach as it called for the openness of global energy markets and for a large role for US energy corporations).

There was also no clear policy on the oil-rich Caspian basin (except dealing with issues on a piecemeal basis), and no clear, unified direction on Afghanistan.[231] There was a contradiction in the US approach to the Taliban, as the Bush administration (like the Clinton administration before it) depended on imposing sanctions on the Taliban, while at the same time trying to woo them to get them to to approve the TAP pipeline and to hand in Osama bin Laden. Also in oil-rich Colombia, the Bush administration withdrew itself from the peace process, instead of finding a solution to the continuous bombing of American pipelines in the country by leftist guerrilla groups. The Bush administration withdrew from the ABM treaty in December 2001, but it was no closer to having a viable NMD system. Before the 11 September attacks, the Bush administration was successful in unilaterally freeing itself from global constraints and commitments, but not successful in taking affirmative action.

Nevertheless, after the 11 September attacks, the Bush administration saw an opportunity on which it sought to capitalize. The attacks provided an opportunity to declare a War on Terror; a war which was used to link all foreign policy strands of anti-terrorism, military preponderance and energy procurement together, with points of intersection seen in the invasion of Afghanistan, followed by the invasion of Iraq.

CHAPTER 3

ANTI-TERRORISM MELDS WITH ENERGY PROCUREMENT

How the 11 September Attacks Affected the Quest for Foreign Oil

Strategically, the 'war on terror' ... reflected traditional imperial concerns over control of Persian Gulf resources as well as neo-conservative desires to enhance Israel's security by eliminating Iraq as a threat.
                                                    Zbigniew Brzezinski.[1]

What {American policy makers} care about is running the world. You lose the major oil resources of the world, and it's finished. And you're not just losing them; you're losing them to ... rising, competing power{s}.
                                                    Noam Chomsky.[2]

The 11 September attacks did not decrease the American commitment to an open global economy, as the Bush administration promoted the global Open Door policy as a means of defeating terrorism.[3] Like George H.W. Bush and William J. Clinton, George W. Bush adhered to the policy of openness, as Bacevich argued that a part of the reason for the War on Terror was that 'terrorism is a threat to openness – essential for American economic expansion and, for that reason, the principle according to which the United States intended to organize the international order'.[4] After the 11 September attacks, Bush administration officials linked a strong American economy to defeating the terrorists, as US trade representative Robert Zoellick wrote nine days after the attacks:

> Economic strength – at home and abroad – is the foundation of America's hard and soft power. Earlier enemies learned that America is the arsenal of democracy; today's enemies will learn that America is the economic engine for freedom, opportunity and development. To that end, US leadership in promoting the international economic and trading system is vital. Trade is about more than economic efficiency. It promotes the values at the heart of this protracted struggle.[5]

Therefore, Bush's War on Terror was at the heart of the project for creating an open and integrated world. 'Terrorists want to turn the openness of the global economy against itself', he told Asian leaders gathered in Shanghai in October 2001. 'We must not let them.' An open economic order, wrote Secretary of State Colin Powell, 'reinforces democracy, growth, and the free flow of ideas' in the world at large. In the War on Terror, trade in itself is a weapon. 'We will defeat them, by expanding and encouraging world trade', Bush said.[6] The National Security Strategy of 2002 echoed the idea that free trade helps fight terrorism:

> The United States will use this moment of opportunity to extend the benefits of freedom across the globe. We will actively work to bring the hope of democracy, development, free market and free trade to every corner of the world ... Poverty does not make poor people into terrorists and murderers. Yet poverty, weak institutions, and corruption can make weak states vulnerable to terrorist networks and drug cartels within their borders ... Free trade and free markets have proven their ability to lift whole societies out of poverty.[7]

Beside the continuation of the Open Door policy, the Bush administration's foreign policy saw some amendments after the 11 September attacks. Before 9/11, the two foreign-policy strands of the Bush foreign policy had been military advancement (as symbolized by the pursuit of a NMD system) and the procurement of energy from foreign sources. Iraq seemed to be the point of intersection between all of these factors: it was a rogue state with ample oil resources, where the Bush administration was considering a form of military action (albeit not yet a full-scale invasion). Anti-terrorism was added as a main strand after the 11 September attacks, and the terrorist threat happened to emanate from countries which were rich in energy sources and/or constituted vital routes to energy pipelines and transportation, namely in Central Asia.[8]

In addition to the three strands, there was a fourth strand highlighted as the main theme of the QDR, released on 30 September 2001: the prevention of the rise of a rival power able to compete with the United States and the promotion of the American 'ability to project power worldwide', especially in 'critical points around the globe'. It was therefore vital, said the QDR, to prevent 'regional powers' from developing 'sufficient capabilities to threaten stability in regions critical to US interests'. The QDR mentioned a region of special interest to the United States: the 'arc of instability that stretches from the Middle East to Northeast Asia', a region which 'contains a volatile mix of rising and declining regional powers', where 'the governments of some ... states are vulnerable to overthrow by radical or extremist internal political forces or movements'.[9] It was obvious that oil-producing regions were included in the expression 'critical points',[10] and that energy resources were among the interests which the United States needed to protect in the 'arc of instability' region. Thus, all four strands of Bush's foreign policy (global power projection, military preponderance, anti-terrorism and energy procurement) were brought together in the QDR:

> The United States and its allies and friends will continue to depend on the energy resources of the Middle East, a region in which several states pose conventional military challenges and many seek to acquire – or have acquired – chemical, biological, radiological, nuclear, and enhanced high explosive (CBRNE)

weapons. These states are developing ballistic missile capabilities, supporting international terrorism, and expanding their military means to coerce states friendly to the United States and to deny US military forces access to the region.[11]

Based on the passage above, which links oil resources to US military access, Michael Klare argued that the NEP was reflected in the QDR and its call for global power projection.[12] Thus, according to Klare, all the foreign-policy strands merged together in one grand strategy with a unified design that governed US foreign policy.[13] To preserve US power, the QDR also called for 'preserving for the President the option to call for a decisive victory in one of these conflicts – *including the possibility of regime change or occupation* [emphasis added]'.[14] Regime change and occupation would be applied in the cases of Iraq and Afghanistan, and with major links to US control over global energy resources.

Taking the QDR's military direction even further, Bush introduced the doctrine of pre-emption in the State of the Union Speech of January 2002, saying: 'I will not wait on events, while dangers gather. I will not stand by, as peril draws closer and closer.'[15] He reiterated his point in his West Point Speech on 1 June 2002, stating that 'containment is not possible when unbalanced dictators with weapons of mass destruction can deliver those weapons on missiles or secretly provide them to terrorist allies' and that 'we must take the battle to the enemy, disrupt his plans, and confront the worst threat before they emerge'.[16] The introduction of pre-emption as a doctrine was also mentioned in the NSS of 17 September 2002. Stating that deterrence would not work 'against leaders of rogue states', it declared that the United States would undertake pre-emptive military action, would not allow its global military strength to be challenged, was committed to multilateral international cooperation but would act alone if necessary, and that the United States would spread democracy and human rights globally, especially in the Muslim world.[17] (Indeed, much of the world's energy resources lie in the Muslim world,[18] which the US wanted to reshape.)

Showing the link between military security and energy security, the National Security Strategy of 2002 also called for the United States to enhance the security of its foreign energy resources. The document summarized:

> We will strengthen our own energy security and the shared prosperity of the global economy by working with our allies, trading partners, and energy producers to expand the sources and types of global energy supplied, especially in the Western Hemisphere, Africa, Central Asia, and the Caspian region. We will also continue to work with our partners to develop cleaner and more energy efficient technologies. Economic growth should be accompanied by global efforts to stabilize greenhouse gas concentrations associated with this growth, containing them at a level that prevents dangerous human interference with the global climate. Our overall objective is to reduce America's greenhouse gas emissions relative to the size of our economy, cutting such emissions per unit of economic activity by 18% over the next 10 years, by the year 2012. Our strategies for attaining this goal will be to:
>
> - remain committed to the basic UN Framework Convention for international cooperation;
> - obtain agreements with key industries to cut emissions of some of the most potent greenhouse gases and give transferable credits to companies that can show real cuts;
> - develop improved standards for measuring and registering emission reductions;
> - promote renewable energy production and clean coal technology, as well as nuclear power – which produces no greenhouse gas emissions, while also improving fuel economy for US cars and trucks;
> - increase spending on research and new conservation technologies, to a total of $4.5 billion – the largest sum being spent on climate change by any country in the world and a $700 million increase over last year's budget; and
> - assist developing countries, especially the major greenhouse gas emitters such as China and India, so that they will have

the tools and resources to join this effort and be able to grow along a cleaner and better path.[19]

Furthermore, Bush hinted to his energy policy and his desire to decrease dependence on foreign oil, urging Congress to 'act to encourage conservation, promote technology, build infrastructure, and ... increase energy production at home so America is less dependent on foreign oil'.[20]

Even though energy has always been linked to national security early in American history, the 11 September attacks highlighted the importance of energy and its links to national security. According to Duncan Clarke, an expert on energy geopolitics, 'the events of September 11 ... may have provoked a new American global activism to enhance oil interests within the US foreign policy agenda'.[21] This may be because the attacks showed the vulnerability of the United States and thus highlighted the need to enhance US security on all levels, including energy. In October 2001, Energy Secretary Spencer Abraham argued that energy security rests on three pillars: diversification, domestic sources and energy infrastructure.[22] He repeated the calls for regional/geographic diversification of America's foreign energy resources:

> Our administration is looking beyond the Persian Gulf. We are expanding our relationships with other energy producers, and building new partnerships to strengthen our overall energy security. This means working with Canada and Mexico and other nations in the hemisphere to develop regional energy cooperation. This means encouraging development of resources in areas as varied as Africa and the Caspian region.[23]

Abraham also called for diversity in the use of fuels, as 'dependence on one type of fuel ... leaves us vulnerable to price spikes, while excessive dependence on one supplier, or one group of suppliers, leaves us vulnerable to reductions or cut-offs'.[24] He reiterated this argument in a statement before the House International Relations Committee in June 2002, as he said that 'energy security is national security'.[25]

With the elevation of the link of energy to security and US foreign-policy tools (such as military action) after 11 September, Iraq was reinforced as a fundamental case. There were other countries where anti-terrorism was linked to military deployment, for instance the Philippines and Colombia, but Iraq was a more demanding case because of energy procurement and resources. Saudi Crown Prince Abdullah was not as pro-American as the ailing King Fahd, and US–Saudi relations were strained due to disagreements over oil prices and the Arab–Israeli conflict. The neo-conservatives were also worried that the US military presence in Saudi Arabia was a factor for instability in the kingdom. However, if the United States could topple Saddam, Bush could pull American troops out of Saudi Arabia and reduce America's strategic dependence on Saudi Arabia.[26] Beyond the specific security concern was the belief in spreading democratic values as a means to decrease the potential for instability and anti-American terrorism in the region (threats which could spread from the Middle East to a wider global scale), as neo-conservatives believed that US interests were best served by 'the aggressive promotion and spread of democracy throughout the globe'[27] and that only by restructuring the 'Arab tyrannies' of the region could US energy supplies and regional security be ensured.[28] Moreover, democracy would stabilize the region (and its oil resources), according to neo-conservative thought. A final geo-strategic goal was that Israel's security would be enhanced, as Bush thought, wrongly, that the road to Jerusalem went through Baghdad, and that the Arabs, awed by America's military victory in Iraq, would be threatened into peace with Israel.[29]

In order to achieve these goals, Iraq had to be rebuilt into a stable, pro-US market democracy, and this could not be achieved without a strong Iraqi economy, which is very dependent on the oil sector. Thus, if Iraq's oil was an end in itself (in order to control this vital oil supply for the consumption of America and the West), then it was also a means for other ends. Iraq could not be a stable pro-American, democratic model without a stable economy, and a strong Iraqi economy is not possible without the development of the Iraqi oil sector. Thus, Iraqi oil was a tool that would help the reconstruction and democratization of Iraq, decreasing dependence on Saudi Arabia, the redrawing of the Middle East map, the stabilization of the region, securing the region's oil resources, the combat of terror, the enhancement of Israel's position and the promotion of the American empire.

But first the Bush administration had to deal with the immediate issue in the post-11 September era, the Taliban and al Qaeda in Afghanistan, a policy which was also vital for energy routes and the New Great Game.

The four strands meet in the invasion of Afghanistan

Afghanistan's significance from an energy standpoint stems from its geographical position as a potential transit route for oil and natural gas exports from Central Asia to the Arabian Sea. This potential includes the possible construction of oil and natural gas export pipelines through Afghanistan.

An Energy Information Administration report
(affiliated to the US Department of Energy),
released a few days before 11 September 2001.[30]

{Central Asia offers} opportunities for investment in discovery, production, transport and refining of enormous quantities of oil and gas resources. Central Asia is rich in hydrocarbons, with gas being the predominant energy fuel. Turkmenistan and Uzbekistan, especially, are noted for gas resources, while Kazakhstan is the primary oil producer.

An *Oil and Gas Journal* report,
published on 10 September 2001.[31]

'How do you capitalize on these opportunities?' That was the question which Condoleezza Rice asked her NSC staff after the 11 September attacks. She wanted to use the attacks to change the fundamentals of American foreign-policy doctrine and thus change the world order in America's favour, as she told journalist Nicholas Lemann in the spring of 2002:

I really think this period is analogous to 1945–1947 [the period when the Containment doctrine took shape], in that events so clearly demonstrated that there is a big threat, and that it's a big global threat to a lot of countries that you would not have normally thought of as being in the coalition. That has started shifting the tectonic plates in international politics. And it's important to try to seize on that and position American interests and institutions and all of that before they harden again.[32]

Rice re-emphasized the 'opportunity' during a speech at the Johns Hopkins School of Advanced International Studies in April 2002, saying that 'this is a period not just of grave danger but of enormous opportunity'.[33]

One could read Rice's statements as an indication that the Bush administration would use the 11 September attacks to promote US power on the global level. She was not the only one who saw an 'opportunity' in the attacks. On the day of the attacks, Bush himself said that this was a 'great opportunity' to improve relations with big powers such as Russia and China. 'We have to think of this as an opportunity', he said.[34] Powell, too, agreed that the attacks were an opportunity to reshape relationships throughout the world.[35] Rumsfeld told the *New York Times* in October 2001 that the attacks created 'the kind of opportunities that World War II offered, to refashion the world'.[36] In fact, on the night of the attacks, and in front of live television, Rumsfeld said to Senator Carl Levin, then Chairman of the Senate Armed Services Committee that the attacks should convince the Democrats to reverse their opposition to increased military spending, especially on the NMD system.[37] Moreover, the NSS of 2002 stated that 'the events of September 11, 2001, fundamentally changed the context for relations between the United States and other main centres of global power, and opened vast, new opportunities'.[38] Linking these 'opportunities' to the US interests in Middle East oil, Bush told business executives in New York on 3 October 2001: 'I truly believe that out of this will come more order in the world – real progress to peace in the Middle East, stability with oil-producing regions.'[39]

Among these 'opportunities' was the American invasion of Afghanistan. Not only did the War on Terror provide the reason for invading Kabul and ousting the Taliban, but also US global power preponderance, foreign energy procurement and control over global energy supplies would be aided by the invasion of Afghanistan. Zbigniew Brzezinski, the architect of the Carter Doctrine of 1980

which called for securing Middle East oil[40] (in addition to being a consultant to US oil company Amoco and a major supporter of the BTC
pipeline, who played a major role in pipeline diplomacy in Central
Asia)[41], wrote in 1997 that 'in a volatile Eurasia, the immediate task
is to ensure that no state or combination of states gain the ability to
expel the United States or even diminish its decisive role ... A benign
American hegemony must still discourage others from posing a challenge [in Eurasia].'[42] Adding that 'a country dominant in Eurasia
would almost automatically control the Middle East and Africa',[43] he
also stated that the American 'imperial geostrategy' in Eurasia should
achieve:

> [T]he twin interests of America in the short-term preservation
> of its unique global power and the long-term transformation of
> it into increasingly institutionalised global cooperation. To put
> it in a terminology that hearkens back to the more brutal age of
> ancient empires, the three grand imperatives of imperial geos-
> trategy are to prevent collusion and maintain security depend-
> ence among the vassals, to keep tributaries pliant and protected,
> and to keep the barbarians from coming together.[44]

This international hegemonic position of the United States rested on
its ability to control the sources and transport routes for crucial energy
and other strategic material supplies needed by other leading industrial
states. This meant that the United States must pursue a superior role
in Eurasia, and to act to ensure that no other state gained the ability
to challenge the it in this vital region.[45] Developing its military presence in Central Asia, the United States could challenge the Russians
and the Chinese and protect the BTC project.[46] The energy resources
of the Caspian region also fitted into the strategy of regional diversification of oil resources because they were 'non-OPEC oil', meaning
that they were less likely to be affected by OPEC price and supply
policies, thus eroding OPEC's ability to maintain high oil prices and
use oil as a political tool,[47] and decreasing OPEC's clout. Furthermore,
Brzezinksi added that large-scale international investment in Central
Asian resources would increase stability, prevent the resurgence of an

imperial Russia, and help post-imperial Russia democratize and integrate in the international economy.[48]

The BTC pipeline was the main US energy project in the region, supported by the US government for five main reasons. First, it would not pass through Russia or Iran. Second, at an expected construction cost of $3.6 billion, the BTC would provide a profitable contract for US construction companies. Third, the pipeline would provide a platform to further promote Washington's political and economic influence in the east, including the Caspian region. Fourth, with Turkey's Ceyhan port only 483 kilometres away from Haifa port in Israel, the BTC would secure oil supplies to Israel. Fifth, it would benefit Turkey's economy.[49]

The United States continued to support the BTC pipeline after the 11 September attacks and the invasion of Afghanistan. During a visit to Kazakhstan in December 2001, US Secretary of State Powell said: 'I see nothing in the post-September 11 environment that leads me to think we should change the US policy on the routing of pipelines from Central Asia.'[50] On 15 December 2001 the *New York Times* reported that 'the State Department is exploring the potential for post-Taliban energy projects in the region.'[51]

The 11 September attacks fostered the militarization of the BTC project, linking it to US military aid to the countries of the region and to plans to integrate the United States' Caucasian, Caspian and Central Asian dependencies (or, as Brzezinski called them, 'tributaries') into a unified military alliance in the War on Terror. Early in October 2001 Bush promised military assistance to Georgia, and on 27 February 2002 Washington announced that it would provide Georgia with $64 million in military aid and send 180 military advisors to train 2,000 Georgian troops. Ostensibly, this was to fight al Qaeda troops in the Pankrisi Gorge bordering Chechnya, but a Georgian Defence Ministry official announced that the United States was training the Georgian rapid reaction force to guard strategic sites, 'particularly oil pipelines'. On 28 March 2002, US Deputy Assistant Defence Secretary Mira Ricardel announced that the United States would provide military aid to Azerbaijan's navy as part of a $4.4 million aid package 'to counter threats such as terrorism ... and to develop trade and transport corridors'. The United States also reversed an earlier decision that it would not intervene militarily to halt incursions by 'Islamic terrorists' such as the Islamic Movement of Uzbekistan.[52] The United States also increased economic and military aid to Uzbekistan from $50 million to $173 million and offered aid of $125 million to Tajikstan.[53]

Moreover, while proposing grants of $51.2 million to Azerbaijan for Fiscal Year 2005, the State Department said that 'US national interests in Azerbaijan centre on our strong bilateral security and counterterrorism cooperation, the advancement of US energy security, [and] progress in free market and democratic reforms'. The State Department emphasized the role of private US firms in securing US energy needs, saying that 'the involvement of US firms in the development and export of Azerbaijani oil is key to our objectives of diversifying world oil supplies, providing a solid base for the regional economy, and promoting US energy security'. Likewise, in requesting $108.1 million for Georgia, the State Department noted that by housing the BTC pipeline, the country would 'become a key conduit through which Caspian Basin energy resources will flow to the West, facilitating diversification of energy sources for the United States and Europe'.[54]

Based on this military–petroleum–private sector link, or what may be called a 'military–petroleum complex', or 'petro–military complex'[55] Klare wrote that it was 'clear from government documents that the war against terrorism intertwined with [America's] Caspian oil policy', not just in Afghanistan, but in the whole Caspian region.[56] 'In fact', he wrote, 'it is getting harder to distinguish US military operations designed to fight terrorism from those designed to protect energy assets. And the [Bush] Administration's tendency to conflate the two is obvious.' Klare also wrote that 'the American military is increasingly being converted into a global oil-protection service'.[57] Furthermore, Anthony Sampson, author of *The Seven Sisters*, said that 'Western oil interests closely influence military and diplomatic policies, and it is no accident that while American companies are competing for access to oil in Central Asia, the US is building up military bases across the region'.[58] Similarly, Noam Chomsky said in 2002 that establishing US military bases in Central Asia after 9/11 was necessary to protect US corporate interests, to ensure US control over energy resources of the Caspian and to encircle the Middle East's oil resources:

> The United States, for the first time, has major military bases in Central Asia. These are important to position US multinationals favourably in the current 'Great Game' to control the considerable resources of the region, but also to complete the encirclement of the world's major energy resources, in the Gulf region. The US base system targeting the Gulf extends from the Pacific to the Azores, but the closest reliable base before the Afghan war [in 2001] was Diego Garcia. Now the situation is much improved, and forceful intervention, if deemed appropriate, will be greatly facilitated.[59]

The invasion of Afghanistan thus supported global power projection not only against individual countries but also against larger potential rivals. For example, following the invasion of Afghanistan, the influence of the Shanghai Cooperation Organization (SCO) decreased on account of the bases that the United States established in the region.[60] The individual member countries have reacted individually to the war in Afghanistan and established independent roles in the US coalition in Afghanistan, rather than acting as a group, and China's influence in the group decreased, as the Central Asian states sought to deepen their relations with the United States.[61]

Beside the challenge for China, the US invasion of Afghanistan also posed a challenge for Russia within the context of the 'New Great Game'. The public rhetoric of Washington and Moscow continued to announce that they would cooperate in the War on Terror, but American military presence was a challenge for the traditional Russian sphere of influence in the Caspian and Central Asia. For example, the United States initially angered Russia by sending US military support and advisors to Georgia in February 2002. After an American explanation that the aid would help Georgia defend pipelines and borders, Russian president Vladimir Putin said that he would accept the US deployment, as long as it was limited to counter-terrorism.[62]

However, empowered by its new military position, the US continued to gain access to the regions' energy resources. A breakthrough to

US energy interests in the Caspian region took place on 15 May 2002, when Hamid Karzai, the interim president of Afghanistan, signed the deal to construct the TAP pipeline, breaking the deadlock on the pipeline negotiations which existed during the days of the Taliban.[63] Thus, the US invasion of Afghanistan, and US military presence in Central Asia, served as a point of intersection of all four foreign policy strands of military enhancement, global power projection, anti-terrorism and foreign energy procurement.

Despite this breakthrough, however, the United States and the West still lacked a clear strategic vision for Central Asia after the 11 September attacks, avoiding the articulation of a more coherent strategic vision for Central Asia, and avoiding getting involved in the domestic politics of the countries.[64] Far from having a plan to redraw the map of the Caspian region (like its ambitions for the Middle East), the Bush administration showed a lack of interest in Afghanistan as preparations for actions against Iraq distracted it from Kabul.

Bush and Saudi Arabia: Oil, Islamic fundamentalism and the democratic exception

The link between anti-terrorism and the procurement of energy resources cannot be discussed without recognizing the complex American relationship with Saudi Arabia. Allegations of a link between Saudi Arabia and the 11 September attacks caused controversy in Washington. This controversy was addressed in the findings of *The 9/11 Commission Report* of July 2004, which said that it found no evidence of a link between the attacks and the Saudi government:

> Saudi Arabia has long been considered the primary source of al Qaeda funding, but we have found no evidence that the Saudi government as an institution or senior Saudi officials individually funded the organization. (This conclusion does not exclude the likelihood that charities with significant Saudi government sponsorship diverted funds to al Qaeda).[65]

Despite these controversies and complexities, the Saudis were traditional allies of the United States. They financed American military operations, supplied the United States with oil and provided strategic support. One immediate reason for the gentle US line with Saudi Arabia after the 11 September attacks was the need to stabilize the price of oil. The price of oil increased immediately after the attacks,[66] but it later fell because of a drop in demand for oil[67] and, more importantly, an increase in supply started by the Saudis on the day after the attacks. The Saudis ignored the quotas they had agreed with fellow producers, and for the next two weeks they shipped an extra 0.5 mbpd to the United States.[68] The price of oil declined from $28 before the attacks to less than $20 a few weeks after.[69]

Despite the friendly Saudi gesture with respect to oil supplies after the 11 September attacks, tensions persisted between the United States and the Saudi royal family, especially regarding the Arab–Israeli question and Islamic fundamentalism which might lead to terrorism. Not just were 15 of the 19 hijackers Saudi nationals, but there were claims that the Saudi government was not fully cooperating on anti-terrorism issues, and claims that Washington was tolerating Riyadh's lack of cooperation in order to protect the oil ties between the two countries.[70]

Ties were further strained by Saudi anger at the lack of sufficient American focus on the Arab–Israeli conflict. Crown Prince Abdullah introduced an initiative in the spring of 2002 which called for the Arab recognition of Israel and normalization of relations with the Jewish state, in return for an Israeli withdrawal to the 4 June 1967 borders, the establishment of a Palestinian state with East Jerusalem as its capital, and a solution to the refugee problem. Bush paid lip service to the Prince Abdullah initiative, saying on 4 April:

> The recent Arab League's support of Crown Prince Abdullah's initiative is promising, is hopeful, because it acknowledges Israel's right to exist. And it raises the hope of sustained, constructive Arab involvement in the search for peace. This builds on a tradition of visionary leadership, begun by President Sadat and King Hussein, and carried forward by President Mubarak and King Abdullah [of Jordan].[71]

Despite the verbal support, however, the Bush administration was not actually supportive of the Saudi initiative. Cheney's trip to the Middle East in March 2002, which sought Arab support for an upcoming war on Iraq, largely failed, because it took no heed of the wishes of Arab leaders who were more interested in a solution to the Arab–Israeli clashes,[72] and a meeting between Bush and Abdullah in April 2002 at Bush's ranch in Crawford, Texas, did not overcome these differences. Bush wanted to discuss terrorism while Abdullah wanted to discuss the Arab–Israeli conflict. According to one American who was present at the meeting, it was 'very rocky and tense'.[73]

The Americans also believed that the Saudis might not be a reliable ally when it came to military action on Iraq. Saudi Arabia feared that an American invasion of Iraq would lead to the exploration of more Iraqi oilfields, promoting a higher supply and increased competition against Saudi oil.[74] With the kingdom struggling with soaring unemployment and falling standards of living, the gush of oil likely to spew from Iraq, the only country whose reserves could rival Saudi Arabia's, could have proved devastating to the Saudi economy.[75]

The second reason for the initial Saudi opposition to the invasion of Iraq was fear of the strong Islamic institutions in the kingdom, which opposed the American-led war on Iraq. The Saudi regime's legitimacy is based on an alliance with the Islamist movements and extremely conservative Sunni sects in Saudi Arabia,[76] and such institutions would not be happy if the royal family showed support for the American invasion of a neighbouring Muslim country. A third reason was the destabilizing effect that the removal of Saddam would have on the region.[77]

On 10 July 2002 a report written by the RAND organization was presented to the Pentagon's Defence Policy Board. It described Saudi Arabia as the 'kernel of evil' in the Middle East, citing that 15 of the 11 September hijackers were Saudi citizens, and that Saudi money was financing the spread of a highly intolerant strain of Islam around the world. The RAND report added that the Saudis were 'active at every level of the terror chain ... from planners to financiers, from cadres to foot soldiers, from ideologists to cheerleaders'. The report recommended that Washington give the House of Saud an ultimatum; the

Saudis must stop backing terrorism or face a seizure of the Saudi oil-fields and financial assets in America.[78] The administration baulked at going that far, as Colin Powell and Donald Rumsfeld assured Riyadh that this report was written by an independent think tank, and that it did not represent the official stance of the Bush administration.[79]

Similarly, the Saudis pulled back from a showdown.[80] In September 2002, Saudi Arabia declared that it would agree to provide bases to the US military if action against Iraq was supported by the United Nations,[81] offering the use of the Prince Sultan Air Base (P-SAB) for any such operation.[82] Also, Riyadh agreed to give its full (albeit secret) support to an invasion if Saudi Arabia was to play a major role in shaping the post-invasion regime in Baghdad, and if Washington promised to genuinely commit to an equitable and lasting Arab–Israeli peace process.[83]

Iraqi oil: An end in itself ... and a means for other ends (Oil makes Iraq a point of intersection of the four foreign policy strands)

The strategic logic for invasion is compelling. It would eliminate the possibility that Saddam might rebuild his military or acquire nuclear weapons and thus threaten the security of the world's supply of oil. It would allow the United States to redeploy most of its forces away from the region afterward.

Kenneth Pollack.[84]

The trouble with diversifying outside the Middle East ... is that this is not where the oil is. One of the best things for our supply security would be to liberate Iraq.

Sarah Emerson, Energy
Security Analysis Inc., 2002[85]

William Appleman Williams has said that economic expansion to foreign markets was 'both a means and an end for American policy-makers' (an end due to its importance to the US economy, and a means to build a US empire).[86] In a similar fashion, this book, too, argues that invading Iraq and seizing its oil was both an end in itself and a means for other, greater ends. Seizing Iraqi oil was an end in itself, since it would achieve the economic interests of US oil companies (and, by extension, boost the US economy). But seizing Iraq's oil was also a means towards other ends, since controlling the region's oil resources would give the US imperial leverage over the oil supplies of other great powers like Europe, Russia and Asia, and since Iraq's oil revenues were necessary to rebuild Iraq in a pro-American model, reshape the region, and secure Israel, thus expand America's neo-imperial influence.

The US invasion of Iraq, although driven by a number of factors, can be seen as part of a long-term drive to perpetuate America's dominance in the Persian Gulf and Caspian region; it can also be read, according to Klare, as a demonstration of America's determination to 'retain control over the spigot of the Persian Gulf oil stream'.[87] Indeed, the invasion of Iraq was an extension of the Carter Doctrine of 1980, which stated that the US would use military power to protect the oil resources of the Persian Gulf.[88] However, this aim was concealed by the official presentation of what Paul Wolfowitz called 'the one issue that everyone could agree on, which was weapons of mass destruction as the core reason' for invading Iraq.[89]

The fallacy of the WMD and Al Qaeda link claims

Events show that the Bush administration built a case for Iraqi WMD despite a clear lack of evidence. For instance, CIA Director George Tenet was pressurized by Cheney, the Pentagon's Office of Special Plans (an office founded in autumn 2002 by Douglas Feith, a corporate lawyer, leading neo-conservative and Wolfowitz's deputy, to plan for invading Iraq) and other members of the Bush administration to produce intelligence reports giving evidence of Iraqi WMD, despite the lack of strong evidence or credible intelligence information.[90] Paul Pillar, the CIA's chief intelligence officer on the Middle East from 2000 to 2005, accused the Bush administration of 'cherry-picking' intelligence on Iraq to justify a decision it had already reached to go to war.[91] Rumsfeld insisted that Iraq possessed a WMD program despite the lack of solid evidence to support this claim, saying that 'the absence of evidence is not an evidence of absence'.[92] Furthermore, retired ambassador Joseph Wilson was sent by the Bush administration to Niger in 2002 to investigate whether Saddam had purchased uranium (yellow cake) from that country. Wilson found that no such transaction had ever occurred. However, Bush claimed in his State of the Union speech of 2003 that 'the British government has learned that Saddam Hussein recently sought significant quantities of uranium from Africa'. Wilson assailed the White House's credibility in a *New York Times* article. Shortly afterwards, his wife's identity as an undercover employee at the CIA's Counterproliferation Division (CPD) was illegally leaked by members of the Bush administration.[93] Hans Blix, head of the UN Monitoring, Verification and Inspection Commission (UNMOVIC) said on 25 August 2002 that there was 'no clear-cut evidence' that Iraq possessed weapons of mass destruction, and that the biological weapons which Saddam possessed in the past may have expired, since they had an average shelf life of five years.[94] Moreover, opposition to this propaganda over Iraq's WMD came from the US intelligence community itself.[95] For instance, Bob Woodward cited various US officials and sources who admitted 'confidentially that the intelligence on WMD was not as conclusive as the CIA and the administration had suggested' and that the evidence was 'shaky' and 'pretty thin'.[96] No WMDs were found in Iraq after the invasion, as they were all destroyed by UN inspections in the 1990s.[97]

After the failure to find any WMD in Iraq after the American invasion, David Kay, the former senior US weapons inspector in Iraq, said in a testimony before the Senate Armed Services Committee in late January 2004 that claims that Iraq possessed WMD were false, and that 'it turns out we were all wrong'.[98] Furthermore, in late March 2005, a report by the Commission on the Intelligence Capabilities of the United States Regarding Weapons of Mass Destruction, chaired by Laurence Silberman, said that 'the intelligence community was dead wrong in almost all its pre-war judgements about Iraq's weapons of mass destruction'.[99] Citing Wolfowitz's statement that the WMD issue was 'the one issue that everyone could agree on', Tenet said that he 'doubted' that the WMD were 'even the principal cause' for the war.[100]

Indeed, Bush's argument for war was not just based on WMD, but also on claims of Saddam's alleged ties to al Qaeda. Bush claimed in his State of the Union speech in January 2003 that the war on Iraq was a necessary step for the elimination of WMD, and a vital part of the global War on Terror due to Saddam's WMD and alleged links to al Qaeda:

> With nuclear arms or a full arsenal of chemical and biological weapons, Saddam Hussein could resume his ambitions of conquest in the Middle East and create deadly havoc in that region. And this Congress and the American people must recognize another threat. Evidence from intelligence sources, secret communications, and statements by people now in custody reveal that Saddam Hussein aids and protects terrorists, including members of al Qaeda. Secretly, and without fingerprints, he could provide one of his hidden weapons to terrorists, or help them develop their own.[101]

Powell reiterated this argument during his presentation to the UN Security Council in February 2003, where he said that 'we've learned that Iraq has trained al Qaeda members in bomb-making and poisons and deadly gases'.[102] However, there was no evidence that Saddam actually had ties to al Qaeda. Shortly after the 11 September attacks, Counter-Terrorism Coordinator Richard Clarke confirmed that there was no link between the attacks and Saddam.[103] Also, FBI director Robert Mueller said in April 2002 that there was also no evidence that Mohamed Atta, the chief hijacker, met an Iraqi agent in Prague in April 2001.[104] On 7 October 2002, Tenet testified before the Senate Intelligence Committee that the CIA could not find any evidence linking Saddam to al Qaeda (which is the opposite of what he said before the same committee on 11 February 2003).[105] In September 2003 Bush admitted that 'we have no evidence that Saddam Hussein was involved with the 11 September attacks'.[106] A 2005 CIA report said that the CIA was aware of several contacts between al Qaeda and Saddam Hussein's regime in the 1990s, but that these contacts never resulted in a serious relationship, as Saddam was 'distrustful of al Qaeda and viewed Islamic extremists as a threat to his regime, refusing all requests from al Qaeda to provide material or operational support'.[107]

Therefore, CIA counter-terrorism expert Michael Scheuer said that 'the US invasion of Iraq was not pre-emption; it was ... an avaricious, premeditated, unprovoked war against a foe who posed no immediate threat but whose defeat did offer economic advantage'.[108] These economic advantages were the control of the global oil supplies (not necessarily for American consumption, but to have influence over rival economies), promoting the interests of US oil companies, and maintaining the US dollar as the main currency for pricing oil.

Control of Iraq's oil

In November 2002, Rumsfeld said in a CBS interview that the invasion of Iraq was not about oil. 'It has nothing to do with oil, literally nothing to do with oil', he insisted.[109] He reiterated these statements in an interview with Al Jazeera in February 2003:

> The United States is not interested in the oil in that region from Iraq. That's just utter nonsense. It is not interested in occupying any country ... We don't take our forces and go around the world and try to take other people's real estate or other people's resources, their oil. That's not what the United States does. We never have and we never will. That's not how democracies operate.[110]

However, other US officials confirmed the role of oil in the invasion of Iraq. For instance, former Federal Reserve Chairman Alan Greenspan said 'I am saddened that it is politically inconvenient to acknowledge what everyone knows: the Iraq war is largely about oil'.[111] General John Abizaid, retired head of Centcom and military operations in Iraq, said in 2007 about the Iraq war: 'Of course it's about oil, we can't really deny that.'[112] Lawrence Lindsey, Bush's White House economic advisor, said in September 2002 that 'Under every plausible scenario, the negative effect [of invading Iraq] is quite small relative to the economic benefits that would come from a successful prosecution of the war. The key issue is oil, and a regime change in Iraq would facilitate an increase in world oil.'[113] 'We will probably need a major concentration of forces in the Middle East over a long period of time', said leading neo-conservative Robert Kagan. 'When we have economic problems, it's been caused by disruptions in our oil supply. If we have a force in Iraq, there will be no disruption in oil supplies.'[114]

Like Ian Rutledge (who said that the invasion did not aim to 'steal' Iraq's oil but to 'control' it),[115] Noam Chomsky said that the Iraq war aimed primarily to seize Iraq's oilfields, stressing that the purpose was not just ensuring American 'access' to Iraqi oil for US consumption, but rather, more importantly, US 'control' of Iraqi and global oil supplies. He agreed with Chalmers Johnson on the importance of having military bases at the heart of the world's largest oil-producing region, as the United States needed to control the world's oil supplies. Since the 1940s, the US used control of oil as strategic leverage against its rivals, and today, three global economic centres are facing off: the United States, Europe and Asia.[116] Japan is a partner whose allegiance to the United States is guaranteed by, among other factors, US control over its oil supplies from the Middle East, and China is a potential threat to US global hegemony, which is also checked by US control of Middle East oil supplies.[117] Control over Iraq's oil would enormously strengthen US control of global energy resources, a crucial lever of world control and establishing an imperial grand strategy.[118] If Washington loses control of Middle East oil supplies, the United States would descend to second-class power status.[119] Clark agreed that access to oil and gas for American consumption was not the ultimate aim of the US invasion of Iraq. Rather, the more ultimate aim was US 'control' over the sources of energy of its economic rivals – Europe, Japan, China and other nations aspiring to be more economically independent from Washington.[120]

'Not only does America benefit economically from the relatively low costs of Middle Eastern oil', wrote Zbigniew Brzezinski, 'but America's security role in the region gives it indirect but politically critical leverage on the European and Asian economies that are also dependent on energy exports from the region'.[121] Chomsky cited Brzezinksi's statement as proof that the invasion of Iraq was about maintaining America's firm hand on the oil spigot to give America 'critical leverage' over Europe and Asia's oil supplies.[122] Brzezinski emphasized the importance of 'domination' in the oil- and gas-producing regions of the Persian Gulf and the Caspian Basin region, in order to have a 'hegemonic asset' over Europe and Asia: 'Since reliable access to reasonably priced energy is vitally important to the world's three economically most dynamic regions – North America, Europe and East Asia – strategic domination over the area, even if cloaked by cooperative agreements, would be a globally decisive hegemonic asset.'[123] (Even though Brzezinksi repeatedly called for US dominance over oil regions, he expressed his concerns over the invasion of Iraq in a *Washington Post* article in February 2003, arguing that 'the preoccupation with Iraq – which [did] not pose an imminent threat to global security – obscure[d] the need to deal with the more serious and genuinely imminent threat posed by North Korea.' He added that 'force may have to be used to [disarm Iraq]. But how and when that force is applied should be part of a larger strategy, sensitive to the risk that the termination of Saddam Hussein's regime may be purchased at too high a cost to America's global leadership.')[124]

David Harvey agreed that control over the 'global oil spigot and hence the global economy' and strengthening America's military and geostrategic position in Eurasia were an aim of the Iraq war.[125] Klare agreed that 'controlling Iraq is about oil as power, rather than oil as fuel ... Control over the Persian Gulf translates into control over Europe, Japan, and China. It's having our hand on the spigot.'[126]

Arguing that Washington was establishing an 'empire of bases' near oil-rich regions,[127] Chalmers Johnson agreed that the invasion of Iraq was a part of a US strategy of building military outposts in oil-rich south-east Eurasia, including Kosovo, Afghanistan, Central Asia and the Persian Gulf, in order to bring the region's oil supplies under US hegemony. The reason for invading Iraq, said Johnson, was not counter-terrorism, or democratization, but control of oil supplies to areas which pose a threat to US power such as Europe or China, ensuring US oil corporations' interests, securing Israel, and fulfilling America's imperialist ambitions, where Iraq would be the 'jewel in the crown' of this grand strategy.[128] Arguing that Bush's War on Terror was a 'variant' of the Open Door policy to expand the US empire,[129] and describing the invasion of Iraq as an 'imperialist power grab',[130] former CIA analyst Stephen Pelletière agreed that the Bush administration saw the invasion of Afghanistan, followed by Iraq, as a plan to establish US bases in an oil-rich region,[131] because in order for the United States to maintain its imperial status, it must control oil resources, and thus it was necessary to invade Iraq.[132] By remaining the dominant power in the Persian Gulf and Central Asia, the United States could achieve more than just the safety of future oil supply; it could also exercise a degree of control over the energy supply of other oil-importing countries.[133] Control of oil is 'the centre of gravity of US economic hegemony',[134] and the United States aimed to control the destination of Iraq's oil exports,[135] thus providing the United States with a potential stranglehold over the economies of potential rivals.[136] Similarly, John Morrissey argued that the Persian Gulf region, a 'geoeconomic pivot vital to US and global economic health', required 'a strong US military presence in key producing areas and in the sea lanes that carry foreign oil to American shores'. The invasion of Iraq fitted into this US military–geoeconomic agenda for Persian Gulf oil, which aimed for 'an overt geoeconomic project' where the US military was used for 'maintaining the conditions for a US-centred global economy, defined by neo-liberal political economic doctrine' and 'defending global capitalism, free markets and [the] neo-liberal order'.[137]

In short, US global dominance required the US military to patrol the major sea lanes used by the world's 3,500 oil tankers, and also to 'physically control the regions with the world's largest remaining hydrocarbons'.[138] The control of Iraq's oil would facilitate this task.

Creating investment opportunities for US energy firms

Ahmed Chalabi, leader of the Iraqi national council, famously said that 'American companies will have a big shot at Iraqi oil'.[139] The major US energy companies are the indispensable means by which the government achieves its energy policy needs,[140] and in pursuit of diversified and expanded oil resources in Iraq, the United States could promote the interests of its own oil companies.[141] The Bush administration believed that Iraq could become a leading oil supplier in the decades ahead, if a stable Iraqi government, willing to open oil reserves for exploitation by US oil companies, was put in power.[142] Thus, Washington needed to install a puppet regime in Baghdad that would align itself with US corporate interests, mainly to award Iraq's oil contracts only to US and UK companies, thus nullifying Saddam Hussein's contracts with French, Russian and Chinese firms, which were revealed by Cheney and Rumsfeld in the 'Foreign Suitors of Iraq's Oil' document.[143] According to Klein, this showed that Iraq fitted the historical pattern of overthrowing regimes, under the claims of over-exaggerated security threats, in order to maintain America's corporate interests.[144] Moreover, if a new regime in Baghdad rolled out the red carpet for the oil multinationals to return, it would be possible that a broader wave of de-nationalization could sweep through the global oil industry, reversing the historic changes of the early 1970s,[145] including oil price shocks and the nationalization of oil assets.

Protecting the petrodollar system from the petroeuro threat

Stressing the role of the US dollar as a main factor in US hegemony, a former UK government official says that 'there were only two credible reasons for invading Iraq: control over oil and preservation of the dollar as the world's reserve currency ... Oil and the dollar were the real reasons for the attack on Iraq, with WMD as the public reason now exposed as woefully inadequate.'[146] Furthermore, one financial expert said that:

> In the real world ... the one factor underpinning American prosperity is keeping the dollar the world reserve currency. This can only be done if the oil producing states keep oil priced in dollars, and all their currency reserves in dollar assets. If anything put the final nail in Saddam Hussein's coffin, it was his move to start selling oil for euros.[147]

Similarly, global energy consultant Alan Simpson said: 'I think that the decision by Saddam Hussein to go for a petroeuro was the last straw. They had to move.'[148]

In November 2000, Saddam decided to sell oil only in contracts denominated in euros, marking the first time an OPEC country attempted to challenge the petrodollar hegemony,[149] threatening to suspend all oil exports (about 5% of the world's total) if the UN's Oil-For-Food program turned down his request. Shortly thereafter, Iraq's oil proceeds went into a special UN account for the Oil-For-Food program, and were then deposited at Baghdad's euro-based account at BNP Paribas. The move did not make economic sense, since the euro was at record lows against the dollar, and thus the move was seen as a largely symbolic political message.[150] (Even the US companies, which purchased around 65% of Iraq's oil exports under the Oil-For-Food program, paid in euros.)[151] However, Saddam's switch to the euro resulted in Baghdad making a profit when the value of the euro later appreciated against the dollar in 2002.[152]

Although Iraq's switch to the euro, in itself, did not have a huge impact on the global oil market, the ramifications regarding further OPEC movement towards a 'petroeuro' system were quite profound. If invoicing oil in euros were to spread, it would create a panic sell-off of dollars by foreign central banks and OPEC oil producers. In the months before the Iraq war, hints in this direction were heard from Russia, Iran, Indonesia and Venezuela.[153] Moreover, OPEC members were considering moving away from the dollar to the euro, as they considered using the euro, as well as the US dollar, when determining price targets for crude.[154] The US dollar was weakened in 2002 due to the economic strain caused by the bursting of the dotcom bubble, and the corporate scandals of Arthur Andersen, Enron, Worldcom and other similar scandals. This caused the euro to appreciate in value in 2002 relative to the dollar, aggravating the fears that OPEC might follow Iraq and move towards a petroeuro.[155] Also, the fear of assets being frozen by the US Patriot Act caused Middle Eastern investors to sell dollar assets.[156] In fact, one former US government employee said that 'one of the dirty little secrets of today's international order is that the rest of the globe could topple the United States from its hegemonic state whenever they so choose with a concerted abandonment of the dollar standard'.[157] (It is unlikely, however, that industrialized countries or OPEC would risk the complete abandonment of the dollar,[158] as such a massive sell-out of dollar assets would cause a sharp decline in the value of the US dollar, and their dollar assets would lose value.)

Therefore, regime change in Baghdad aimed to prevent further movement within OPEC toward the euro as an alternative oil transaction currency. In order to pre-empt OPEC, the US Government needed to gain control of Iraq and its oil reserves, and the currency to price it, in order to preserve the petrodollar system.[159]

Weakening OPEC's influence in the global oil market

The conventional wisdom on the relation between OPEC and the Iraq invasion was that invading Iraq would help roll back OPEC (which supplied 40% of the world's global oil production), which was a major objective for the US Government.[160] The neo-conservatives wanted to cripple OPEC which they considered 'evil'.[161] For instance, Claudia Rosett, influential journalist at the *Wall Street Journal* and member of the neo-conservative think-tank The Foundation for the Defence of Democracy (FDD), famously called OPEC a 'purely evil cartel ... a gang of price-fixing, oil-rich thug regimes that meet and reinforce assorted terrorist-sponsoring tyrants and gouge consumers'.[162] Rosett's statement has two weaknesses. First, whether OPEC is a cartel is a much-disputed point among economists because, strictly speaking, OPEC lacks the key ingredients of a true cartel: sufficient control over supply, and the discipline to enforce its rules.[163] (Indeed, price wars did occur between OPEC members, most notably between Saudi Arabia and Venezuela in 1998.) Second, there is no definitive evidence that petrodollars fund global terrorism. Anti-terrorism expert G.I. Wilson argued that oil money does not fund global terrorism, that the conflation of terrorism and oil is a 'contrivance', and that 'terrorists don't need oil money. For terrorists, the money flow doesn't come from oil. It comes from drugs, crime, human trafficking, and the weapons trade.'[164]

There was disagreement in Washington whether it would be in America's favour to destroy OPEC. The neo-conservatives favoured a sell-out of Iraq's oilfields, in order to use Iraq's oil to destroy OPEC through massive overproduction above OPEC quotas. The US oil industry, on the other hand, favoured state control of Iraq's oil reserves, according to Amy Jaffe (one of the authors of the April 2001 *Strategic Energy Policy* report), since the industry feared that over-production would decrease the oil prices to hazardous levels[165] (as in the 1980s). Former Saudi Petroleum Minister Zaki Yamani agreed that Washington would not actually want to see the end of OPEC.[166] So did Rutledge, who said that since the 1980s, the US had never actually wanted to 'bust OPEC', since the American oil industry depended on prices within an optimum level, high enough for the US oil industry to make profit, and low enough not to hurt the US economy.[167] America's objective, instead, was to 'control' OPEC.[168] Yergin, too, did not believe that the new Iraqi government would want to bust OPEC by flooding the market in oil, since Baghdad's need for money would make it more interested to sell oil at $20 or $25 per barrel[169] (which was within the optimum oil price range in March 2003).[170]

The Bush administration resisted a US Congress resolution to sue OPEC for anti-trust practices and price manipulation. OPEC member states enjoy immunity from prosecution under US law, which states that 'courts of one country will not sit in judgment on the acts of the government of another, done within its own territory'. In early 2005, a group of senators led by Senator Mike DeWine (R-Ohio) introduced the 'No Oil Producing and Exporting Cartels Act', or NOPEC, to amend US law to allow the US Department of Justice and the Federal Trade Commission to bring suits against OPEC for monopolistic practices.[171] In April and May 2007, committees in both the Senate and House of Representatives voted to approve versions of NOPEC.[172] In May 2008, the House of Representatives passed NOPEC in a 324–84 vote, enough to override a presidential veto. However, the White House opposed the bill, saying that targeting OPEC investments in the United States as a source for damage awards 'would likely spur retaliatory action against American interests in those countries and lead to a reduction in oil available to US refiners'.[173] The White House always maintained that it would veto the bill, arguing that it could stimulate retaliatory action from OPEC governments to limit US access to global oil supply, thus push crude oil and gasoline prices higher and damage American business interests abroad.[174]

While Bush wanted to increase US control over global oil supplies, and weaken any power that might challenge the US, including OPEC, he saw that the NOPEC bill was not the correct way to weaken OPEC. Rather, his way of weakening OPEC was to invade an OPEC member (Iraq) and use its vast oil resources to gain global influence.

Decreasing dependence on Saudi Arabia for strategic and economic assistance

Former Saudi Petroleum Minister Sheikh Zaki Yamani said that oil was *the* major objective for the United States in seeking to occupy Iraq, because the US was aiming to secure its oil supplies by reducing dependence on oil from Saudi Arabia, adding that the 11 September attacks accelerated the American direction on the regional diversification of oil resources.[175] Indeed, Lawrence Kaplan and William Kristol said that invading Iraq could make it 'replace Saudi Arabia as the key American ally and source of oil in the region'.[176] Kristol specifically called for an invasion of Iraq to decrease dependence on Saudi oil, saying that removing Saddam would 'reduce the Saudis' leverage' and that 'returning Iraqi oil fully to market can only reduce the Saudi ability to set oil prices, and make the US bases there superfluous'.[177] Kristol reiterated this argument in a testimony to the House of Representatives in May 2002, saying that Riyadh could use the oil weapon against the US, that removing Saddam would be 'a tremendous step towards reducing Saudi leverage,' and that 'bringing Iraqi oil fully into world market would improve energy economics. From a military and strategic perspective, Iraq is more important than Saudi Arabia. And building a representative government in Baghdad would demonstrate that democracy could work in the Arab world. This, too, would be a useful challenge to the current Saudi regime.'[178] (In the first half of 2001, 8.9% of the United States' oil came from Saudi Arabia and 3.2% came from Iraq.)[179]

With Saddam gone, thought the neo-conservatives, Bush would pull American troops out of Saudi Arabia and decrease Washington's dependence on Riyadh for oil and strategic assistance.[180] Iraq's oil might provide both a major addition to world reserves and (therefore) a means of reducing Saudi Arabia's central role as the sole effective swing producer.[181] Also, the presence of US military bases in Saudi Arabia, the land of Mecca and Medina, was a main cause of al-Qaeda's attacks inside and outside Saudi Arabia.[182] In an interview with *Vanity Fair* in May 2003, Wolfowitz said that the US military deployment in Saudi Arabia 'over the last 12 years has been a source of enormous difficulty' for Riyadh, and 'a huge recruiting device for al Qaeda. In fact if you look at bin Laden, one of his principal grievances was the presence of so-called crusader forces on the holy land, Mecca and Medina. I think just lifting that burden from the Saudis is itself going to open the door to other positive things.'[183] Thus, Cary Fraser argued that seeing that Saudi Arabia was willing to distance itself from the United States (as seen in the exchanges between King Abdullah and Bush before 11 September), that 15 of the 19 hijackers on 11 September were Saudis and that the US military presence in Saudi Arabia failed to secure US interests against militant Islamist attacks, the Bush administration decided to invade Iraq where one of the aims was to replace US military presence in Saudi Arabia with US military presence in Iraq.[184] There was also a fear of Islamist extremists taking over Riyadh and, by extension, Saudi oil supplies, which would blackmail the West by cutting off its energy supplies.[185]

The Saudis were initially against the war on Iraq, fearing that removing Saddam would open Iraq's oilfields as a competitor to Saudi oil, that removing Saddam would cause disturbance in the region by encouraging the Kurds and the Shiites to try for their own independent states, and that the strong Islamist current in the Kingdom would cause trouble if the Saudi royal family supported the war. However, following the RAND report of July 2002, which described Saudi Arabia as the 'kernel of evil' in the region, the Saudis felt compelled to cooperate in the war on Iraq, albeit secretly.[186]

Rebuilding the Iraqi economy and democratizing Iraq

In implementing the Bush Doctrine's call for spreading democracy in the Middle East as a means to fight terrorism, and redrawing the region's political map in Israel's favour, Bush saw Iraq as a potential model for democratization, as it had an educated population and a strong economic and strategic potential that could set an example for democratization. Richard Perle thought that Iraq provided the best hope for the democratic experiment in an Arab country. 'I think there is a potential civic culture in the Arab countries that can lead to democratic institutions, and I think Iraq is probably the best place to put that proposition to the test because it's a sophisticated educated population that has suffered horribly under totalitarian rule.'[187] Similarly, Wolfowitz said that Iraq was the best place to test democracy.[188] The Bush administration believed that democratization would help fight terrorism and stabilize the region. Richard Haass, the State Department director of policy planning, said in December 2002:

By failing to help gradual paths to democratization in many of our important relationships – by creating what might be called a 'democratic exception' – we missed an opportunity to help these countries become more stable, more prosperous, more peaceful, and more adaptable to the stresses of a globalizing world. It is not in our interest – or that of the people living in the Muslim world – for the United States to continue the exception. US policy will be more actively engaged in supporting democratic trends in the Muslim world than ever before.[189]

For decades, however, democratization in the Middle East was complicated by the Democracy Conundrum; the need to maintain despotic but pro-American regimes to stabilize the region and its oil resources.[190] The United States did not want to take the risk of full democratization, lest it brought to power a regime that would not be as lenient towards US interests. Charles Krauthammer acknowledged that Washington ignored Middle East democratization, due to fear of disrupting oil supplies: 'Oil is why America kept its distance from the region for so long. Ever since Franklin Roosevelt made alliance with Saudi Arabia, the US chose to leave the Arab world to its own political and social devices so long as it remained a reasonably friendly petrol station. The arrangement lasted a very long time.' In other words, the two regions that remained exempt from the democratizing impulse of the United States were Africa and the Middle East, 'because of oil and apparent benignity'. The reason for the war on Iraq was that, after 11 September, the United States realized that 'the old offshore, hands-off, over the horizon policy' towards the Middle East would not work, that it needed to get more involved in the Middle East, and that 'Iraq [was] the beckoning door' to this hands-on approach.[191] The neo-conservative rationale was that only by restructuring the 'Arab tyrannies' of the region could US energy supplies and regional security be ensured,[192] in addition to the fact that democracies would be more willing to fight terrorism and secure Israel.

After the removal of the Saddam regime in Iraq, next on the neo-conservative list of regime change were Syria, Iran, Libya, Sudan, Yemen, Saudi Arabia, the Gulf States, and perhaps even Egypt.[193] As Richard Perle pointed out, 'if a tyrant like Saddam is brought down, the others will start to think'.[194] Similarly, Kaplan and Kristol argued that democratization in the Middle East became a matter of national security after 11 September, since 'Arab repression [fueled] Islamist terror and anti-American sentiments'.[195]

To achieve a stable, strong and democratic Iraq, Iraq's oil exports would help rebuild the country's economy, as well as build up Iraq's market democracy and fulfil America's objectives. Paul Wolfowitz testified before the House of Representatives on 27 March 2003:

> There's a lot of money to pay for this, that doesn't have to be US taxpayer money, and it starts with the assets of the Iraqi people ... and on a rough recollection, the oil revenues of that country could bring between $50 and $100 billion over the course of the next two or three years ... We're dealing with a country that can really finance its own reconstruction, and relatively soon.[196]

Thus, Iraq's oil became more than just a resource to be controlled; oil exports were seen as the 'cash lifeline' or the 'cash crop'[197] that would provide funds for economic and political development in Iraq, thus trigger a democratic domino effect in the region.

However, a February 2003 report by the US State Department's Bureau of Intelligence and Research, titled *Iraq, the Middle East and*

Change: No Dominoes, said that the invasion of Iraq would not likely lead to democracy in Iraq or the Middle East, as Iraq was made up of several ethnic groups deeply hostile to one another. Even if some form of democracy took place in Iraq, said the report, it would probably be manipulated by anti-American Islamic groups, and it would not lead to the spread of democracy throughout the Middle East. The US State Department declined to comment on the report, saying that it did not reflect the Department's opinion. Also, George Tenet was sceptic about a democratic domino effect, saying at a hearing in Capitol Hill: 'I don't want to be expansive in, you know, a big domino theory about what happens in the rest of the Arab world.'[198] Moreover, Paul Pillar, who was the CIA's chief intelligence office on the Middle East from 2000 to 2005, said that intelligence assessments before the invasion indicated that a post-war Iraq 'would not provide fertile ground for democracy', that Iraq would need 'a Marshall Plan-type effort' to restore its economy despite its oil revenue, and that Sunnis and Shiites would fight for power, thus undermining the argument for a democratic domino effect and the argument that oil revenues would cover the costs of reconstruction.[199]

In any case, democratization continued to be an official justification for the war, as the neo-conservatives argued that democratization would stabilize the region, and that democratic governments would be more willing to cooperate on the War on Terror and make peace with America's ally in the region: Israel. Given that Bush and the neo-conservatives had staunch pro-Israel views, the security of Israel and scaring the Arabs into peace with Tel Aviv was part of the justification for invading Iraq.

Advancing the Arab–Israeli peace process, and securing Israel

Bush said in June 2002 that, by removing Saddam, 'our ability to advance the Israeli-Palestinian peace process would be advanced'.[200] Kaplan and Kristol argued that removing Saddam and installing a pro-American regime would improve Baghdad's relations with Israel.[201] Bush thought, wrongly, that the road to Jerusalem went through

Baghdad, and that awed Arabs would be threatened into peace with Israel,[202] where it would be easier to impose on the Palestinians a peace agreement acceptable to the neo-conservatives' Israeli allies, Ariel Sharon and the Likud. Thus, remaking the whole region on a pro-American model would strengthen Israel.[203] Furthermore, Israel hoped that the invasion of Iraq would next lead them to Iran and Syria. 'The war in Iraq is just the beginning', former Prime Minister Shimon Peres said in February 2003.[204]

Therefore, US control over Iraq's oil was an end in itself (due to the economic and business benefits to the US) and a means for other greater ends (first, it would advance the US imperial grip on global oil supplies, and, second, Iraq's oil revenues would help rebuild Iraq according to Washington's mould, serve as a model for the Arabs, secure Israel and help fight terrorism). All four of the Bush foreign policy priorities (energy procurement, military advancement, War on Terror and global power projection) were seen in the arguments for the invasions of Iraq (although, obviously, the War on Terror as a motive to invade Iraq was only false rhetoric, since there were no links between Saddam and al Qaeda, and since he had no WMD). Therefore, it was Iraq's oil which made Iraq the point of intersection of the four strands of foreign policy, as military advancement and global power projection were linked to Iraq's oil, and War on Terror rhetoric was used to justify the invasion.

Iraq policy after 11 September

Before 11 September the Bush administration was still undecided on how to pressurize or overthrow Saddam, considering the three options of regime change through an armed coup by Iraqi opposition groups, enforcement of the UN economic sanctions, and enforcement of the NFZs to protect the Kurds in the north and the Shiites in the south. Bush empowered three working groups to study each of these options.[205] However, the smart sanctions approach, favoured by Secretary of State Colin Powell, faded away in July 2001, amidst UN Security Council disagreement.

It took 24 hours for the proponents of regime change to cite 9/11 as the foundation for military action. During the NSC meeting on 12

September 2001, Rumsfeld said that Afghanistan did not have decent targets for bombing, but Iraq did, so the United States should bomb Iraq. Richard Clarke, US Counter-Terrorism Coordinator, thought that Rumsfeld was joking, but was surprised to see that he was actually serious. Bush replied that the Americans had to change the government in Iraq, not just hit it with missiles.[206] Rumsfeld pressed ahead, instructing General Richard Myers (who was the acting Chairman of the JCS because JCS Chairman Henry Shelton was en route to Europe at the time) to find out as much as he could about Saddam Hussein's possible responsibility for the attacks.[207]

Bush repeated the wider possibilities of a post-9/11 response to his advisors. On 16 September Bush told Rice that the focus would be on Afghanistan, although he still wanted plans for Iraq, in case Iraq somehow took advantage of the 11 September attacks, or in case it turned out that it was involved in the attacks.[208] During an NSC meeting on 17 September he said: 'I believe that Iraq was involved, but I'm not going to strike them now. I don't have the evidence at this point.' He added that he wanted the NSC to keep working on plans for military action in Iraq but indicated that there would be plenty of time to do so.[209] He thus ordered the Pentagon to be ready to deal with Iraq if Baghdad acted against US interests, with plans to include possibly occupying Iraqi oilfields,[210] while working on Afghanistan.[211] On 20 September Bush told British Prime Minister Tony Blair that, although Iraq was not the immediate problem at the time, he was seriously considering attacking Iraq, saying: 'Iraq, we keep for another day.'[212]

'There must be a Phase Two', wrote Richard Perle, head of the Defence Policy Board in November 2001. 'At the top of the list for Phase Two is Iraq.'[213] Eventually, Iraq was marketed to the public as 'Phase Two' of the War on Terror.[214] Other possible targets were eliminated.[215] Somalia, Sudan, Yemen, Georgia and the Philippines were often mentioned, and in some instances Washington dispatched Special Forces to help train local armies in anti-terror campaigns.[216] On 16 November Bush privately asked Rumsfeld to create a secret plan for Iraq.[217] On 21 November Bush ordered Rumsfeld to update the war plans for Iraq.[218] Then, in January 2002, Bush gave his State

of the Union speech, where he falsely accused Saddam of supporting terrorism and seeking WMD, claiming that 'the Iraqi regime has
plotted to develop anthrax, and nerve gas, and nuclear weapons for
over a decade', in order to justify the neo-conservative plans to invade
Iraq, reshape the Middle East and protect oil supplies. He also claimed
that Iraq, Iran and North Korea formed an 'Axis of Evil' in alliances
with 'terrorists', saying that 'states like these, and their terrorist allies,
constitute an Axis of Evil, arming to threaten the peace of the world.
By seeking weapons of mass destruction, these regimes pose a grave
and growing danger.'[219] According to journalist Julian Borger, who
interviewed David Frum, the author of the speech in January 2003,
the main aim of the Axis of Evil speech was to put the case for 'going
after Iraq', as Michael Gerson, Bush's main speechwriter, expressed it
to Frum.[220]

In March 2002 Senator John McCain was in Rice's office at the
White House with two other senators, briefing her about their meetings with European allies, when Bush unexpectedly stuck his head in
the door. 'Are you all talking about Iraq?' Bush asked, his voice tinged
with schoolyard bravado. Before McCain and the others could answer,
Bush waved his hand dismissively and said 'F**k Saddam! We're taking him out!' McCain was appalled. Despite his support for the war
on Iraq, McCain saw that this encounter with Bush was more evidence
of Bush's shallow intellect and dangerous self-regard. Later, McCain
retold the story, saying 'Can you believe this guy? He's the President
[instead of me]!'[221]

However, Bush could not attack Iraq in the spring of 2002, due
to political and military obstacles. First, Osama bin Laden had not
been captured yet (even though Bush said in March 2002 that he was
'truly ... not that concerned about him'[222]). Furthermore, military
action against Iraq would have been a large-scale operation, so Bush
needed time to mobilize the military and move troops and equipment to the region and obtain international support. Moreover, there
was a major conflict going on in the Middle East in the spring of
2002: the Arab–Israeli military clashes. These needed to be calmed
down in order to ensure Arab support for an American invasion of
Iraq. (Cheney's trip to the region in the spring of 2002 failed to reach

such agreement, as the Arab rulers were more interested in solving the ongoing Arab–Israeli problem than in a military action against Iraq.) Finally, there was another global conflict in the spring of 2002: the Indian–Pakistani military clashes which almost led to a nuclear war between the two countries. Bush was hoping to invade Iraq in the autumn of 2002, but he decided to delay to March 2003 in order to build up America's petroleum reserves to counter the expected oil price shock, and to re-stock the hi-tech weapons that had been used up in the Afghanistan war.[223]

Despite these obstacles, the Bush administration's focus remained upon Iraq. Richard Haass, Director of Policy Planning at the State Department, met with Rice in July 2002 in one of their regular meetings, and asked her whether Iraq should really be front and centre in the administration's foreign policy, given the War on Terror and other issues. Haass recalls Rice saying 'that decision's been made, don't waste your breath'.[224] The signal for the renewed American campaign against Iraq came in late August 2002, when Cheney delivered a speech saying that if Saddam built a nuclear weapon, the Iraqis would 'seek domination of the entire Middle East' and the world's oil supplies. Cheney also opposed the inspectors' return to Iraq, saying that it 'would provide false comfort that Saddam was somehow back in the box'. He also warned that Saddam could use control over oil to threaten US interests, as he could 'take control of a great portion of the world's energy supplies, directly threatening America's friends throughout the region, and subjecting the United States or any other nation to nuclear blackmail' if he was not stopped, and that Saddam would continue to develop WMD using his oil wealth.[225]

By the end of August 2002, the final and irrevocable decision to attack Iraq was made, with the signing of the necessary documents on 29 August.[226] However Bush continued to pursue a diplomatic route, at least in appearance.[227] In an attempt to expand the coalition for the Iraq war, he told the UN General Assembly on 12 September 2002 that the UN should pass a resolution authorizing the American war on Iraq. He said 'we will work with the UN Security Council for the necessary resolutions', but stated clearly that the United States was ready to act unilaterally if necessary.[228]

Saddam initially forestalled military action by suddenly allowing the inspectors into Iraq in mid-September.[229] But the UN Security Council adopted Resolution 1441 anonymously on 8 November 2002, which specified 'serious consequences' if Iraq was found in material breach of the UN terms.[230] The first inspections started on 27 November, under the supervision of the United Nations Monitoring, Verification and Inspection Commission (UNMOVIC) and the International Atomic Energy Agency (IAEA). Both agencies reported in early 2003 that they had found no 'smoking gun' or evidence for Iraq's WMD program,[231] and that they needed six more months to complete their investigation.

Thus, the die was cast, as the US rejected the pursuit of a second UN resolution, and on 19 March 2003 the United States started its invasion of Iraq.

Conclusion

US commitment to the Open Door policy of informal imperialism did not wane after the 11 September attacks, as the Bush administration promoted trade and global economic openness as a means to fight terrorism. In fact, the practice of using the military to protect US corporate and economic interests increased significantly after the attacks, as the US set up more and more military bases close to the world's energy regions. The wars in Afghanistan and Iraq were part of a global strategy to reassert US dominance in the international system[232] and maintain the Open Door empire, (except replacing China and the Far East with Central Asia and the Persian Gulf,[233] and the fact that President Karzai and US ambassador to Afghanistan Zalmay Khalilzad were former Unocal employees was a sign of corporatism). After the 11 September attacks, the Bush Doctrine, based on the idea that America was in a War on Terror, called for taking the war to the enemy, pre-emptive strikes, and spreading democracy as a way to fight terror. The Bush administration saw Iraq as an optimum starting point for its democratization programme. It saw Iraq as a the key point of intersection that linked the four strands of US foreign policy, bringing together US military power, the control of oil supplies and the prospect of democratization with the extension of US presence in the Persian Gulf, making Iraq (rather than other rogue states) the focal point of US foreign policy. Iraq's oil had become an end in itself, and a means for the wider ends of rebuilding Iraq as a democratic model for the Middle East, securing Israel, substituting Saudi Arabia as a strategic ally and main oil supplier, fighting terrorism, and extending US empire.

Nevertheless, there were differences between Iraq as a point of inter-section of the four strands, and the Caspian region as another point of intersection. A first difference was that, in the case of Iraq, it was Iraqi oil which made Iraq a point of intersection, while in the case of the Caspian, it was not oil, but rather the War on Terror, which brought all the strands together after the 11 September attacks. In other words, in the case of Afghanistan, it was the 11 September attacks which triggered the American invasion. But in Iraq there were no links to 11 September, so it was oil which brought the strands together, and oil was seen as an end in itself, and a means to achieve other greater ends in the Middle East. (The War on Terror did exist in the Iraq case, but only as rhetoric, or as a false claim, if not an outright lie.) A second dif-ference was that, in the case of the Caspian, the Bush administration did not seem to have a vision to redraw the map of the Caspian region, as was the case with Iraq and the Middle East region (where the US intervention was meant to be a demonstration to establish and confirm US power, and extend it to the rest of the region, in effect redrawing the map of the entire Middle East).

The 11 September attacks also facilitated the conversion of the Bush administration's ambitions into actions, as the US invaded Afghanistan, established bases in Central Asia and prepared for regime change in Iraq. Subsequently, success or failure in Iraq was not only significant for the fulfilment of the four strands in that country. Rather, it would mark whether the US would be able to defend and extend its influence elsewhere because, at that point, the Bush administration made US foreign policy and strategic vision too dependent on success in Iraq.

Following the invasion of Afghanistan, the Caspian region was ignored, in favour of the Persian Gulf. Even after 9/11, the United States and the West still lacked a clear strategic vision for Central

Asia.[234] The American foreign policy refocused on the Persian Gulf, and by the spring of 2002, the four strands of US foreign policy came together in the Persian Gulf, where invading Iraq would help all four strands (the link between Saddam and al Qaeda being false, of course). Iraq's oil (being an end in itself and a means for other ends) was the factor which turned the country into the point of intersection of all foreign policy strands, as Iraq's oil revenues were meant to help the economic reconstruction, and thus democratization, of Iraq, turning Iraq into a model after which the whole region should be reshaped in accordance to US interests.

However, the US invasion of Iraq failed to exploit Iraq's oil resources as it planned, and failed to come up with a reliable Phase IV plan for a stable political and economic structure for Iraq,[235] and thus failed to redraw the Middle East as desired. And as Iraq moved after April 2003 not towards democracy and order, but towards instability and ethnic political and military clashes, it would put not only US control of energy, but also its pursuit of preponderance of power into doubt.

CHAPTER 4

'A BIG SHOT' AND
'A LOT OF MONEY'

Operation Iraqi Freedom and
US Foreign Energy Policy

American oil companies will have a big shot at Iraqi oil.

Ahmed Chalabi, leader of the Iraqi
National Council.[1]

If oil was one of the main reasons for the invasion of Iraq, then it was both an end in itself and a means to other ends. Iraq's vast resources made it a candidate to be a starting point for the neo-conservative plan to trickle down democracy in the Middle East. Democracy would lead to more stability and thus more security for the region's energy supplies, while decreasing the threat of global terrorism. This would not be possible without Iraq's oil, as Iraq cannot be a stable, pro-American democratic model without a strong and stable economy, and a stable Iraqi economy is not possible without Iraqi oil and its revenues.

Donald Rumsfeld said in an interview with *Fortune* magazine in November 2002 that the cost of Iraq's reconstruction would be 'a very different situation from Afghanistan' since 'Iraq has oil'.[2] He said in testimony before the Senate Appropriations hearing on 27 March 2003 that he did not believe that the United States had the responsibility for reconstruction, since reconstruction funds could come from 'various sources', including 'frozen assets, oil revenues, and a variety of other things including the Oil-For-Food program'.[3] His deputy, Paul Wolfowitz, had over-optimistic expectations for Iraq oil revenues, as he testified before a House of Representatives committee on the same day that 'there's a lot of money to pay for this' from the Iraqi oil revenues, not American taxpayers' money.[4] Also on the same day, Deputy Secretary of State Richard Armitage testified to a House of Representatives committee: 'This is not Afghanistan. When we approach the question of Iraq, we realize here is a country which has a resource. And it's obvious, it's oil. And it can bring in and does bring in a certain amount of revenue each year ... $10, $15 or even $18 billion ... this is not a broke country.'[5] State Department official Alan Larson added in a Senate Foreign Relations hearing on 6 April 2003: 'On the resource side, Iraq itself will rightly share much of the responsibilities. Among the sources of revenue available are $1.7 billion in invested Iraqi assets, the found assets in Iraq ... and unallocated Oil-For-Food money that will be deposited in the development fund.'[6] Andrew Natsios, administrator for the Agency for International Development, reiterated the $1.7 billion figure in an interview with Ted Koppel on 23 April 2003, saying that the cost for the US of rebuilding Iraq would be only $1.7 billion, and that the rest of the cost would be paid by international pledges and by Iraqi oil revenues, which he estimated would reach $20 billion/year, when the Iraqi oil sector was 'up and running':

> The rest of the rebuilding of Iraq will be done by other countries who have already made pledges, Britain, Germany, Norway, Japan, Canada, and Iraqi oil revenues, eventually in several years, when it's up and running and there's a new government that's been democratically elected, will finish the job with their own revenues. They're going to get $20 billion a year in oil revenues. But the American part of this will be $1.7 billion. We have no plans for any further ... funding of this.[7]

When Koppel recapped by saying 'As far as the reconstruction goes, the American taxpayer will not be hit for more than $1.7 billion no matter how long the process takes?' Natsios replied 'That is our plan and that is our intention'.[8]

Similarly, the importance of Iraq's oil was evident when Wolfowitz said at an Asian security summit in Singapore in June 2003 that the 'difference between North Korea and Iraq is that we had virtually no economic options with Iraq because the country floats on a sea of oil. In the case of North Korea, the country is teetering on the edge of economic collapse.' He meant that the US had no economic options by means of which to achieve its objectives in North Korea, while in Iraq the US could use oil revenues to rebuild Iraq into a pro-American democracy, according to the neo-conservative vision. [9]

These aspirations were evident in the fact that the Iraqi Oil Ministry was the only government building heavily guarded by US troops during the looting which took place after the occupation,[10] (along with the Interior Ministry which which contained secret dossiers incriminating Saddam, the US and Israel).[11] The relation between Iraqi oil and US foreign policy was thus more than just a war to seize Iraq's oil resources. Oil was an end in itself, and a means for other ends.

These statements by US officials were based on ambitions which went back to the start of the administration in 2001, as seen in the documents used by the Energy Task Force with detailed descriptions of Iraq's 'super giant oilfields', oil pipelines, refineries, and tanker terminals.[12] Tom Fitton, President of Judicial Watch (a public interest law firm that investigates government corruption) said that the presence of such maps and lists on Iraq's oil projects showed that 'Iraq was on the minds of at least some of the members of the [Energy] Task Force long before the war [on Iraq]' and that one could not talk about the Middle East oil situation without talking about Iraq.[13] The declassified documents also included maps of Saudi and UAE oilfields, refineries, pipelines and tanker terminals, with supporting charts detailing major oil and gas development projects. The fact that the same maps and lists used by Secretary of Defence Rumsfeld (when he was arguing for his 'dissuade and demonstrate' doctrine) were used by the National Energy Policy Development Group showed the linkage between the strands of military preponderance and foreign energy procurement.[14]

Shortly after the invasion, the US took steps to pave the way for control of Iraq and its oil resources. On 22 May 2003 the UN Security Council adopted Resolution 1483, which recognized the USA and the UK as occupying powers and ended UN sanctions on Iraq. The Resolution approved the transfer of 95% of revenues on Iraqi oil to the Development Fund for Iraq (DFI), a bank account which held Iraq's oil revenues, in addition to $1 billion in cash that US soldiers had found hidden in the homes of Saddam and his cronies, and money from frozen bank accounts. In total, the DFI, which was controlled by the United States until authority for the fund was transferred to the Iraqi government on 28 June 2004, contained $20 billion which the Coalition Provisional Authority (CPA) was free to spend as it wished. Resolution 1483 added that the money would be spent in the interests of the Iraqi people and that all accounts related to the oil revenues should be independently audited. However, the DFI was abused by the CPA; it was a source of easy money to hand out and pay for reconstruction contracts handed to US companies without the hassle of accounting and contracting regulations. But it did not matter for the CPA, because the DFI was Iraqi money, not American money.[15] UNSCR 1483, adopted by 14 votes to zero, with Syria abstaining, provided the legal framework for Iraqi oil exports to resume.[16] (On 15 December 2010, the UN adopted Resolution 1956 to formally terminate DFI on 30 June 2011, thus allowing Iraq to take full responsibility for its oil revenues.)[17]

On the same day, 22 May 2003, the Bush administration issued Executive Order 13303, which gave oil contractors lifelong exemption from lawsuits and legal proceedings in the US. The order applied to Iraqi oil products 'in the United States, [that] hereafter come within the United States, or that are or hereafter come within the possession or control of United States persons'.[18] According to Jim Vallette, an analyst with the Sustainable Energy & Economy Network of the Institute of Policy Studies in Washington, the Executive Order 'reveals the true motivation for the occupation: absolute power for US interests in Iraqi oil'.[19] He adds that the Executive Order gave US companies complete immunity from legal accountability:

Anything that happened before with US oil companies around the world – a massive tanker accident, an explosion at an oil

refinery, the employment of slave labour to build a pipeline, mur-
der of locals by corporate security, the release of billions of tons
of carbon dioxide into the atmosphere, or lawsuits by Iraq's current creditors or the ... Iraqi government demanding compensation – anything at all, is immune from judicial accountability.[20]

UNSCR 1483 and Executive Order 13303 were complemented by
Order 39, the official order concerned with foreign investment, issued
by Paul Bremer, head of the Coalition Provisional Authority (CPA) on
19 September 2003. The Order included the following provisions:

1) Privatization of Iraq's state-owned enterprises (except the oil
   reserves);
2) 100% foreign ownership of Iraqi business;
3) 'National Treatment', which meant that it was not necessary to
   give Iraqis preference over foreigners;
4) Unrestricted, tax-free remittance of all profits and other funds,
   authorizing investors to transfer abroad all funds associated
   with their investment, including shares, profits, dividends, proceeds ... etc., thus giving no obligation to re-invest in Iraq;
5) 40-year ownership licences: Baghdad could deny foreign investors
   the right to own real property (land, buildings, etc). However, it
   granted foreigners 40-year licences with unlimited renewal options
   to lease Iraqi real estate;
6) Dispute settlement: Foreign companies have the right to reject Iraq's
   domestic courts and turn to international tribunals instead.[21]

Order 39 was a practical application of neo-conservative theories which
believed in the elimination of the state-run economic sector and 'shock
therapy' economic revolution according to Milton Friedman's teach-
ings.[22] However, Order 39 had important exclusions, namely in Iraq's
energy sector, as it stated that extraction and initial processing of oil
in Iraq were excluded from the provisions of Order 39:

Foreign investment may take place with respect to all economic
sectors in Iraq, except ... the natural resources sector involving

primary extracting and initial processing remains prohibited. In addition, this Order does not apply to banks and insurance companies.[23]

Thus, Bremer changed Iraq's laws to implement an economic model preferred by the Bush administration that, according to its 'corporate globalization' agenda, could ensure increased profits for the corporations without consideration of the welfare of the Iraqi economy and people.[24] Vital to this agenda was getting rid of the Baathists in order to eliminate the opponents of the liberalization of Iraq's national economy and remove all remnants of state economic control, implementing privatization to open Iraq for foreign investment, and bringing in foreign companies and an Iraqi bourgeoisie whose prosperity would inspire Syrians, Iranians and others to seek the same. Thus, Coalition Provisional Authority Order No. 1, 'De-Baathification of Iraqi Society', signed by Bremer on 16 May 2003, was necessary for this economic agenda. The top four rings of the Baath Party's organizational structure (about 120,000 people) were to be dismissed from their posts and barred from taking up employment in the public sector in the future. A de-Baathification committee was established in November 2003, under the leadership of Ahmed Chalabi. De-Baathification also aimed to eliminate Chalabi's political opponents, as he was a major supporter of American business in Iraq. However, these people were experienced professionals whose experience was needed for the reconstruction effort, and they had joined the Baath Party under Saddam's reign not out of ideological commitment, but out of necessity, as they would not have been able to advance their government careers without being members of the party.[25] The provisions of Order 39 were emphasized on 21 September 2003, when Iraq's Finance Minister Kamil al-Kilani confirmed during an IMF–World Bank meeting in Dubai that Order 39 had opened up all non-oil sectors in Iraq to foreign investors, who could own 100% of any enterprise.[26]

The awarding of new contracts was not a new strategy. It followed months of discussion on the economic model for the oil industry after the invasion, where the US Government, the Iraqis in exile and the US companies all agreed, in principle, that Production Sharing

Agreements (PSA) would be the type of contract or form of privatization used in Iraq's oil sector, instead of full privatization of Iraqi oil
resources. US officials decided to keep the Iraqi oil sector under state
control, since privatization would give an impression of imperialism,
would cause widespread opposition among the Iraqi people, and would
sharpen opposition to the occupation.[27] Furthermore, Iraqi advisors
warned that any move to privatize the state oil company or to lay
claim to untapped reserves before an Iraqi government was in place
would be seen as an act of war.[28] Thamer Ghadban, an Iraqi oil expert
who worked with the US-controlled Office of Reconstruction and
Humanitarian Assistance (ORHA, which was formed by Jay Garner,
and later replaced by the CPA), was sceptical about privatization of the
state oil company, saying during the World Economic Forum meetings in June 2003 that 'privatisation does not ring a bell inside of Iraq.
The economy and people's livelihood revolve around the oil industry.
Privatisation is not only premature, it is unacceptable.'[29] Furthermore,
Philip Carroll, a former Shell executive who was appointed by the
Bush administration to run Iraq's oil industry, opposed privatization,
telling Bremer: 'There was to be no privatisation of Iraqi oil resources
or facilities while I was involved.'[30] Bremer's Order 39 also avoided
the privatization of oil resources or facilities because Grand Ayatollah
Ali al-Sistani, Iraq's most respected Shiite cleric, held that minerals
belonged to the 'community', meaning the state.[31] Thus, the Americans
decided to use Production Sharing Agreements (PSAs) as the favoured
form of privatization, instead of full privatization.[32]

Production Sharing Agreements differ from full privatization
because of two distinctive features. The first feature of a PSA is that
privatization is not necessary; the state would retain ownership of the
oilfields, and the foreign company would not actually own the *in situ*
oil reserves. This is a crucial political feature of PSA contracts, which
enables the host country to maintain the principle of national sovereignty over its petroleum resources. However, accounting principles
and regulations allow the company to book the oil reserves of the host
country, to which they will eventually become entitled as they extract
the oil, in the company's accounts, even though the company does not
actually own these reserves. This feature is particularly important to

the company, since stock markets have the propensity to value the company according to the size of its proven reserves.[33]

The second distinctive feature of PSAs is that oil is divided into 'cost oil' and 'profit oil' as follows: the company is compensated for its costs by the 'cost oil' which it takes and sells in the international market, or perhaps back to the state oil company. When the sales recover the company's costs, the amount of oil left in the field, which is the 'profit oil', is divided between the company and the country according to an agreed proportion; for instance 40% to the company and 60% to the country.[34]

Thus, foreign firms would not be able to purchase existing fields but would be allowed to collaborate with the state-owned oil company on joint rehabilitation and development projects.[35] In other words, they could participate in upstream activities without actually owning the fields.[36]

Nevertheless, the broad arrangement to link private enterprise to Iraq's oil industry was fundamentally flawed. Numerous Iraqi commentators and oil experts, like Fouad Qassem al Amir, Tariq Shafiq and others, criticized PSAs, saying that they would give too much leverage to the foreign oil companies vis-à-vis the national Iraqi oil company. They argued that such agreements were useful only for countries where the extraction of oil is very costly and/or difficult, or where the state did not have the funds necessary for the extraction operations, whereas extraction cost in Iraqi oilfields did not exceed $2/barrel.[37] Furthermore, they argued that Iraq did not actually need foreign investment to develop its oil resources, and that it could obtain the necessary funds through international loans, without the need for foreigners.[38] Essam al Jalabi, Iraqi oil minister under Saddam Hussein, argued that it was all right to bring in foreign investment if it could provide technology, know-how and other services, but not for production sharing.[39] Experts also noted that Saudi Arabia, Iran, Kuwait and the UAE never used PSAs, and that Venezuela and Russia were gradually getting rid of their own PSA deals.[40] Moreover, PSAs favour private companies at the expense of the exporting government, therefore only 12% of world oil reserves are subject to PSAs, compared to 67% developed solely or primarily by national oil companies.[41]

PSAs would provide Iraq with higher revenues than the concessions provided to foreigners during the first half of the twentieth century. However, these PSAs did not provide Iraq with as much revenue as the technical services contracts which were used following the nationalization of Iraq's oil in 1972. Furthermore, using PSAs with American companies could lead to contractual conditions which would not be beneficial to Iraq's interests.[42] Companies could force the Iraqi government to guarantee very high compensation to the foreign companies for their costs and high risks, thus decreasing the sum allocated to the state, and the more unstable the security situation became, the more the companies would pressurize the state for a higher ratio of profits to compensate for risks and costs.[43]

These criticisms did not (initially) prevent the calls for PSAs. Before the invasion, the US State Department organized a number of meetings known as the Future of Iraq Project, involving Iraqi exiles, administration officials and corporate executives and consultants.[44] The project was supervised by Thomas Warrick of the Department's Office of Northern Gulf Affairs, with the help of the Working Group on Oil and Energy, which was filled with expatriate Iraqi oil officials sympathetic with Washington's war aims.[45] In September 2002 a State Department official said that oil was not on the agenda of the Future of Iraq meetings.[46] In fact, the State Department and Vice President Cheney convened two 'Oil and Energy Working Group Meetings'; the first in December 2002 and the second in January–February 2003.[47] In the first meeting the State Department convened the Iraqi opposition leaders to discuss the future of Iraqi oil 'to meet the needs of the Iraqi people'.[48] State Department officials also said that Iraq 'should be opened to international oil companies as quickly as possible after the war'.[49]

The Future of Iraq Oil and Energy Working Group's report, *Considerations Relevant To An Oil Policy For A Liberated Iraq*, dated 27 January 2003, stated that 'the political, social and economic well-being of a liberated Iraq depend on Iraq rapidly and substantially increasing its volume of oil production'.[50] Arguing for the decentralization of Iraq's oil industry, the document said that 'war and sanctions are not the reason that Iraq's oil industry has chronically failed to achieve its potential output',[51] and that 'we can reject any suggestion that Iraq's low level of oil

production is due primarily to war and sanctions'.[52] The report went on
to say that war and sanctions had reduced Iraq's oil production by only a
'modest' amount: from an average of 3.5 mbpd (in 1979) to an average of
2.4 mbpd, while the actual potential for Iraq is around 8 mbpd.[53] Rather,
said the document, the reason why Iraq had not reached its potential of
8 mbpd was because Iraq was using the wrong policy by favouring the
monopoly of a national oil company instead of decentralization:

> It may be relevant to note that five of the six countries with the
> highest production yields have highly decentralized oil indus-
> tries. By contrast, five of the six countries with the lowest pro-
> duction yields have highly centralized oil industries dominated
> by politically controlled, monopolized, nationalized oil indus-
> tries. Formulation of an oil policy that has as its goal the delivery
> of the full benefits of Iraq's enormous oil reserves to the Iraqi
> people, must involve a critical analysis of the role, consequences
> of and alternatives to the monopoly rights of the centralized
> national oil company (NOC).[54]

It was this 'monopoly status', argued the document, which resulted in
'underproduction', and it was necessary to increase production to alleviate
Iraq's economic impoverishment, especially as it was 'inconceivable' that
the non-oil sector in Iraq could generate the economic growth necessary
to alleviate Iraq's poverty within a politically acceptable timeframe.[55]
The oil industry was 'the one industry in Iraq that has the possibility
of inducing the enormous amount of foreign direct investment that is
necessary to alleviate the impoverishment of the Iraqi people', and it was
thus necessary to 'restructure' the oil industry in order to raise production above 3.4 mbpd.[56] 'The regime change provides the opportunity to
liberate not only the country but also the economy', added the report.[57]

Iraqi oil and the American business corporations

*We know where the best reserves are {and} we covet the opportunity to
get those some day.*

Chairman of ConocoPhillips,
referring to Iraqi oil, February 2003[58]

*I think this is very promising to the American investors and to American
enterprise, certainly to oil companies.*
                    Adel Abdul Mahdi, Interim Finance Minister,
                         Washington, DC, December 2004[59]

As with Central Asia, the US Government encouraged US private
companies to invest in Iraq and its oil to help reconstruct the Iraqi
economy, leading to the realization of the US agenda for Iraq and
the region. To further facilitate investment for American companies,
Wolfowitz announced on 10 December 2003 that companies from
countries opposed to the war on Iraq would not be allowed to bid
for new Iraq rebuilding contracts worth $18.6 billion, and that this
step was necessary to protect America's 'essential security interests'.
(The Bush administration hinted, even before the war, that countries
opposed to the war would pay a price during post-war reconstruction.)[60]
But on 13 January 2004, at a meeting with Canadian Prime Minister
Paul Martin in Mexico (during negotiations over an Americas summit
over trade and a proposed Free Trade Area for the Americas, FTAA),
Bush said that Canada would be 'eligible to bid' for reconstruction
projects in Iraq, even though Canada opposed the war.[61]

Philip Carroll (former head of Shell's US operations, appointed by
the Bush administration to head a team of experts to run Iraq's oil
industry)[62] hired more American oil executives to devise an 'American-
style' corporate management structure that would oversee the Iraqi oil
industry.[63] (Carroll quit his post in September 2003, due to the chaos
in Iraq, and was replaced by Robert McKee III, Chairman of Enventure
Global Technology, a company owned jointly by Halliburton and
Shell.)[64] Working closely with US defence contractors like Halliburton,
these executives established Team RIO (Restore Iraqi Oil) to undertake the rehabilitation of Iraq's infrastructure.[65] The RIO program
had two goals. The first, the re-opening of old refineries and facilities,
has been accomplished. The second goal, which is to increase Iraq's
ability to export fuel,[66] would be harder to achieve given the security
situation in Iraq.

The RIO program was complemented by the Logistics Civil
Augmentation Program (LOGCAP). Halliburton's work in rebuilding

Kuwait and putting out Kuwaiti oil-well fires, in the aftermath of the 1991 Gulf War, impressed the Defence Department (which was then headed by Secretary of Defence Dick Cheney). In 1992, Kellogg, Brown & Root (KBR), a subsidiary of Halliburton, signed a cost-plus LOGCAP contract with the Department of Defence to provide services to the US army in Kuwait, Somalia, Haiti and the Balkans. (The cost-plus arrangement meant guaranteed profit for KBR.) In December 2001, KBR renewed the cost-plus LOGCAP contract with the Department of Defence. The contract stipulated that KBR would 'put out well fires and assess the facilities, clean up oil spills or other environmental dangers at the sites, repair or reconstruct damaged infrastructure and operate facilities and distribute products'.[67]

In autumn 2002, Douglas Feith (a leading neo-conservative and former corporate lawyer) established the Office of Special Plans (OSP) to plan for the invasion of Iraq. Feith appointed Michael Mobbs as leader of the OSP's Energy Infrastructure Group. Mobbs concluded that the Iraqi oil system was in worse shape than previously thought, and that no US Government agency had the experience necessary to extinguish well fires and re-launch Iraq's oil industry. Oil executives, military and government agencies kept saying that Halliburton and KBR, who held the LOGCAP contract, could repair the Iraqi oil sector. Thus, Mobbs pressed for contracts to be awarded to KBR. On 11 November 2002 the US Army officially awarded KBR a $1.9 billion contract to repair Iraq's oil infrastructure.[68]

The RIO and LOGCAP programs were not competitively bid; they were given to Halliburton 'out of convenience' because they already had plans to extinguish oil fires. Bob Faletti, spokesperson of the US Army Corps of Engineers (USACE), the department of the army overseeing RIO, said that 'it made sense' to award the contract to KBR, since awarding without competition helped 'save time and money'.[69]

But this led to accusations of favouritism in the Bush administration towards certain companies. An iconic case was that Halliburton (the US oil services company whose CEO was Dick Cheney between 1995 and 2000, and in which the Bush family has business interests)[70] and Bechtel (which had business connections to some members of the

Bush administration),[71] benefited from the war and won non-bid contracts. Many company officials who benefited from post-war Iraq contracts were also members of the Committee for the Liberation of Iraq (CLI), which was founded in 2002 as an extension of the Project for the New American Century (PNAC) and supported the Iraq invasion. Its members included George Shultz of Bechtel[72] (who had been Secretary of State under Ronald Reagan). The CLI disbanded after the invasion.[73] (Congressman Henry Waxman (D- California) was the leader of the Democratic opposition to Halliburton, as he was concerned about the government issuing non-bid contracts to a company that had clear ties to Cheney,[74] saying that the terms of the vague and open-ended LOGCAP contract 'would allow Halliburton to profit from virtually every phase of the war'.[75])

For the USACE, KBR was 'the only source capable of developing a contingency support plan' to extinguish oil well fires if the Iraqi army were to set them on fire, in addition to the fact that 'KBR, by virtue of its familiarity with the Centcom operational plans, [and] related Logistics Support Plans[76] ... is the only contractor that can satisfy a requirement for immediate execution of the contingency support plan'. A USACE memo dated February 2003 also mentioned the time factor to justify the no-bid contract to KBR. 'Because the Iraqi economy depends on oil revenue, the well-being of the Iraqi people depends on continuity of oil production and distribution. Any delays in providing support required by US Centcom will result in significant negative international repercussions,' stated the memo. Thus, it explained, 'no delay can be tolerated', since 'these specialized services can be available only from KBR in the time available'. Furthermore, the USACE was instructed to 'plan for competition for the execution effort at the earliest reasonable opportunity consistent with the needs of the mission'. The memo also said that this decision was agreed 'w vp', i.e. with the Vice President's office.[77] Wolfowitz also defended the no-bid procedures, saying that 'it is necessary in the public interest to use other than competitive procedures' for the contracts awarded by the Department of Defence for the rebuilding of Iraq.[78] Bunnatine Greenhouse, senior contracting official in the USACE, said in February 2003 that she opposed the contract to KBR. She understood why the contract was issued in secret, as a war was taking place. She also understood that bidding would be a waste of time. But she did not understand why it needed to be a five-year contract. Why not make it for one year, and extend it later if necessary? She also failed to understand why it was a cost-plus contract, which gave KBR a guaranteed 2% profit plus a 5% bonus.[79]

On 8 March 2003 the USACE awarded KBR a $7 billion contract to put out oil fires. The contract was later expanded to include fuel delivery.[80] A month later, Bechtel was awarded a contract to rebuild Iraqi infrastructure.[81]

Halliburton, KBR's parent company which contributed $1,146,248 to the Republican Party between 1998 and 2003, received at least $19.3 billion in Iraq's single-source contracts, and Halliburton's stock price increased by 229% after the war began.[82] (Of course, the Iraq war was not the only reason why Halliburton's stock prices rose; the rise in global oil prices was another reason.)[83] Furthermore, in late October 2003, a report by the Centre for Public Integrity (CPI) said that many of the US firms which won Iraqi reconstruction contracts were major donors to George W. Bush's political campaigns, especially the 2000 campaign. However, the USACE and USAID, the bodies jointly responsible for allocating the contracts, rejected any suggestions of favouritism. 'We have made sure that it is not politically motivated', a USAID spokesperson said.[84]

But thanks to the Iraq contracts, Halliburton was saved 'from the brink of bankruptcy.'[85] In December 2003 two of Halliburton's divisions (KBR and DII) declared bankruptcy due to lawsuits filed by people harmed from inhaling asbestos, although the KBR unit did not include the government contract division working on Iraq reconstruction because it did not have any asbestos liabilities.[86] (In June 2004, Democratic Senator Patrick Leahy and Vice President Cheney had a 'brief but fierce exchange' in the US Senate where Leahy criticized Halliburton's no-bid contracts in Iraq and Cheney's ties to Halliburton. Cheney's reply to Leahy was an obscenity: 'F**k yourself!')[87]

Naomi Klein argued that 'the lack of competition on awarding contracts has been one of the distinguishing features of the Bush years',[88] (not just in Iraq, but also in US domestic security contracts).[89]

This leads us back to Klein's analysis of Milton Friedman's economic 'shock treatment',[90] and 'disaster capitalism',[91] where crisis was needed to promote his fundamentalist form of capitalism,[92] showing the convergence of national and corporate interests.[93]

Democratic developments in Iraq

Despite the agreement between the State Department and the Department of Defence on using PSA contracts in Iraq's oil sector, there was tension between the development of the Iraqi oil industry and the establishment of a post-invasion political system.[94] Jay Garner, the first US administrator in post-war Iraq, said in a BBC interview in April 2004 that he was sacked partly because he wanted to hold elections whereas his superiors wanted first to privatize Iraq's oil.[95] Garner stated: 'My preference was to put the Iraqis in charge as soon as we can, and do it with some form of elections … I just thought it was necessary to rapidly get the Iraqis in charge of their destiny.'[96] Garner also disagreed with many of the Bush administration's economic proposals, such as full privatization of Iraq's 192 state-owned enterprises, arguing that the Iraqis should be in charge of determining their own economic fate.[97]

So there was a contradiction in the Bush administration's political and economic strategy. It held that Iraq should be democratized but, despite calling for democracy, the administration was initially against elections and paid little attention to political development in Iraq. The call by al-Sistani for elections was a serious challenge to Bush's vision for Iraq, as they could lead to an Iraqi constitution based on Islamic Shari'a law and so threaten the secularization of Iraqi governance. In addition, the United States was worried that Islamic Shiite leaders in Iraq might form an anti-American partnership with the Mullahs in Iran,[98] and undermine US plans for controlling Iraqi oil.[99] Bremer was also opposed to holding elections, fearing that elected Iraqis would produce a constitution that would not separate the mosque from the state, might not provide equal rights to women, or might not follow any of the other principles supported by the White House, which wanted Iraq to be a model for democracy in the Arab world.[100]

The United States wanted an orderly political transition that Washington could control; a system that would be a model to the Arab world[101] (or what could be called a 'controlled, low-intensity democracy').[102] Moreover, polls showed that if Iraqis were allowed to vote, they would vote for candidates who would limit the powers of US troops and the US multinationals.[103]

On 28 June 2003 al-Sistani issued a *fatwa* calling for elections, saying that the constitution should be written by officials elected by the Iraqi people, not a council appointed by the US occupation as Paul Bremer planned.[104] Al-Sistani also questioned the legitimacy of a security agreement between an un-elected Iraqi government and the United States. However, the Bush administration decided that early elections would be a recipe for civil war. If elections were held before political parties had time to establish themselves, well-organized Baathists and Islamic parties with ties to Iran would dominate the outcome. Al-Sistani's demands were thus seen as impractical, as Iraq lacked a population census and an electoral law, and it would take months to lay the infrastructure for a nationwide ballot.[105]

Justifying his opposition to early elections, Bremer explained why he was more concerned with privatization than with elections:

> We're not going to rush into elections because Iraq simply has none of the mechanisms needed for elections – no census, no electoral laws, no political parties, and all the related structure we take for granted. We've also got to get this economy moving and that's going to be a helluva challenge. A stable Iraq will need a vigorous private sector.[106]

Bremer said elsewhere that he was 'not opposed' to elections, but that he wanted the elections to be held 'in a way that takes care of our concerns. Elections that are held too early can be destructive. It's got to be done very carefully. In a situation like this, if you start holding elections, the people who are rejectionists tend to win.'[107] In late September 2003 he argued against early elections during his testimony before the Senate Armed Forces Committee, saying that elections should come after the constitution was ratified, not vice versa.[108]

By the autumn of 2003, however, the Bush administration was forced to reconsider. With Iraqi support for the occupation declining, the transition of power to the Iraqis had to be speeded up,[109] especially as scores of angry Iraqis were joining the insurgency in opposition to the occupying American forces on Iraqi territory and anger over the lack of security, jobs and electricity. Mindful of the growing anti-American sentiment, the members of the Governing Council called for the end of the American occupation, saying that the council should assume sovereignty and administer the country until elections could be held.[110]

In October 2003, Bush and Rice told Bremer that his seven-step plan for the restoration of Iraqi sovereignty was untenable.[111] They suggested that he think of a way to speed up the handover process and speculated on whether the open-ended occupation was fuelling the insurgency, especially that they did not want American forces to be involved in a bloody guerrilla war as the American people voted in the presidential elections of November 2004. The State Department and the Pentagon also pressed Bremer to speed up the handover process, as the Pentagon told Bremer to hand over sovereignty to an expanded version of the Governing Council by the following spring. Bremer initially resisted.[112] But by mid-November 2003, under pressure from both the White House and Ayatollah al-Sistani, Bremer abandoned his seven-step plan.[113]

Bush decided on 12 November 2003 that the United States should pursue early sovereignty in Iraq.[114] The 15 November agreement, which the CPA wrote and then obliged the Governing Council to sign, set 30 June 2004 as the day when sovereignty would return. Both the CPA and the council, which most Iraqis correctly saw as unrepresentative and ineffective, would then dissolve. An interim assembly, chosen through a complex process of nationwide caucuses, would govern Iraq until elections in early 2005. The elected government would then write the new constitution.[115]

Even with this adjustment in US policy, Bush's hopes that democracy in Iraq would reshape the region failed to materialize, either at the national or the regional level. The root failure lay in post-invasion planning and operations. According to Major Isaiah Wilson III, former war planner in Iraq, 'there was no Phase IV plan', with no one producing a strategy to deal with the situation after major combat operations ended.[116]

Bush and his team began to focus on the post-war phase only shortly before the war started in March 2003, and with little attention compared to the attention given to the invasion plan.[117] Jack Keane, one of the top US military commanders in Iraq, said that the Americans failed to plan for the insurgency, and admitted that this was not just an intelligence failure, but also a failure of senior military leaders, including himself.[118] The primary cause of this haphazard planning was the Bush administration's disdain for nation-building.[119] David Kay (an ex-CPA official and security expert who led the CIA effort to investigate Iraq's missing WMD) said that 'there was a reluctance to divert resources to post-conflict maintenance of law and order' in Iraq.[120] Rumsfeld did not want the Iraqis to develop a 'culture of dependency', where they would rely upon the United States to take care of their security, as had allegedly happened in Kosovo. Therefore, the focus of post-war planning was not how to build a democratic Iraq, but how to get the US troops out of Iraq as soon as possible,[121] since, according to a senior Pentagon planner, Rumsfeld wanted to 'do war on the cheap' so that a small, lightly armed force could take over Iraq.[122] (In May 2003 there were only 150,000 US troops and 18,000 UK troops in Iraq.)[123] Moreover, the Pentagon saw post-war planning as a block standing in the way of war, since it attracted attention to costs and potential problems that might have weakened the case for launching the war.[124] In fact, the CIA conducted several war games in late spring 2002, concluding that civil disorder would likely occur after Saddam's removal, but the Pentagon was not pleased with these results and ignored them.[125] According to Tenet:

> There was precious little consideration, that I'm aware of, about the big picture of what would come [after the invasion]. While some policy makers were eager to say that we would be greeted as liberators, what they failed to mention [was] that the intelligence community told them that such a greeting would last for only a limited period.[126]

Secretary of State Colin Powell did raise the issue of insufficient troops and the difficulties that would face the Americans in the post-war phase. If these concerns were taken more seriously, then perhaps more efficient post-war planning would have taken place.[127] Instead, a small group of Pentagon planners took over the planning and ignored the State Department's Future of Iraq Project.[128] Another sign of departmental strife was that Rumsfeld wanted the Department of Defence (DoD) to overtake the reconstruction process, in part, it appears, in order to spite Powell.[129] Rumsfeld rejected the Future of Iraq Project because he wanted the DoD, not the State Department, to lead Phase IV. (The State Department was working with one group of Iraqis, and Wolfowitz with another.)[130] The situation was compounded by the absence of healthy debate between the State Department and the DoD. Rumsfeld and Commander in Chief of Centcom Tommy Franks dominated planning, marginalizing JCS Chairman Richard Myers and keeping away those who disagreed with them.[131]

Failure to rely on the Future of Iraq Study had deprived Jay Garner and his team of a vital blueprint at a critical moment. David L. Phillips, former senior advisor to the US State Department, who worked on the Future of Iraq Project, wrote that 'the ... Project did not have all the answers. However, its findings, had they been heeded, could have provided useful guidance enabling the US to fulfil its promise of liberation [sic]. Ignoring it was just the first in a series of mistakes.'[132]

The Bush administration did have a Phase IV plan, but it was neither practical nor realistic enough, as the goals of the invasion (imposing a pro-US government in Iraq, demonstrating American power to threaten Iran, Syria and the militant groups which they support, and transforming the region into a pro-American mould) were over-optimistic and extraordinarily ambitious.[133]

In order to achieve these goals, Washington had to rebuild a strong and democratic Iraq, using Iraq's oil revenues to rebuild the country. However, Paul Bremer later wrote that the development of the Iraqi oil sector suffered from problems such as sabotage, Washington's contradicting bureaucracy, the slow Iraqi bureaucracy, and smuggling, adding that Iraq had 4,000 miles of pipelines, and it was impossible to protect them all.[134] Also, even though the Oil Ministry was guarded by US troops during the looting which followed the invasion, other oil facilities were not safe. A coalition assessment in 2003 found that looters stole or damaged more than $900 million worth of equipment at Iraqi oil facilities.[135] Furthermore, the civilian leadership at the Pentagon was aware before the war that Iraqi oil was not a ready source for reconstruction, despite the opposite impression given by Congress testimonies. Amy Myers, one of the authors of the April 2001 *Strategic Energy Policy* report, was consulted by the government before the war, and she and her group concluded in December 2002 that oil revenues would certainly not be sufficient to cover the cost of reconstruction, due to years of deterioration and sanctions.[136] Even though members of the Bush administration (such as Alan Larson from the State Department and Andrew Natsios from USAID) testified that rebuilding Iraq would cost American taxpayers only $1.7 billion, Bush signed Public Law 108–106 on 6 November 2003, setting $18.4 billion for the reconstruction of Iraq.[137] US Special Inspector General for Iraq Reconstruction (SIGIR) Stuart Bowen said that, by mid-2008, $117 billion had been spent on the reconstruction of Iraq, including about $50 billion in US taxpayers' money, that the US Government never developed a sound plan for stabilizing Iraq, due to 'blinkered and disjointed pre-war planning for Iraq's reconstruction', that US reconstruction efforts in Iraq were 'grossly overburdened' by wasteful spending, and that the rebuilding effort never did much more than restore what was destroyed during the invasion and the looting that followed.[138]

Bremer wrote that 'reality on the ground made a fantasy of the rosy pre-war scenario under which Iraq would be paying for its own reconstruction through oil exports within weeks or months of "liberation"'.[139] Generally speaking, the CPA lacked the resources necessary to perform good post-war management, and it also lacked an understanding of the Iraqi social/political apparatus.[140] This failure in Iraq had repercussions on Bush's plan to remake the Middle East.

The Middle East in May 2003:
A narrow window of opportunity

After the removal of Saddam Hussein under the guise of democratization and the free entry of US business corporations into Iraq, Bush was hoping to spread the democratization and free enterprise programme to the rest of the region. He called for democratization in Iraq and the Middle East in a major speech at the AEI on 26 February 2003[141] in which he argued that the 'liberation', rebuilding and democratization of Iraq would lead to the democratization of the whole region:

A liberated Iraq can show the power of freedom to transform that vital region, by bringing hope and progress into the lives of millions. America's interests in security, and America's belief in liberty, both lead in the same direction: to a free and peaceful Iraq ... Iraq, with its proud heritage, abundant resources and skilled and educated people, is fully capable of moving toward democracy and living in freedom ... There are hopeful signs of a desire for freedom in the Middle East. Arab intellectuals have called on Arab governments to address the 'freedom gap' so their peoples can fully share in the progress of our times. Leaders in the region speak of a new Arab charter that champions internal reform, greater politics participation, economic openness, and free trade ... A new regime in Iraq would serve as a dramatic and inspiring example of freedom for other nations in the region.[142]

On 6 November 2003 Bush reiterated these ideas at the National Endowment for Democracy (NED), where he announced that accepting dictatorship in the Middle East had not yielded any positive results, and that his administration had 'adopted a new policy, a forward strategy of freedom in the Middle East'. He cited the lack of democracy as a cause for the spread of WMD and of terrorism:

Sixty years of Western nations excusing and accommodating the lack of freedom in the Middle East did nothing to make us safe because in the long run, stability cannot be purchased at the expense of liberty. As long as the Middle East remains a place where freedom does not flourish, it will remain a place of stagnation, resentment, and violence ready for export. And with the spread of weapons that can bring catastrophic harm to our country and to our friends, it would be reckless to accept the status quo. Therefore, the United States has adopted a new policy, a forward strategy of freedom in the Middle East … We believe that liberty is the design of nature; we believe that liberty is the direction of history.[143]

The fact that he gave this speech before the US Chamber of Commerce (a major participant in the NED programmes) and that the event was sponsored by AT&T and AIG, showed that US business interests backed the type of democracy that Bush and his administration were promoting.[144] (AT&T was awarded an exclusive contract by the US DoD to provide pre-paid calling cards for military personnel stationed in Iraq. However, the soldiers were complaining about the high prices which they had to pay to make calls to their loved ones in the United States, giving rise to allegations of over-charging, especially as AT&T was the only card that could be used by US troops in Iraq.[145] AIG offered insurance to American contractors working in Iraq, but there were allegations that AIG was overcharging them.[146])

Of course, this drive for democratization and openness for business started with Iraq, with Bremer declaring in late May 2003 that Iraq was 'open for business'. Arguing for the imposition of the free market formula in Iraq, he said 'history tells us that … resources protected by private property [and] private rights are the best protection of political freedom. Building such prosperity in Iraq will be a key measure of our success here.'[147] Indeed, Joseph Stiglitz said that the invasion of Iraq was 'one of the last attempts to impose neo-liberal doctrines', as the invasion pushed for 'instant liberalization, instant privatization, even if it was against international law'.[148] Neo-conservatism provided the impetus for the invasion, while neo-liberalism called for free markets, as Bush associated political freedom with market freedom.[149]

Opening Iraq's oil sector for private foreign investment was a major part of this plan, as the link between Iraq's oil and the corporate agenda for Iraq was evident in Bremer's statement to the US DoD in June 2003, where he said that 'Now that [our] sanctions have been lifted, it's important for Iraq to re-enter the world economy.

The most obvious example of that is the sale of Iraqi oil.'[150] (Of course, the sale of Iraqi oil was switched back to the US dollar shortly after the invasion.)[151] This shows, according to Rutledge, that 'US policymakers and oil executives seemed to be working as closely together ... as they had sixty years earlier'.[152]

Even during his final weeks in office, Bush's desire to establish an 'empire of bases' to control Iraq's oil was evident. In the 2009 Defence Authorization Bill, issued in October 2008, Bush exempted himself from a provision which banned funds '(1) To establish any military installation or base for the purpose of providing for the permanent stationing of United States armed forces in Iraq', or '(2) to exercise United Sates control of the oil resources of Iraq'. Experts viewed this expansion of Bush's powers in Iraq as an attempt to expand US control over Iraq's oil.[153]

Furthermore, Bush wanted to spread his 'model of imperial-style corporate globalization' from Iraq to the rest of the Middle East,[154] since Bush's speeches associated democracy in the Middle East with the elimination of trade barriers and promoting the Open Door policy, free trade, and the corporate globalization policy. On 9 May 2003 he announced that he would work on establishing MEFTA 'within a decade', and he put Cheney's daughter Liz in charge of the project.[155] Robert Zoellick, the main architect of MEFTA,[156] (and a former advisor to Enron)[157] wrote in the *Wall Street Journal* in October 2004 that trade would help spread democracy and fight terrorism:

> America has been attacked by those who want us to retreat from world leadership. Let there be no misunderstanding: the United States will continue to advance the values that define this nation – openness, opportunity, democracy and compassion. Trade reinforces these values, serving as an engine of growth and a source of hope for workers and families in the United States and the world.[158]

Robert Lawrence (international trade expert and former economic advisor to Bill Clinton) agreed with the Bush administration's argument that trade could help stabilize the economically vital Middle East and its oil resources:

> US interest in the MEFTA reflects geopolitical and security considerations based on concerns about the Middle East's central role in global oil markets, the Arab–Israeli conflict, and violent clashes between the West and militant Islamic fundamentalists. The United States has an interest in a stable and friendly Middle East because the American economy is highly dependent on imported oil. It is vulnerable to high prices and disruptions caused by political turmoil, wars and terrorist attacks ... The MEFTA initiative reflects that US interests cannot be advanced through purely military or political initiatives. To be effective in the battle for hearts and minds in the region, the policy needs an economic component.[159]

He stressed that these goals would not be achieved if the United States resorted to mercantilist policies, or failed to lift trade barriers as stipulated by the MEFTA.[160] However, there is no guarantee that the United States would not resort to mercantilist policies (as it has repeatedly resorted to such tools with its other free trade agreements with other regions).[161]

US business coalitions showed interest in the MEFTA, notably the National Foreign Trade Council (NFTC), arguably the most powerful US corporate lobbying association on international trade, whose members include Bechtel, Chevron, Halliburton and others.[162] Another business coalition interested in the MEFTA was the Business Council for International Understanding (BCIU).[163] Charles Dittrich of the NFTC said that 'the [Bush] administration looks at the MEFTA as a key component of its Middle East policy'.[164] Thus, democratization was Bush's method of promoting the Open Door policy.

There was indeed a brief wave of democracy in the Middle East in early 2005. In Iraq, free parliamentary elections took place in January. In February, Egyptian president Mohammad Hosni Mubarak announced that he would allow multi-candidate presidential elections in September. In Saudi Arabia, elections for local councils were held for the first time. In Kuwait, the parliament passed a law to allow women the same political rights as men. In Palestine, presidential elections took place (claimed to be free elections despite Israeli pressure on candidates other than Abbas). In Lebanon, the Cedar Revolution, following the assassination of Rafiq Harriri, led to the withdrawal of Syrian forces from the country. It was, according to analysts, the 'Arab Spring'. The Bush administration claimed that its policies, especially the invasion of Iraq, were behind this movement to democracy. In his Inauguration Address in January 2005 Bush argued for his new policy, the Freedom Agenda, saying that 'it is the policy of the United States to seek and support the growth of democratic movements and institutions in every nation and culture, with the ultimate goal of ending tyranny in our world'. He said that Washington would 'encourage reform in other governments by making clear that success in our relations will require the decent treatment of their own people. America's belief in human dignity will guide our policies'. He asserted that 'America's vital interests and our deepest beliefs are now one'.[165] The idea which he wanted to communicate was that the future of America and the security of America depend on the spreading of liberty.[166] He reaffirmed these ideas in his State of the Union address in February 2005, as he argued that democratization in the Middle East would help eliminate the causes of global terrorism:

> In the long term, the peace we seek will only be achieved by eliminating the conditions that feed radicalism and ideologies of murder. If whole regions of the world remain in despair and grow in hatred, they will be the recruiting grounds for terror, and that terror will stalk America and other free nations for decades. The only force powerful enough to stop the rise of tyranny and terror, and replace hatred with hope, is the force of human freedom. Our enemies know this, and that is why terrorist Zarqawi recently declared war on what he called the 'evil principle' of democracy.[167]

American pressure was indeed a factor in the wave of democracy which took place in the Middle East in early 2005. However, the connection between the invasion of Iraq and the steps towards democracy in parts of the region was tenuous at best,[168] as the 'Arab Spring' was caused mainly by internal factors in Saudi Arabia, Egypt, Kuwait, Palestine and Lebanon, not by the invasion of Iraq. Furthermore, this 'Arab Spring' was a short-lived phase that did not provide any real change in the region; nor did it substantially help fight terrorism or promote US business interests there. Bush had no coherent policy to use the narrow window of opportunity, which the Arab Spring provided, to promote US interests in the Arab world.

On the Arab–Israeli peace process, too, a narrow window of opportunity came to the Bush administration in the form of Yasser Arafat's death on 11 November 2004 (only a few days after Bush's re-election) which Bush hoped would pave the way for a new Palestinian leadership, and a new hope for peace.[169] However, even after Arafat's death and the election of Abbas as president, Bush did not offer any real plan to promote Arab–Israeli peace (beyond the Annapolis conference of 2007, which did not lead to any real progress).

Conclusion: Iraqi oil, energy security and the Arab spring: The collapse of the Bush vision

If the Bush administration managed to impress the world with the easy victory over Iraq, it squandered the narrow window of opportunity to devise a realistic policy to remake the Middle East. The invasion of Iraq was meant to hold the four US foreign-policy strands together, by transforming the country into a model for a transformed Middle East, and Iraq's oil sector was central to rebuilding Iraq into a model for the region. When Iraq collapsed, the ambitions of the Bush administration collapsed with it, as it was too dependent on success in Iraq, and the failure in Iraq made the administration more dependent on Saudi Arabia for strategic assistance and swing production of oil.[170]

Due to the lack of political stability and control in Iraq, al Qaeda flourished in Iraq after the invasion, and Bush repeatedly argued that failure in Iraq or premature US withdrawal would give terrorists access to Iraq's oil revenues, enabling them to finance terrorist attacks against the West. 'An emboldened al Qaeda with access to Iraq's oil resources could pursue its ambitions to acquire weapons of mass destruction to attack America and other free nations', he said.[171] He also said that 'failure in Iraq would be a disaster for the United States' because radical Islamic extremists in Iraq 'would be in a better position to topple moderate governments, create chaos in the region, and use oil revenues to fund their ambitions'.[172]

The Bush administration also saw oil as a factor that should help keep Iraq united, as it called for oil revenues to be divided equally among Iraq's population (not according to each region's share of the oil reserves, where the Shiite south contains about 65% of Iraq's oil reserves and the Kurdish north about 10%) in order to prevent Iraq from splitting into three Sunni, Shiite and Kurdish regions. Bush repeatedly echoed this argument, rejecting the partition of Iraq. 'I happen to believe the oil belongs to the people and to the extent that they understand that, it'll help unify the country', he said.[173] Rice, too, said that oil should be a resource for the Iraqi people as a whole, 'not one that would help to make the country less unified'. She said that Iraqi leaders must settle their differences, and that Kurdish independence, especially if it meant taking control of the northern oilfields, was something that Washington would not support.[174]

An Iraqi oil law was originally drafted in 2006 to settle the disputes between the regions over oil revenues.[175] But the passage of the law was delayed due to disagreement between the government in Baghdad, the Kurdish government in Irbil and various Iraqi factions and parties over control, revenue distribution and foreign companies' involvement in Iraq's oil sector. Due to these disagreements, it may be years before the law is passed.[176] Thus, in November 2007, Iraqi oil minister Hussein Shahrestani announced, to Washington's relief, that the government would work with foreign oil companies without the Iraqi oil law.[177] The first major foreign oil deal came in August 2008, when Iraq signed a $3 billion oil service contract with China's CNPC. The fact that China was the first country to sign an oil deal with Iraq, even though it opposed the American invasion, shows that America's agenda for control of Iraq's oil did not go according to plan.[178]

Despite its oil riches, Iraq is going through a sensitive phase. Oil provides about 95% of the government revenue[179] and about 50% of the GDP,[180] and this over-dependence on the oil sector is a cause of severe economic problems.[181] The collapse in global oil prices in late 2008 and early 2009 made it even harder to rebuild the Iraqi economy. Not only is there too little money for reconstruction; there is also not enough to maintain and expand the oil industry, the government's main source of revenue. The government needs oil revenues to hold the country together, as Prime Minister Nouri al Maliki has squared many opponents by handing out well-paid government jobs. Without a rapid increase in oil revenues, the rising budget deficit could create a fiscal crisis and this fragile political settlement could shatter.[182] Indeed, there are signs that al Qaeda is rising again in Iraq,[183] and that members of the Awakening Councils, a group of Sunni fighters who used to fight against the US troops but have now joined the Baghdad government, are rejoining the insurgency.[184] Maliki blamed these developments and the recent rise in bombings on budgetary problems.[185] Security improved after the US troop surge in 2007, but these gains are described by US officials as 'fragile' and 'reversible'.[186]

Iraq has 80 known oilfields, of which only 15 have been partly developed.[187] But equipment in many oilfields is 30 years old, and is in dire need of development.[188] Shahrestani said that his goal was to increase production from the current level of about 2.4 mbpd to 6 mbpd by 2014, for which the Iraqi oil sector needs $52 billion in investment on top of the $8 billion the Iraqi government has spent during the past several years.[189] (Out of Iraq's current 2.4 mbpd production, 1.8 mbpd are exported.) However, international companies keen to move into Iraq are still wary of political instability, legal uncertainty, inadequate security and corruption.[190] The oil sector is also suffering from a brain drain; under Saddam's regime, many oil-sector employees who held technical jobs were members of the Baath Party, and many of them fled abroad, or were arrested or killed after the American invasion, leaving the ranks severely depleted.[191] The insurgency also took its toll on Iraq's security and oil sector, as the insurgency is financed by revenues from oil smuggling.[192] (The Bush administration assumed that Iraq's physical and administrative infrastructure would be largely intact after the war.)[193]

Oil production remains at about 2.4 mbpd, which is not very different from the pre-invasion level.[194] Repairs to Iraq's oil infrastructure by KBR were botched, due to cutting corners, involvement in legal disputes, missing deadlines, accumulating costs over and above the original budgets, which caused delays in oil production.[195] Also, Russian,[196] Chinese[197] and French[198] companies are investing in Iraq's oil sector, even though Washington aimed to prevent their investments because these countries opposed the war.

Thus, Baghdad is calling for foreign oil companies to invest in Iraq, to provide expertise and funding. But the contracts offered by Baghdad are not PSA contracts, which the companies prefer, as there is strong opposition in Iraq to such contracts. The foreign companies would not take over the operations entirely, or share in the oil produced. Instead, Baghdad offers technical service contracts, where Iraq would retain ownership of the oilfields, a joint venture would be established between the foreign company and whichever Iraqi state company is responsible for the oilfield in question, where Iraq would retain a 25% share in the project. For each field, the ministry would specify a minimum production level, and the companies would not be paid for anything up to the minimum production level. After the minimum production level, the company would be paid a fee of, for example, $2/barrel.[199]

Initially, the international oil companies believed that Baghdad's conditions were harsh (which showed that the government was under great domestic pressure to demonstrate that it was not giving away Iraq's oil cheaply).[200] In an auction which took place in Baghdad in June 2009, the first since the US invasion, only one company, a consortium led by BP and CNPC, agreed to the oil ministry's terms to run the 17-billion-barrel Rumaila field, for only $2/barrel, instead of the $4 which the consortium initially demanded (although it would also be able to charge the ministry for the costs of repairing the production facilities).[201]

However, international oil companies could not afford to stay out of Iraq. Investment in Iraqi oilfields remains very attractive to international oil companies, as proven global reserves are dwindling and new oil finds are becoming scarcer.[202] Another reason why Western oil firms are prepared to go back to Iraq despite the unfavourable terms of contract and the uncertain investment climate is fear that Chinese or Russian oil firms might acquire Iraqi oil contracts, making the risk of staying out of Iraq greater than the risk of getting into Iraq.[203] Alex Munton, a consultant at energy and mining consultancy firm Wood Mackenzie, said that Iraq's oil is very attractive despite the pitfalls.[204] Indeed, in November 2009, California's Occidental and Italy's Eni agreed to Baghdad's $2/barrel fee and signed a deal on the 4.4-billion-barrel Zubeir oilfield in southern Iraq, followed by a deal with ExxonMobil and Royal Dutch/Shell on the West Qurna-1 oilfield, for $1.9 per barrel.[205] During the auction which took place in Baghdad in December 2009, a joint venture between Shell and Malaysia's Petronas won the right to develop the 13-billion-barrel Majnoon oilfield in the south for a fee of only $1.39 per barrel, a consortium led by China's CNPC (and including Malaysia's Petronas and France's Total) won the contract for the 4.1-billion-barrel Halfaya oilfield in the south, for only $1.40 per barrel, and Lukoil got the right to the 12.9-billion-barrel West Qurna-2 field in the south, in addition to other deals for oilfields in the north and south of Iraq.[206]

The one place where foreign companies were able to reach PSA deals was in Kurdistan,[207] which contains almost 10% of total Iraqi oil reserves,[208] as Irbil signed about 30 contracts of its own with international oil companies. Baghdad refuses to recognize these contracts, however,[209] arguing that Kurdish PSA contracts waste Iraq's oil, and that contracts signed by the Kurdish Regional Government (KRG) are illegal because they were made independently from Baghdad. (The Kurds give foreign companies 11–18% of the 'profit oil', while Baghdad, although rejecting PSAs and preferring technical service agreements, says that giving foreigners 3–16% would be more appropriate.)[210] Furthermore, while the central government in Baghdad failed to pass a national oil law, the Kurdish parliament in Irbil passed its own oil and gas law for the KRG, separate from Iraq's national oil law, on 6 August 2007.[211]

The KRG's most controversial oil deal took place in September 2007, when the KRG signed a PSA with American oil firm Hunt Oil Company of Dallas, the first oil agreement signed between the KRG and a foreign oil company after the approval of the KRG's oil law. Iraqi oil minister Hussein Shahrestani condemned the agreement, saying that it was illegal because the central government in Baghdad had not approved it, despite the KRG's insistence that the Iraqi constitution gave it the right to strike oil deals.[212] A senior US State Department official acknowledged that the KRG contract with Hunt Oil contradicted the US policy of strengthening Iraq's central government.[213] However, it later emerged from leaked letters and e-mails that the US State Department and Commerce Department actually knew in advance about the KRG deal with Hunt, and even supported it, even though it undermined Washington's peace efforts in Iraq. This raised suspicions of Bush cronyism, given that Ray Hunt, Hunt Oil's chief executive, was a close friend of the Bush family and a major financial sponsor of Bush's presidential campaign. Bush denied any prior knowledge of the deal.[214]

Furthermore, Arab–Kurdish tensions over the oil-rich Kirkuk province resulted in military clashes between predominantly Arab Iraqi army units and the Kurdish Peshmerga forces.[215] Even more dangerous for Washington's influence in Iraq, Tehran's influence is evident in Baghdad, through Maliki's pro-Iranian government and various pro-Iranian militants.[216] Armed skirmishes still occur between Sunnis and Shiites, as they battle for influence in Baghdad.[217] For years, militias in Basra (including the Mahdi Army, the Badr Brigade, Fadhila and others) have been competing for control over neighbourhoods, oil revenues, electricity, the ports and even local universities.[218]

This shows that the Iraq war provided few advantages for the US oil industry. In most cases, it was non-American companies that won oil deals, for example France's Total, China's CNPC, and other companies from Malaysia, Vietnam, Angola, Norway, Britain and Russia. Only two US oil giants managed to secure contracts for oilfields: Exxon and Occidental. The US spent more than $700 billion on Iraq, but now Iraq's oil profits are going to non-American companies.[219] By mid-2008, $117 billion had been spent on the reconstruction of Iraq, including about $50 billion in US taxpayers' money.[220]

Only one US company truly profited: Halliburton. Most foreign oil firms are hiring US engineers, usually from Halliburton. Halliburton's chief in Iraq told US diplomats in December 2009 that 'it is really Halliburton [to whom] all the oil majors are turning to extract, process, and deliver the oil'.[221] American private security firms also made good money in Iraq, as Washington could not have functioned in Iraq without them due to the lack of sufficient US troops.[222] US companies working in the security field and in the reconstruction field made profit (although Bechtel had to pull out of Iraq in November 2006 due to the deteriorating security situation).[223] The corporatist political economy was clear in the pre-war planning phase, and in the dependence on private corporations to run Iraq in the post-invasion phase.

The contracting system by which these firms got their business was not without its flaws. Former senior official at the US Departments of Defence, State and Energy Anthony Cordesman criticized the contracting system in Iraq, telling Congress in November 2004 that 'US economic aid has lagged behind the need for urgent action, has wasted vast resources on an impractical contracting effort'.[224] Stuart Bowen Jr., the US SIGIR, confirmed in October 2007 that there were long-standing contract-administration problems within the US State Department bureau that awarded contracts, as a result of a 'lack of controls' and 'serious contract management issues'.[225] Similarly, in November 2007, an independent panel strongly criticized the US Army's management of contracts to supply its troops in Iraq and Afghanistan, saying that there were high levels of fraud and waste in relation to contracts worth $4 billion a year, that only about half the army's contracting staff were properly qualified, and there was a lack of oversight.[226] Furthermore, wrote Cordesman, US economic aid was 'made worse by an almost completely dysfunctional reporting system within the US Government that does not tie plans to realistic requirements'.[227]

Iraqi oil, as the 'cash lifeline' or the 'cash crop',[228] did not help democratize the region, due to the failures in post-war planning. The elections of 2005 obviously would not have taken place without the American invasion. However, elections in themselves are not democracy. In order for a country to become a genuine democracy, first, major ethnic or religious strife has to be avoided, or else elections would exacerbate group tensions. Second, an oil-based economy has to be avoided, because easy oil revenues could make the government

'Rumsfeld "wanted cheap war"', BBC News, 30 March 2003, http://news.bbc. co.uk/1/hi/world/americas/2899823.stm [accessed 26 June 2008]

Stiglitz, Joseph and Linda Blimes, 'The Three Trillion Dollar War', *Times Online*, 23 February 2008, http://www.timesonline.co.uk/tol/comment/columnists/ guest_contributors/article3419840.ece [accessed 10 May 2010]

— 'The $3 Trillion War', *Vanity Fair*, April 2008, http://www.vanityfair.com/ politics/features/2008/04/stiglitz200804 [accessed 10 May 2010]

— '$3 trillion may be too low', *Guardian*, 6 April 2008, http://www.guardian. co.uk/commentisfree/2008/apr/06/3trillionmaybetoolow [accessed 10 May 2010]

— 'Ask a question: The Three Trillion Dollar War', *McClatchy*, 15 April 2008 http://www.mcclatchydc.com/qna/forum/three_trillion_dollar_war/index. html [accessed 10 May 2010]

Stiglitz, Joseph, 'It's not the economy, stupid', *Financial Times*, 28 September 2004, http://www2.gsb.columbia.edu/faculty/jstiglitz/download/opeds/ Economy_Stupid.pdf [accessed 10 May 2010]

— 'The Three Trillion Dollar War', *Project Syndicate*, 7 March 2008, http://www. project-syndicate.org/commentary/stiglitz97/English [accessed 10 May 2010]

— 'War at any cost? The total economic costs of the war beyond the federal budget: Testimony before the Joint Economic Committee', Columbia University, 28 February 2008 http://www2.gsb.columbia.edu/faculty/ jstiglitz/download/papers/Stiglitz_testimony.pdf [accessed 10 May 2010]

Yergin, Daniel, 'Energy security will be one of the main challenges of foreign policy', Cambridge Energy Research Associates (CERA), 19 July 2006 http://www.cera.com/aspx/cda/public1/news/articles/newsArticleDetails. aspx?CID=8248 [accessed 10 February 2010]

— 'The fight over Iraq's oil', BBC News, 14 March 2003, http://news.bbc.co.uk/1/ hi/business/2847905.stm [accessed 14 March 2003]

Television programmes

'Empire – Running on empty?' Al Jazeera English, 1 July 2010, http://www. youtube.com/watch?v=JlE-i4QzT04 [accessed 20 December 2010]

Lauer, Matt, 'Views of America from the streets of Iran – Matt Lauer', MSNBC, YouTube, 13 September 2007 http://uk.youtube.com/watch?v=7Z6cIA9B_ Mw) [accessed 30 June 2008] Also see http://www.msnbc.msn.com/ id/20746548/

Reeve, Simon, 'Conspiracy Theories: Iraq', Sky One, 10 January 2007, 22:00 GMT.

Salameh, Mamdouh, 'Oil prices forecasted to rise again – 08 September 2008', *Al Jazeera English*, 9 September 2008, http://www.youtube.com/ watch?v=G4B0YwGBqoc [accessed 10 September 2008]

reluctant to create a real economy, and this would result in a government detached from the real needs of the people. Third, the rule of law has to be established, which is not thoroughly established in Iraq, as seen in the unstable security situation.[229] Iraq thus failed as a demonstration case to become a stable, democratic, pro-American country, and the Bush vision was so reliant on success in Iraq that it had no back-up plan on what to do in case Iraq failed.

After the failure in Iraq, the Bush administration thought that Iran (the next member of the Axis of Evil) would replace Iraq as the next terror-sponsoring state, as it linked Tehran to the major threats in the Middle East: Tehran's nuclear program, terrorism (due to Iran's support to Hamas and Hezbollah) and regional influence in Iraq, Afghanistan, Lebanon, Syria and the Palestinian Territories.[230] Condoleezza Rice described Iran as 'the single most important, single-country challenge ... to US interests in the Middle East and to the kind of Middle East that we want to see'.[231] But Bush had limited options on Iran. He could not use the military option for fear that Iran might retaliate by closing the Strait of Hormuz, through which 14 million barrels of oil (a fifth of total global production, or 43% of total global exports) pass every day, making it 'the world's most important oil checkpoint' according to the US Department of Energy.[232] In fact, Israeli Prime Minister Ehud Olmert asked Bush in May 2008 to give him a green light to attack Iran, but Bush refused the request. He feared that the costs would outweigh the benefits, because of the possibility of Iran's retaliation on US interests in the Gulf region, and anxiety that Israel would not succeed in disabling Iran's nuclear facilities in a single assault (Israel could not mount a series of attacks over several days without risking full-scale war).[233]

Regime change was not a realistic option, either, as there was no substantial opposition group in Tehran to receive Washington's support.[234] On the diplomatic side, Bush needed Tehran's cooperation to stabilize Iraq, where Tehran's influence in has increased since the invasion.[235]

On a wider global level, Iran used its energy resources to cut oil, gas and pipeline agreements and spread its influence and gain the friendship of US rivals like Russia and China,[236] and even US allies

like India, Pakistan,[237] Turkey,[238] Malaysia[239] and some European countries.[240] Had Washington's power not been consumed by Iraq, the Bush administration could have had more influence over Iran. (Similarly, Sudan, another state on Washington's terror list, is using Chinese investments in Sudan's oil sector to evade Washington's sanctions and influence.)[241]

Another consequence of the failure in Iraq was the Bush administration's increased dependence on Saudi strategic support and swing oil production.[242] Saudi Arabia is no longer the number one supplier of US oil imports.[243] (In fact, Saudi Arabia is currently the fourth largest exporter of oil to the United States, after Canada, Mexico and Nigeria. The fifth largest is Venezuela, and the sixth is Iraq.[244]) But it is still the most important US ally in the Gulf region (because of its strategic assistance and its role as a swing producer of oil). But again, Russian and Chinese influence in Saudi Arabia is rising. Moscow is increasingly selling arms to Riyadh,[245] and Saudi Arabia is supplying China with 17% of its imported oil needs, making it China's largest foreign oil supplier.[246] These shifts are causing what Daniel Yergin called 'a significant re-orientation' of global oil politics.[247] But even after the invasion of Iraq and the US troop withdrawal from the kingdom, Saudi Arabia still has a potential for instability. In May 2003 a terrorist bomb struck a foreign workers' compound in Riyadh, killing many. The US completed the withdrawal of its troops from Saudi Arabia (to station them in Qatar instead)[248] in August 2003, but this did not ease the situation, as another bomb struck Riyadh in November 2003. In late May 2004, a group affiliated to al Qaeda attacked the Al-Waha business compound in Khobar, killing 22 people, wounding 25, and taking hostages. The hostages were released after a Saudi security offensive. Furthermore, al Qaeda attempted several terrorist attacks on oil facilities in the Kingdom, causing more concern about the safety of Saudi Arabia and its oil resources. In late February 2006, for instance, the Saudi authorities foiled a plot by al Qaeda to bomb the world's largest oil-processing plant at Abqaiq, which handles about two-thirds of Saudi Arabia's oil production. Al Qaeda advised followers to attack oil processing facilities in Saudi Arabia, but not the oil wells, since al Qaeda would need the oil revenues if they took over

the country.[249] Furthermore, in April 2007 the Saudis foiled a plot to carry out air attacks on oil installations and military bases, arresting 172 suspects linked to the plot.[250]

Moreover, Riyadh was not eager to raise oil production as much as the Bush administration wanted in January and May 2008 when global oil prices were escalating to record highs,[251] and Riyadh continued to reject foreign investment in the Saudi upstream oil sector.[252] During the period between 1999 and 2009, total oil production in the Persian Gulf states still averaged around 20 mbpd,[253] showing no sign of rising to 44.5 mbpd by 2020, as the NEP of May 2001 desired.[254] Had Bush succeeded in reshaping Iraq to replace Saudi Arabia as the main ally and oil source in the region, he would not have faced these difficulties in Riyadh. The Bush administration tried to use the Arab–Israeli peace process (and the $30 billion arms sales to the Gulf states in July 2007) to rally Arab support against Iran.[255] Again, Bush depended mainly on Saudi (not Iraqi) support in the Arab–Israeli peace process. The Annapolis summit of November 2007 promised Arab–Israeli peace and a Palestinian state before Bush left office, but such promises were not met, as the Bush administration had no serious policy to follow up on these promises.

However, the good news for the US is that a Saudi oil boycott similar to that of 1973 is very unlikely. Any exporter attempting to discriminate against the US would hurt itself as much as the United States.[256] Back in the 1970s, oil-producing states sold directly to customers, and could punish them individually. Today, oil is sold on an international market mediated by thousands of middlemen and futures exchanges around the world, a system that has done much to undermine OPEC's clout. According to Professor William Hogan of Harvard, 'They can't cut off oil supply. It is a very blunt political weapon, so they stopped using it as a weapon.'[257] 'Even if major oil-producing states tried to raise the price level by cutting production, how long would it be before others rushed in to profit from the market vacuum?' wrote Graham Fuller in 1997.[258] Riyadh ruled out the use of the oil weapon during the Israeli war on Lebanon in 2006,[259] and during the Israeli offensive on Gaza in December 2008–January 2009, saying that the concept of an oil cut-off would be futile.[260] Also,

even though 'Saudi Arabia is *the principal enabler* of the petrodollar recycling system, thus holding the sword of Damocles over the dollar supremacy',[261] and even though a CIA memo warned that OPEC, and in particular Saudi Arabia, could use its 'accumulated wealth as a political weapon [if it] became displeased with US policy',[262] neither Saudi Arabia or OPEC would drop the 'sword of Damocles', fearing economic repercussions.[263]

The tragedy of the Bush Doctrine

Bush was a follower of the Open Door policy, where he used over-exaggerated threats to justify military intervention in Iraq to promote America's strategic, economic and corporate interests. According to Williams, an American foreign policy based on the Open Door policy could have hazardous effects on US interests, causing a 'tragedy of American diplomacy', where America's policies in the quest for economic openness (policies which are usually made by an elite few without accountability, and which usually involve the betrayal of America's own democratic and human rights values) could lead to the rise of nationalist, anti-American movements frustrated with America's interference in their national affairs and economic benefit.[264] Kolko agreed, arguing that by dragging the United States into such dangerous conflicts, US foreign policy would always be in 'perpetual crisis', and the 'traditional US solutions to global problems will be increasingly futile' in the face of 'the uncontrollable nature of the international situation'.[265] (Indeed, Ikenberry argued that Bush's neo-imperial agenda to invade Iraq would lead to imperial overstretch.)[266] Such was the tragedy of the Bush Doctrine. Bush's invasion of Iraq was planned by elite neo-conservative groups, military officials and oil-industry members, who presented the public with false information on an over-exaggerated Iraqi threat with almost no accountability, thus betraying the US principles of transparency at home, and international law and human rights abroad. The invasion therefore led to more anti-American feelings and the rise of more anti-American militant groups. The cost of the war and its financial strain was also a factor in the decline of US power.

The Bush Doctrine was different from George H.W. Bush's New World Order and from William J. Clinton's Engagement and Enlargement because while the democratization goals of the New World Order and Engagement and Enlargement were limited and restrained by the need for Middle East oil, the Bush Doctrine stated that democratization would take place despite the need for Middle East oil, and that democratization would actually stabilize the region. In other words, the New World Order and Engagement and Enlargement were limited by what could be called the 'democracy conundrum'; the contradiction between American calls for democratization and American support of the non-democratic regimes of its allies.[267] Unlike the New World Order and Engagement and Enlargement doctrines, the Bush Doctrine tried to surpass and defeat the democracy conundrum by invading Iraq and imposing democracy on it, thus hoping to start a democratic domino effect in the region and reshape it in America's favour.

However, the Bush Doctrine failed in several ways. First, it depended too much on success in Iraq. When intervention in Iraq failed, the Bush Doctrine (and its plans for democratization) collapsed with it. Second, the Bush Doctrine could not defeat the democracy conundrum, as the failure in Iraq forced the Bush administration to abandon its regional agenda and continue supporting undemocratic regimes in the Middle East in return for oil and strategic cooperation.[268] Third, Iraq's oil became one of the financial sources of different anti-American Iraqi militias, whether Sunnis, Shiites or groups affiliated to al Qaeda. The threat of global terrorism has increased after the war on Iraq, due to the failure of democratization, the establishment of the invasion of Iraq as a cause célèbre for terrorists, and the effect of the Iraq war in diverting Western efforts away from fighting terrorism.[269] William Appleman Williams would have called this increase in anti-Americanism as a result of the invasion of Iraq a 'tragedy'.

Due to the lack of political stability, rule of law, and basic security in Iraq, the US oil companies are struggling to get the 'big shot' that they expected. And everyone is still waiting for the 'lots of money' from the Iraqi oil sector to pay for the reconstruction of Iraq that would, according to neo-conservative hopes, give it the strong economy needed to

turn it into a stable, pro-American democratic model, as advocated by the Bush Doctrine's democratization drive. In fact, by the time of the Republican losses in the Congressional elections of November 2006, Bush had effectively abandoned the 'freedom agenda', had no functioning foreign-policy doctrine, and followed a haphazard foreign policy.[270] This haphazard, ad hoc policy would be reflected in the decline of US power and the Bush administration's lack of real solutions to this decline.

CONCLUSION

Overstretch, the Decline of Energy
Security and the Fall of the American Empire

*You and your predecessors in the oil and gas industry played a large role
in making the twentieth century the 'American Century'.*
> Secretary of Energy Spencer Abraham, remarks
> before the American Petroleum Institute,
> Texas, June 2002[1]

Bush, the Open Door and oil capitalism

The concept of US imperialism is far from new. But the policies of the
Bush administration confirmed Williams' and Kolko's writings on an
informal American empire based on the Open Door policy, deploying
the US military to protect America's economic and corporate interests,
promote the political economy of the large corporation, and protect US
access to global energy resources.[2] This was seen in America's increased
military involvement in the world after the 11 September attacks to make
up for relative economic decline, to pursue the renewed Open Door policy, and to protect oil-rich regions in the Middle East and Central Asia,[3]
as Washington (especially under the Bush administration) replaced
access to Asian (and Chinese) markets with access to Persian Gulf oil by
building military bases in and around Afghanistan and Iraq:

> Call it variations on a theme. Instead of access to Asian markets,
> access to Persian Gulf oil had become the main issue. The Open
> Door Notes of 1899–1900 found their functional equivalent in

the Carter Doctrine [of] 1980. In place of China, US policymakers [under George W. Bush] were soon to fixate on Iraq.[4]

George W. Bush distinguished himself from previous administrations by giving special attention to energy resources from his first
days in office. Bush sought the foreign-policy strands of US military enhancement and global power projection at the same time that
he endorsed the energy strand, with a strategy that would lead to
increased oil supplies from unstable regions. Although arising from
different sets of concerns – one energy driven, the other security/mili-
tary driven – 'these two strategic principles have merged into a single,
integrated design for American world dominance in the twenty-first
century'[5] (or, as the NSC document discussed by Jane Mayer would
say, these two priorities were 'melded').[6] Bush used the 11 September
attacks (which raised the foreign-policy strand of anti-terrorism) as
an 'opportunity' to advance this agenda and take the Carter Doctrine
to another level: invading Afghanistan (thus establishing US bases
in Central Asia for the first time) and Iraq (hoping to strengthen the
US position in the Gulf) to secure energy routes and supplies. The
11 September attacks thus facilitated Bush's geopolitical, corporate,
military and economic agendas.[7] Had William Appleman Williams
been alive, he would have stressed the centrality of US control of
global oil resources (especially Middle East oil) to maintain US global domination over supplies to Europe and Asia, citing Iraq as an
example of deploying US military in order to secure the Open Door
for US participation in Iraq's oil and banking enterprises as a form of
'pell-mell privatisation'. He would also have cited the importance of
the dollar–euro competition in the international oil market.[8]

Bush was indeed committed to the Open Door model, which 'found
a new lease on life in the Long War [the War on Terror]' as it pursued
the interests of US corporations,[9] especially the 'petro–military complex', which was particularly powerful under George W. Bush, as it
sought to expand with the US imperial order.[10] The no-bid contracts
to favoured corporations showed that 'crony capitalism' went hand
in hand with US empire-building under the Bush administration,[11]
and the Bush agenda of promoting corporate globalization (especially

in the field of oil) through war showed that military and economic interests went hand in hand to further America's interests, especially in expanding US corporate access to the Middle East.[12] US military dominance is thus linked to 'oil capitalism':

> The United States is using its military power to fashion a geopolitical order that provides the political underpinning for its preferred model of the world economy: that is, a relatively open, liberal international order. US policy has aimed at creating an oil industry in which markets, dominated by large multinational firms, allocate capital and commodities. State power is deployed not just to protect present and future consumption needs of the United States and the profits of US 'oil capitalism', but also to guarantee the general preconditions for a world oil market. So, to the [considerable] extent that the openness and stability of the international oil market are premised on American geopolitical and military commitments ... the military power and geopolitical influence of the United States provide the necessary and sufficient conditions for the stable operation of the international oil market. This system has, of course, been designed for US interests.[13]

The role of the military in maintaining US control of global energy resources was especially evident under the Bush administration and its plans for Iraq. However, Bush failed to reshape Iraq in America's image. This failure had consequences beyond the Middle East, according to Zbigniew Brzezinski: 'Beyond destabilizing the Middle East, the Iraq war had a further, much more important consequence. It made the success or failure of US policy in the Middle East the test case of American global leadership.'[14] This 'test case' failed to bring democracy to Iraq or to the region, and now the main aim of Washington in Iraq is stability (instead of democratization).[15]

Nevertheless, in pursuing the energy strand, Bush prioritized preponderance and empire over economic and energy interests in cases where the goals conflicted. This was evident in the fact that Bush never allowed US oil investments in anti-American rogue states like Iran, Sudan and Libya (before 2003/4) and Cuba,[16] despite pressures

from the oil lobby. Thus (despite the strong link between the national interest, economic interests and energy interests) the national interest had priority over economic interests in cases of conflict between energy and security in US foreign policy. This showed that Bush's priority was the global preponderance of power. (Allowing the Hunt Oil deal in Iraq's KRG was an obvious exception.)

Beyond Iraq, Bush's quest for a global empire was primarily centred on redrawing the map of the vast region extending from the Caucasus of Europe to central and south-western Asia, including the Persian Gulf, North Africa and the Middle East, projecting massive military power in this region.[17] He also endeavoured to remake the societies and politics of Muslim states in America's image, in order to spread the neo-liberal model and promote the Open Door for US investments.[18] Bush believed that he had the mission of spreading freedom and democracy in 'the darkest corners of our world', and, according to Bacevich, 'the world's darkest corners coincided with the furthest reaches of the Islamic world, which, not coincidentally, contained the world's most significant reserves of fossil fuels'.[19]

The democratization project failed. Due to the instability and corruption in Iraq, Iraq's oil sector failed to help Iraq develop economically or democratically, to act as a model for other Islamic states to reshape their political systems and liberalize their economies for US investment. Bush's implementation of the Open Door policy by invading Iraq went wrong, as his administration presented false justifications for the war to the American public and failed to come up with a sound post-war plan for the management of Iraq, leading to the loss of the hearts and minds of the Iraqi people (and the Arab and Islamic world). He ignored serious warnings that the invasion would not result in democratization, and that instead it might result in chaos and ethnic violence. He also ignored the needs of the Iraqi people and focused on the needs of the US corporations. Bush's corporate globalization agenda was hazardous for the Iraqi people. First, Iraqi companies were excluded from the rebuilding operation, even though they had the necessary skills, in favour of foreign companies. Iraqi companies were hired by US companies, but only as subcontractors, and for short-term projects, thus making very little money compared to US companies. Furthermore, US contractors did not hire Iraqis in

reconstruction jobs; instead, they imported workers from India, Nepal and Indonesia, because they were cheaper and because the Iraqis were considered a security threat. The reconstruction process is not succeeding as planned, because, first, the US companies are putting their profits first, rather than the immediate needs of the Iraqi people, and, second, because of the insurgency fuelled by hatred towards the invaders.[20] Unemployment among Iraqis has reached 25–40%.[21] This has led to more violence and the deterioration of the security situation, making international oil companies hesitant to invest in Iraq at times, despite their eagerness to invest in Iraq's oil. [22]

The deteriorating situation in Baghdad absorbed the Bush administration's efforts and made the job of implementing the American agenda in Iraq and the region even harder. This made Washington less powerful *vis-à-vis* its challengers (Iran, Russia, and China) who were trying to expand their influence in the region.

Furthermore, the United States today is less secure regarding its energy resources. As Iraq's oil resources are still underdeveloped, the Middle East is in turmoil, the War on Terror is more difficult to win due to the invasion of Iraq, energy facilities in the Middle East and elsewhere are under the constant threat of militant attacks from al Qaeda and other groups (or even threatened by war, as in the 2008 war in Georgia), and the US is facing tough competition over energy from China and Russia.

Failure in Iraq leads back to ad hoc policy

As a nation, America has accepted for too long an ad hoc approach to energy and foreign policy.

Former Secretary of Energy
Bill Richardson, 2005.[23]

For the first time in three decades, energy has become a central problem in foreign policy ... A foreign policy weakened by oil dependence is a national security problem.

James Schlesinger and
John Deutch, 2006.[24]

We have a serious problem: America is addicted to oil, which is often imported from unstable parts of the world.

George W. Bush, State of the
Union, January 2006.[25]

For too long our nation has been dependent on foreign oil. And this dependence leaves us more vulnerable to hostile regimes, and to terrorists – who could cause huge disruptions of oil shipments, and raise the price of oil, and do great harm to our economy.

George W. Bush, State of the
Union speech, 23 January 2007[26]

Due to the failure to reshape Iraq, and due to the military quagmire in Iraq, the Bush administration failed to use the invasion as a platform to link the four strands of US foreign policy together, or to reshape the region in America's image. Global terrorism has increased because of the invasion of Iraq,[27] and the oil sector in Iraq was not reshaped according to Washington's wishes (as technical-services contracts were used instead of PSAs, and companies from France, Russia and China were allowed to invest in Iraq's oil sector and other economic sectors). The quagmire in Iraq also took its economic and military toll on Washington, leading to the decline of Washington's regional and global influence, and the empowerment of actors rival to the United States, namely Iran, Russia, China and (in the Latin American region) Venezuela.

Bush's foreign policy aimed to secure America's vital oil supplies, not only in the Middle East, but also on a global level through the regional diversification of oil resources away from the Persian Gulf. One of the alternative regions was Central Asia, where the West is unable to make a serious start on the TAP pipeline project because of the increasing insecurity in Afghanistan,[28] where the Taliban still pose a serious threat to pro-American interests. The Bush administration could not pay the necessary attention to Afghanistan or Central Asia, because Iraq consumed most of the administration's efforts and attention, and because of the lack of a clear foreign-policy direction in the Bush administration. (Ahmad Rashid said that Washington and the West still lacked a clear strategic vision for Central Asia, even

after the American invasion of Afghanistan.[29] Former EU envoy to Afghanistan Francesc Vendrell agreed, saying in September 2008 that the Bush administration, and the West in general, did not have a coherent plan to stabilize Afghanistan.)[30] Even the BTC pipeline, inaugurated on 25 May 2005, was not without risk, as it runs through the volatile Caucasus and will require constant surveillance to protect it from attacks[31] which may result from the Chechnyan conflict and the ethnic conflict in the breakaway regions of Georgia,[32] as seen during the war in Georgia in 2008. Moreover, Caspian oil will not last like Middle East oil, as the Caspian contains only 4–5% of global oil reserves,[33] compared to 65% in the Middle East. Energy and mining consultancy firm Wood Mackenzie forecast that Caspian oil production might reach a peak of around 2.5–2.8 mbpd by 2015, then decline,[34] and most experts indicate that the Caspian's energy production will probably last no more than a quarter of a century.[35] Caspian oil is also three or four times more expensive to extract than Middle Eastern oil,[36] and is of lower quality than Middle East oil as it is 20% sulphur, thus more expensive to refine than Middle East oil.[37] Nor was Washington assured of cooperation from Central Asian states like Uzbekistan, Kazakhstan and Kyrgyzstan, whose loyalties tend to swing between Washington and Moscow. In August 2005 the US troops in Uzbekistan were given six months to leave the Khanabad base in Uzbekistan, after Washington condemned the Andejan massacre. The US Senate voted to block a $23 million payment for the use of the base to Uzbekistan,[38] prompting Uzbekistan to tilt away from Washington towards Russia and China. In February 2009, due to disagreement over rental fees, the Kyrgyz government decided to close the Manas airbase, the only remaining US base in Central Asia.[39] Bishkek reversed the decision only after Washington agreed to raise the annual rent from $17.4 million to $60 million.[40] The rising influence of Russia and China showed the increasing clout of the SCO/vis-à-vis Washington's declining influence.[41]

Also, with the US military mired in Afghanistan and Iraq, and unwilling to enter a military conflict with Russia on its border, the Bush administration had few options for dealing with the Russian offensive on Georgia in August 2008, apart from sending Condoleezza

Rice to Georgia and promising aid.[42] This war had repercussions for the region's energy routes, as Georgia was a main transport route for several major oil and gas pipelines, including the BTC pipeline, the Baku–Supsa pipeline and the Baku–Erzurum pipeline (all managed by BP). Before the war, Georgia was championed as a reliable country through which pipelines could avoid both Russia and Iran. Even though the pipelines themselves were not attacked, the claim for Georgia's reliability as an energy-transport route was put into doubt by the Russian offensive, since the BTC pipeline is only 35 miles from the South Ossetian border. The conflict also called into question America's efforts to diversify its oil supplies away from the Middle East or gain access to more Central Asian energy resources,[43] and showed the emergence of a 'New Cold War' between Russia and the United States[44] as a part of the New Great Game over oil resources. (However, Russia avoided directly bombing the pipelines. Mindful of the need for good relations with Azerbaijan and Turkey, Russia was careful to point out that this was not a war for oil. Nevertheless, the vulnerability of the BTC pipeline, which transports 850,000 barrels of oil per day, or 1% of the world's total oil supplies, was evident when the Kurdish militant group PKK bombed the Turkish part of the pipeline in early August 2008, in an attack unrelated to the war in Georgia. There were unconfirmed reports that Russian jets targeted the pipeline, but missed, and BP said that the pipeline was not hit.)[45]

The offensive was seen as a part of Russia's pipeline competition with the West. Washington is promoting oil routes which do not pass through Iran or Russia, like the BTC. Another project is Nabucco, a pipeline project (named after the famous opera by Verdi) supported by the EU and the United States, to bring gas from the Caspian region, mostly from Azerbaijan, going through Bulgaria, Romania, Hungary, Austria and Turkey, to reduce dependence on Russian gas.[46] Moscow is promoting other pipelines which pass through Russian-friendly territories. For example, Germany and Russia have started work on building a new north European gas pipeline, Nord Stream, a joint venture between Gazprom (which owns a 51% stake), German firms E.On and BASF (both sharing a 40% stake) and Dutch energy firm Gasunie (9% stake), and managed by former German Chancellor Gerhard

INDEX

Schroeder. Nord Stream would pass beneath the Baltic Sea to transfer gas directly from Siberia to Germany. Another Russian-backed pipeline is South Stream, which would pass under the Black Sea to transfer Russian gas to Bulgaria, Greece, Italy, Romania and Austria.[47]

Moscow has also repeatedly used Gazprom, Russia's biggest and most important company, which controls a fifth of the world's gas reserves,[48] as its most brutal and effective geopolitical weapon, as well as a tool for bullying Russia's neighbours, since the company provides 25% of Europe's gas (or 42% of the EU's gas imports). Moscow has repeatedly threatened to cut off gas supplies to European importers unless they pay higher prices for Russian gas.[49]

In Latin America, another energy-rich region that could provide an alternative to the Middle East, Bush's efforts were marked by the failure of the Washington-backed coup in Venezuela in April 2002 (a coup partially caused by fears over oil boycotts, fears that Chavez might price his oil in euros, and opposition to a new hydrocarbon law which imposed higher royalties on US oil companies).[50] Chavez used his oil money to improve relations with China,[51] Russia[52] and Iran,[53] and spread his socialist, anti-Washington agenda in Latin America.[54] He also imposed partial nationalization on US oil firms in Venezuela.[55] And even though the Venezuelan economy is heavily dependent on oil exports to the US (Venezuela supplies 12–14% of US oil imports,[56] or 1.5 mbpd, which amounts to approximately 50% of Venezuela's oil)[57] the Bush administration did consider adding Venezuela to the list of terror-sponsoring states in September 2008 (accusing Caracas of supporting FARC in Colombia), but decided against it, fearing that this might affect oil imports from Venezuela.[58]

Even cases like Canada show the limits of US foreign policy, as Canada's tar sands cannot replace Middle Eastern oil, due to the expenses of extracting oil from the sands, and its low EROEI (Energy Return on Energy Invested, which is the ratio of the energy spent on extracting a barrel of oil, compared to the energy provided by that barrel of oil). The EROEI of Canada's tar sands is only 1.5 or 3, which is very low compared to the average EROEI of Middle East oil, which is 30.[59] According to a Council on Foreign Relations report, Canadian tar sands cannot free the United States from its dependence on Middle East oil.[60]

The Canadian energy sector is also being penetrated by Chinese invest-ment.[61] Therefore, Canada's tar sands, despite the billions of barrels in reserves, will never be a panacea for US energy needs.[62]

There were small success stories in Bush's foreign policy. One example was Libya, which possesses the largest oil reserves in Africa (the ninth largest in the world), close to 39 billion barrels, or 3% of the world's oil reserves. It is Africa's third-largest oil producer (after Nigeria and Angola). The oil and gas industry contributes 70% of Libya's GDP,[63] and Libyan oil has a production cost of just $1 a bar-rel. Moreover, most Libyan oil is low sulphur, sweet crude, making it easy to refine. Libya's coming in from the cold (after Qaddafi gave up his WMD program in December 2003), and allowing in US energy investments after decades of sanctions,[64] was a rare and small success for Bush, but it was mostly due to the decade of economic sanctions, not due to the invasion of Iraq.[65]

Also, Bush's relatively successful visit to Africa in March 2008 man-aged to improve relations with oil-rich African countries.[66] Bush also set up a Pentagon command centre for Africa, known as Africom, to oversee US military activities in all of the continent, except Egypt,[67] to help fight terrorism in Africa, protect US interests in African oil and confront the growing Chinese influence in Africa.[68] However, no African country is willing to host Africom's headquarters, for fear of being seen as too pro-American,[69] and fear that US facilities could become targets for terrorism.[70] Thus, Africom decided to scale back its ambitions of finding a base in Africa and settle for its current headquar-ters in Stuttgart, Germany.[71] Furthermore, despite providing as much foreign oil to the US as the Persian Gulf (indeed, in 2005 the United States imported more oil from Africa than from the Middle East, and it imported more oil from the Gulf of Guinea than it did from Saudi Arabia and Kuwait combined[72]), Africa would not be a reliable oil sup-plier in the long run, as it suffers from instability and corruption, and its resources are not as massive as the Persian Gulf's.[73]

Based on these experiences, the US discovered that energy resources outside the Middle East were neither sufficient in amount, nor cheap to extract and transport.[74] Even if the non-Middle East regions raised production as the NEP hoped, they would still be vulnerable to instability and conflict, just like the Persian Gulf.[75] None of these alternative regions are safe from instability.[76] Thus, regional diversification would not avoid dangers and would not decrease dependence on the Persian Gulf,[77] forcing Washington to return to the Middle East as its main test case for the success of US foreign policy, and as its main source of energy supplies. As Cheney said as CEO of Halliburton in the autumn of 1999, 'While many regions of the world offer great oil opportunities, the Middle East with two thirds of the world's oil and the lowest cost, is still where the prize ultimately lies'.[78] Sarah Emerson of Energy Security Analysis Inc. echoed this sentiment in 2002 when she said 'the trouble with diversifying outside the Middle East ... is that this is not where the oil is'.[79]

As seen with Iran, Sudan, Canada, the Central Asian states and the Gulf States, oil producers are using their oil wealth to acquire the friendship of great powers other than the United States, most importantly Russia and China, which are competing with the US over control of global energy supplies. This has decreased Washington's influence over these oil-producing countries. Also, rising global oil prices in 2007–2008 increased the incomes and the clout of oil-rich countries with anti-American agendas, like Venezuela, Iran and Russia, empowering them to crack down on opposition and/or spend their oil money on anti-American policies.[80]

Due to the failure to reshape the Middle East through Iraq, and the fact that other regions do not provide enough hope for emancipation from Middle East oil, Bush was forced to deal with global issues, including energy security, on a piecemeal, case-by-case basis without a unified strategy. The failure of the Bush Doctrine to enhance US influence led to the decline of American power worldwide.

Decline of US empire and energy security

During the past eight years, our energy policy has been directed by the two oil men in the White House {Bush and Cheney}. Their failed policy has increased our dependence on foreign oil, damaged our economy, and left consumers paying record prices at the pump.

Nancy Pelosi, August 2008[81]

This might be the beginning of the end of the American empire.
                                    Nouriel Roubini, economist,
                      New York University, August 2008[82]

US power has declined not only because of the failure in Iraq, but also because of the recent global financial crisis. Like Paul Kennedy, Chalmers Johnson forecast that the 'overstretched American empire' will probably first begin to unravel in the economic sphere.[83] The 2008 financial crisis proved these commentators correct. The financial meltdown and the economic drain caused by the invasion of Afghanistan and Iraq have weakened US power, costing the US much of its unipolar advantage.[84] John Gray wrote that the 2008 global financial crisis was 'more than a financial crisis'; it was a sign of a 'historic, geopolitical shift', resulting from the financial strain of the Iraq war and an American ideology of financial deregulation.[85] Like Paul Kennedy, Gray linked the fall of empires to a combination of wars and financial problems:

> The fate of empires is very often sealed by the interaction of war and debt. That was true of the British Empire, whose finances deteriorated from the First World War onwards, and of the Soviet Union ... Despite its insistent exceptionalism, America is no different. The Iraq War and the credit bubble have fatally undermined America's economic primacy. The US will continue to be the world's largest economy for a while longer, but it will be the new rising powers that, once the crisis is over, buy up what remains intact in the wreckage of America's financial system.[86]

Economists Joseph Stiglitz and Linda Blimes agreed that the Iraq war, despite not causing the Great Recession of 2008–2009 by itself, did contribute to the recession by exacerbating the US deficit and helping raise the global oil price. They argued that the Iraq war, which, according to their calculations, cost the US economy at least $3 trillion and helped erode the global leadership of the United States, was the most expensive war for the United States since World War II. Since the Iraq war was entirely financed by borrowing, mostly from abroad (mainly from China and the Gulf states), it helped raise the national deficit and the national debt (both of which were further exacerbated by Bush's tax cuts). Thus, the Iraq war strained the US economy, weakened the US dollar, and diverted money away from domestic projects which may have stimulated the US economy (refuting the argument that war is good for the economy). Pointing out that global oil prices rose from $25/barrel before the war to more than $100/barrel in early 2008, Stiglitz and Blimes 'conservatively' assumed that the war has contributed $5 or $10 to this rise in global oil prices (although they point out that the actual figure may be closer to $35). This raised the cost of US oil imports, and helped divert money away from the US economy.[87] (In fact, US oil imports rose from an average of 11.8 mbpd in 2001 to an average of 12.9 mbpd in 2008.)[88] Analysing the role of rising oil prices in the Great Recession, Stiglitz wrote that 'the burden on monetary policy was increased when oil prices started to soar after the invasion of Iraq in 2003. The United States spent hundreds of billions of dollars on importing oil – money that otherwise would have gone to support the US economy.'[89]

Oil economist and World Bank consultant Mamdouh Salameh gives greater emphasis to the effect of the war on oil prices, as he said that if the invasion of Iraq never took place, 'oil today would have been no higher than $50/barrel'.[90] Similarly, financial expert Loretta Napoleoni agreed that the costs of the wars of Afghanistan and Iraq, and the fact that they were both funded by external debt (through the sale of US treasury bonds to China, Japan and the Gulf states), raised US debt and deficit, contributing to the credit crunch and the Great Recession. The fact that Greenspan's Federal Reserve followed a policy of cutting interest rates, partially to make treasury bonds (necessary for war funding) more attractive to foreigners, encouraged the irresponsible wave of loans and mortgages. The rise in oil prices, too, played a role in the credit crunch, as a one-cent rise in the price of fuel wipes $1 billion from the pockets of US consumers, depriving the nation of extra liquidity that could be spent on necessary investments.[91] Economist James D. Hamilton focused on the effect of oil prices on the Great Recession. As oil prices doubled between June 2007 and June 2008, he argued that 'in addition to housing ... oil prices were an important factor in turning that slowdown into a recession'. In fact, 'had there been no increase in oil prices between 2007:Q3 and 2008:Q2, the US economy would not have been in a recession over the period 2007:Q4 through 2008:Q3'. Due to 'an interactive effect between the oil price shock and the problems in housing ... oil prices indisputably made an important contribution to both the initial downturn and the magnitude of the problem'. Thus, 'the economic downturn of 2007–08 should be added to the list of recessions to which oil prices appear to have made a material contribution'.[92] Nouriel Roubini, too, acknowledges that the rise in oil prices in 2008 'played a role' in the crisis and helped push oil importers like the US, Europe and Japan into recession.[93]

Due to the Great Recession, the United States lost its role as the sole engine of growth in the global economy, because of the rise of new growth engines, particularly in Asia.[94] Paul Kennedy agreed that 'the real threat to US power in the future ... is the steady rise of Asia and, in particular, China'.[95] Nouriel Roubini agreed that 'recent economic, financial and geopolitical events suggest that the decline of the American Empire has started', citing three factors which 'suggest that the US has squandered its unipolar moment': excessive reliance on hard military power in Iraq and Afghanistan, the rise of the BRIC countries, the EU, South Africa and Iran as economic rivals and/or regional powers, and most importantly, the fact that 'the US squandered its economic and financial power by running reckless economic policies, especially in its twin fiscal and current-account deficits'. This twin deficit, financed by strategic rivals like Russia and China, or 'unstable petro-states' like Saudi Arabia and the Gulf States,[96] led to a 'balance of financial terror', since these creditors could 'pull the plug' and sell US debts en masse, leading to the collapse of the US dollar. And even though it was unlikely that these creditors would 'pull the plug' (since they themselves would suffer huge losses by such a move), it would still be risky for Washington to depend on them to finance its twin deficit.[97] George Soros agreed that the invasion of Iraq and Bush's policies decreased US global power and contributed to the rise in oil prices and the weakness of the US dollar in 2008–2009.[98]

On the global level, Paul Kennedy agreed that Krauthammer's 'unipolar moment' was over, as US power declined due to 'inconsiderate and sometimes arrogant diplomacy, an obsession with the War on Terror, and reckless fiscal policies'. (This decline was evident in Washington's passive reaction to the Russian offensive in Georgia in August 2008.)[99] Richard Haass, President of the Council on Foreign Relations, agreed that 'what some dubbed the "unipolar moment" is history'.[100] Therefore, argued Charles Kupchan, managing the competition with rising powers will be a greater challenge than terrorism.[101]

The US intelligence community agrees with this analysis. 'The US will remain the pre-eminent power, but ... American dominance will be much diminished', forecast Thomas Fingar, the US intelligence community's top analyst, in September 2008. America's influence was diminishing as economic powerhouses such as China asserted themselves on the global stage. Washington would no longer be in a position to dictate what the global structures would look like. Nor would any other country, but the world would shift to a less US-centric system.[102] According to *Global Trends 2025*, a report prepared by the National Intelligence Council (NIC) in November 2008, the US would suffer 'relative' decline in economic and military power over the next decade or so:

> The United States will have greater impact on how the international system evolves over the next 15–20 years than any other international actor, but it will have less power in a multipolar world than it has enjoyed for many decades. Owing to the relative decline of its economic, and to a lesser extent, military power, the US will no longer have the same flexibility in choosing among as many policy options. We believe that US interest and willingness to play a leadership role also may be constrained as the economic, military and opportunity costs of being the world's leader are reassessed.[103]

The Global Trends 2025 report paid special attention to energy supplies and other raw material, because it regarded 'energy scarcity as a driving factor in geopolitics',[104] as 'continued global growth – coupled with 1.2 billion more people by 2025 – will put pressure on energy, food and water resources'.[105]

> Resource issues will gain prominence on the international agenda. Unprecedented global economic growth – positive in so many other regards – will continue to put pressure on a number of **highly strategic resources**, including energy, food, and water, and demand is projected to outstrip easily available supplies over the next decade or so. For example, non-OPEC liquid hydrocarbon production – crude oil, natural gas liquids, and unconventionals such as tar sands – will not grow commensurate with demand.[106]

Energy security would also become a major issue as India, China and other countries join the United States in seeking oil, gas and other energy sources. And since China and Europe receive a large portion of their energy supplies from Iran, US options on Iran would be limited. 'So the turn-the-spigot-off kind of thing, even if we could do it, would be counter-productive', said Fingar.[107] Also, Russia would continue to use energy as a tool of coercion and influence, al Qaeda and other terrorist groups would continue to target Persian Gulf oil facilities, and ethnic conflict in oil-producing regions would continue,[108] especially considering that oil countries currently host about a third of the world's civil wars, up from one-fifth in 1992.[109] Therefore, market forces would not be the only solution to the energy situation, due to rising demand and the political agendas of oil-producing countries:

> Unlike earlier periods when resource scarcities loomed large, the significant growth in demand from emerging markets, combined with constraints on new production – such as the control exerted now by state-run companies in the global energy market – limits the likelihood that market forces alone will rectify the supply-and-demand imbalance.[110]

Indeed, experts agree that the power of the major oil corporations of the US (and the West in general) is falling relative to the power of the national oil companies (NOCs) of the producing countries, as NOCs now control 81% of global oil reserves, and their management and production are highly politicized. Therefore, the US empire of oil is weakening due to the rise of other players. There will be no single empire of oil in the twenty-first century, but there will be many players.[111]

Furthermore, the world today is taking the concept of Peak Oil (the point where global oil production would reach its maximum, then start to decline) more seriously. There are two camps on Peak Oil theory. The pessimist camp says that Peak Oil has already been reached (in 2006), and global oil production has been in decline ever since. The more optimistic camp, which includes the International Energy Agency (IEA), believes that the world has not reached Peak Oil yet, but it will in a decade or two. Fatih Birol, the chief economist at the IEA, believes that global oil production will peak around 2020, a decade earlier than most governments have estimated. Even if demand remained steady, the world would have to find the equivalent of four Saudi Arabias (i.e., 40 mbpd) to maintain production, and six Saudi Arabias if it is to keep up with the expected increase in demand between now and 2030, according to Birol.[112]

Moreover, Klare has argued that the era of 'easy oil' (the oil which is easy to find on shore or close to shore, in safe and friendly locations) is over, and that we are now entering the era of 'tough oil'; oil buried far offshore or deep underground, or scattered in small, hard-to-find reservoirs, or oil that must be obtained from unfriendly, politically dangerous or hazardous places. This is evident as oil companies are finding less and less oil.[113] Energy expert Robert Bryce, who does not believe that Peak Oil will arrive soon, agrees that 'easy' oil has ended, and oil is harder and more expensive to replace, as new discoveries are becoming rarer while the costs associated with finding and producing oil are rising.[114] More significantly, Jeroen van der Veer, former CEO of Royal Dutch/Shell, has said that the world has ran out, or is about to run out, of 'easy oil'.[115]

These fears over global energy security (arising from fears of dwindling oil supplies globally, shift of production from global North to the more dangerous, unstable and risky global South, and the targeting of oil facilities by terrorists)[116] have led the major oil consuming countries to think of energy procurement as a zero-sum game.[117] Therefore, the global competition over energy resources has led to a security-dilemma mentality, where the actions of one state to increase its security are perceived as a threat to other states' security, leading them to take similar actions, which may be perceived by others as a threat to their security, leading to a vicious cycle of competition.

On the other hand, the current talk about the replacement of US power by Asian power may be premature. Roubini acknowledged that the erosion of the US empire will not occur overnight, and that it would take 'a couple of decades'.[118] Furthermore, despite its economic rise, Asia is not ready yet to assume economic leadership.[119] Therefore, it would be too simplistic and premature to assume that the US will lose its power in the short term, because, despite the signs that the 'unipolar moment' is over, and signs that other powers will rise to challenge US power, the United States will remain the world's most powerful nation for at least a few years to come, and the decline of the US empire will take a decade or two.

It would also be premature to think that the Bush's corporate globalization agenda will simply disappear because Bush has left office, since this agenda actually predates him, being the work of a group of powerful politicians supported by the world's most powerful corporations.[120] According to energy expert Terry Lynn Karl, the National Energy Policy, or the Bush–Cheney energy plan, 'remains in effect' in the US. The major implication of this is that the US is in a race with China for increasingly scarce oil resources, thus the US must secure resources in as many places as possible.[121] Under the Barack Hussein Obama administration, the United States is still seeking to maintain and expand its military bases in the Middle East, Central Asia, the Far East, Africa and Latin America.[122] In 2009, US Army Chief of Staff General George Casey said that plans had been drawn up in case American fighting forces had to remain in Iraq for another decade, despite the written agreement with Baghdad to pull all troops out by the end of 2011.[123] Richard Haass has said that in Iraq and Afghanistan, the US 'is counting on Iraqis and Afghans to do more so that Americans can do less, but in neither country is it obvious, or even likely, that this will turn out to be the case', making it likely that US troops would stay.[124] Obama's visit to Africa in July 2009 and his support for the Africom force (at a time when US imports from Africa are increasing) marked the expansion of the 'empire of bases' near Africa's oil riches.[125] (It is telling that Washington's FY 2010 budget doubled the funds allocated to Africom.)[126] Obama's deal with Bogotá to use Colombia's military bases, too, shows the expansion of the empire of bases near Latin America's oil riches.[127]

Although similar to Bush on continuing the 'empire of bases', Obama is different in that he gives more focus to green solutions to energy problems and wider US economic problems. Obama has repeatedly said that climate change and dependence on foreign oil were threats to America's economic and national security, proposing alternative energy as a solution to these two threats. During his 2008 election campaign, Obama pledged to lessen addiction to foreign oil, describing it as 'one of the most dangerous and urgent threats this nation has ever faced' and pledging to 'eliminate our imports from the Middle East and Venezuela within 10 years' by developing alternative sources of energy and encouraging conservation.[128]

During his 2008 campaign, Obama initially opposed reversing a 27-year-old ban on offshore drilling in the United States to counter rising fuel prices (while President Bush and Republican candidate John McCain supported lifting the ban). But he later reversed his position, saying that this shift in position was a necessary compromise with the Republicans to achieve his broader goals of energy independence and moderate prices.[129]

Like Bush, Obama addressed the problem of energy during his first week in office, linking it to national security and the threat of terrorism, saying that 'no single issue is as fundamental to our future as energy. America's dependence on oil is one of the most serious threats that our nation has faced. It bankrolls dictators, pays for nuclear proliferation, and funds both sides of the struggle against terrorism. It puts the American people at the mercy of shifting gas prices.'[130] Furthermore, renewable energy/green technology (along with other similarly strategic industrial sectors like health care, broadband and infrastructure) are the main focus of Obama's strategy in dealing with the recent economic crisis and the decline in America's global standing; a strategy which Obama calls the 'new energy economy'.[131] He repeatedly vowed to spend $150 billion on green technology to stimulate jobs and conserve energy,[132] strongly endorsing ethanol as an alternative fuel, as he had advisors and prominent supporters with close ties to the ethanol industry[133] (in a continuation of the business–government ties that are a feature of US policy). On 17 February 2009, Obama signed into law the American Recovery and Reinvestment Act, a $790 billion stimulus bill which included $60 billion to be spent on clean energy, scientific research, setting new fuel-efficiency standards and addressing greenhouse-gas emissions.[134] In June 2009, under Obama's support, the House of Representatives passed the American Clean Energy and Security Act of 2009 to reduce gas emissions. Obama called it a 'jobs bill' that would transform the US economy.[135] Also, like Bush who called for 'energy independence' in his State of the Union address in January 2003,[136] Obama said in his Democratic nomination acceptance speech in August 2008 that 'government must lead on energy independence', promising that 'in ten years, we will finally end our dependence on oil from the Middle East'.[137]

However, critics argue that although the green-technology sector does help create jobs, it would not be sufficient to create jobs and stimulate growth on the desired level.[138] The concept of US energy independence is also unrealistic. Bryce has argued that the United States cannot be energy independent for three reasons. First, America's demand for energy is too large, and growing. Second, increased energy efficiency, although it is vital, does not lead to less energy consumption because each household purchases more and more electric appliances which consume more energy. Third, alternative energy sources (such as biofuels, solar energy, etc.) do not necessarily produce as much energy as fossil fuels, and are not necessarily less polluting or less expensive.[139] Biofuels, for instance, are as expensive and as polluting as fossil fuels, they can cause deforestation and food shortage, and they do not produce as much energy as fossil fuels. The reason why biofuels are being promoted is political pressure from the biofuel industry, especially corn producers. Even if the US switched its entire grain crop to ethanol, it would only replace one-fifth of US gasoline consumption.[140]

Furthermore, a CFR report said in October 2006 that energy independence was a 'myth' and a 'chimera':

US energy policy has been plagued by myths, such as the feasibility of achieving 'energy independence' through increased drilling or anything else. For the next few decades, the challenge facing the United States is to become better equipped to manage its dependencies rather than pursue the chimera of independence.[141]

The report added that 'the voices that espouse "energy independence" are doing the nation a disservice by focusing on a goal that is unachievable over the foreseeable future', and that the US cannot be energy independent 'because liquid fuels are essential to the nation's transportation system … The United States will depend on imported oil for a significant fraction of its transportation fuel needs for at least several decades.'[142]

Similarly, a National Petroleum Council report released in July 2007 said that the concept of 'energy independence' is unrealistic, and that US energy security is best achieved through moderating consumption and global cooperation on energy:

'Energy Independence' should not be confused with strengthening energy security. The concept of energy independence is not realistic in the foreseeable future, whereas US energy security can be enhanced by moderating demand, expanding and diversifying domestic energy supplies, and strengthening global energy trade and investment. There can be no US energy security without global energy security.[143]

Moreover, a Congressional report said that energy independence would not protect the US economy from global price fluctuations: 'US suppliers of energy participate in the world energy market. So long as prices are determined in that market, energy independence will not free the United States from oil price shocks.'[144]

Therefore, green technology is unlikely to significantly decrease dependence on oil, especially Middle East oil, any time soon, and Middle East oil will remain America's main source of energy for a long time.[145]

* * *

Bush raised the place of foreign energy procurement as a US foreign-policy goal, and linked it to other foreign-policy strands on a level never seen before. The failure to secure the oil resources of Iraq led to failure in Bush's overall foreign policy. Before Bush came to office, energy had always been a vital part of the informal Open Door American empire, but Washington dealt with energy resources on an ad hoc basis. Bush elevated the level of energy in US foreign policy to an unprecedented level, as seen in the NEP of May 2001, and he tried to solve America's energy problems, reshape the Middle East and establish a renewed American empire by invading Iraq. This was a part of the Open Door policy which Bush tried to implement. However, the application of the Open Door failed in Iraq, as Bush stuck to false claims over an Iraqi threat which did not exist, and to assumptions that Iraq would be intact after the invasion. When Bush failed to remake Iraq into a model for the Islamic world, his foreign policy vision collapsed as it was too dependent on Iraq. Also, the invasion (and the consequent violence and chaos in Iraq) led to the rise of more anti-Americanism, instead of turning the Middle East into a pro-American region, resulting in the rise of terrorism and militant assaults on American and Western interests in the area. The issue at stake was not just the security of oil resources, but also the stability of the US position in the Middle East, the Caspian, Latin America, Africa and other regions. Thus, the Bush administration failed to advance its foreign-policy agenda, failed to make the world a safer place for the United States, and failed to improve the energy security of the United States.

NOTES

Introduction

1. Antonia Juhasz, *The Bush Agenda: Invading The World, One Economy At A Time* (London: Duckworth, 2006): 162, from Leslie Gelb, 'Kissinger means business', *New York Times*, 20 April 1986.
2. Ian Rutledge, *Addicted to Oil: America's Relentless Drive for Energy Security* (London: I.B.Tauris, 2005): 30. The Petroleum Industry War Council was an advisory group that served as a link between the US Government and the oil industry (Rutledge, *Addicted to Oil*: 30).
3. Simon Bromley, *American Hegemony and World Oil* (Pennsylvania State University Press, 1991): vii.
4. George W. Bush. 'President promotes energy efficiency through technology', *White House*, 25 February 2002 [accessed 2 March 2008] http://www.white-house.gov/news/releases/2002/02/20020225-5.html.
5. Zbigniew Brzezinski, Second Chance: Three Presidents and the Crisis of American Superpower (New York: Basic Books, 2008): 20–21.
6. Zbigniew Brzezinksi, *The Grand Chessboard: American Primacy and its Geostrategic Imperatives* (New York: Basic Books, 1997): 215. Brzezkinski also said that America is 'the world's first truly global power' (Brzezinski, *Second Chance*: 20–21).
7. Brzezinksi, *Second Chance*: 3.
8. Susan Strange, 'The future of the American empire', in Richard Little & Michael Smith (eds.) *Perspectives on World Politics* (London: Routledge, 2006): 353.
9. Bill Emmott, *20:21 Vision: The Lessons of the 20th Century for the 21st* (London: Penguin, 2003): 28–29.
10. Vassilis Fouskas & Bülent Gökay, *The New American Imperialism: Bush's War on Terror and Blood for Oil* (Connecticut: Praeger Security International, 2005): 13, 74, 231–232.

11. William Appleman Williams, *The Roots of the Modern American Empire: A Study in the Growth and Shaping of Social Consciousness in a Marketplace society* (New York: Vintage Books, 1970): xiv; Joyce Kolko and Gabriel Kolko, *The Limits of Power: The World and United States Foreign Policy, 1945–1954* (New York: Harper and Row Publishers 1972): 2, 8, 709; Gabriel Kolko, *The Politics of War: Allied Diplomacy and the World Crisis of 1943–1945* (London: Weidenfeld and Nicolson, 1969): 294.

12. William Appleman Williams, *The Tragedy of American Diplomacy* (New York: Norton: 1984): 45–46, 229; William Appleman Williams, *The Contours of American History* (Chicago: Quadrangle Books: 1966): 349, 368, 369, 370, 372 and passim and William Appleman Williams, 'The Large Corporation and American Foreign Policy', in David Horowitz (ed.) *Corporations and the Cold War* (London: Monthly Review Press, 1969): 71–104.

13. Williams, *The Tragedy of American Diplomacy*: 47, 96, 97 and Williams, *The Contours of American History*: 369.

14. Williams, *The Roots of the Modern American Empire*: 246–247 and Williams, *The Tragedy of American Diplomacy*.

15. Williams, *The Roots of the Modern American Empire*: xii.

16. Ibid: 450.

17. Williams, *The Tragedy of American Diplomacy*: 59 and Williams, 'The Large Corporation and American Foreign Policy': 81.

18. Williams, *The Contours of American History*: 346, 347.

19. Bradford Perkins, 'The Tragedy of American Diplomacy: Twenty Five Years After' in Williams, *The Tragedy of American Diplomacy*: 324.

20. Kolko, *The Politics of War*: 244–248, 254, 265, 280 and Gabriel Kolko, *Main Currents of Modern American History* (New York: Harper and Row Publishers, 1976): 223, 350. Williams agreed that Hull was 'a vigorous advocate of expanded exports and raw material imports' (Williams, *The Contours of American History*: 455).

21. Williams, *The Tragedy of American Diplomacy*: 129.

22. Gabriel Kolko, *The Roots of American Foreign Policy* (Boston: Beacon Press, 1971): 52, 55 and Kolko, *Main Currents in Modern American History*: 205.

23. Paul Kennedy, *The Rise and Fall of the Great Powers: Economic Change and Military Conflict From 1500 to 2000* (London: Fontana Press, 1989): 315–316, 318.

24. Andrew Bacevich, *American Empire: The Realities and Consequences of US Diplomacy* (Massachusetts: Harvard University Press, 2002): 2, 3, 4 and Andrew J. Bacevich, 'Tragedy renewed: William Appleman Williams', *World Affairs*, Winter 2009 [accessed 22 April 2009] http://www.worldaffairs.org/2009%20-%20Winter/full-Bacevich.html.

25. Naomi Klein, *The Shock Doctrine: The Rise of Disaster Capitalism* (London: Penguin, 2008): 3–21. This policy was promoted by the late famous economist Milton Friedman.

26. Stephen Kinzer, *Overthrow: America's Century of Regime Change from Hawaii to Iraq* (New York: Times Books, 2006): 34.

27. Ibid: 321.

28. William Appleman Williams, 'The Large Corporation and American Foreign Policy': 71–104, and Williams, *The Contours of American History*: 368, 369, 370, 372 and passim.

29. Gabriel Kolko, *The Triumph of Conservatism: A Reinterpretation of American History, 1900–1916* (London: The Free Press of Glencoe, 1963): 3; Kolko, *The Roots of American Foreign Policy*: 5, 16, 17, 18, 24–25; David Painter, *Oil and the American Century: The Political Economy of US Foreign Oil Policy, 1941–1954* (London: The John Hopkins Press Ltd, 1986): 1, 2 and Joseph R. Stromberg, 'The Political Economy of Liberal Corporatism,' *Center for Libertarian Studies, The Memory Hole*, 1977 [accessed 15 February 2009] http://tmh.floonet.net/articles/strombrg.html

30. Kolko, *The Roots of American Foreign Policy*: 26.

31. Ibid: 53, 55, 79, 85.

32. Ibid: xvi.

33. Kinzer, *Overthrow*, 3–4 and Klein, *The Shock Doctrine*: 309–310.

34. Bromley, *American Hegemony and World Oil*: 53, from Ernest J. Wilson III, 'World Politics and international energy markets', *International Organization*, 41, 1987: 125–149.

35. Daniel Yergin, *The Prize: The Epic Quest for Oil, Money and Power* (New York: Free Press, 1992): 13.

36. Ibid: 410.

37. Bromley, *American Hegemony and World Oil*: 9.

38. Ibid: 46, from Susan Strange, *States and Markets* (London: Pinter, 1988): 191, and Susan Strange, *States and Markets* (2nd edn) (London: Pinter, 1997): 195.

39. Bromley, *American Hegemony and World Oil*: 46.

40. Strange, *States and Markets* (2nd edn): 194.

41. Ibid: 195.

42. Michael T. Klare, 'Energy Security', in Paul D. Williams (ed.) *Security Studies: An Introduction* (Oxford: Routledge, 2008): 484.

43. Klare, 'Energy Security': 484, 485, 487–488 and Daniel Yergin, 'Energy security will be one of the main challenges of foreign policy', *Cambridge Energy Research Associates (CERA)*, 19 July 2006 [accessed 10 February 2010], http://www.cera.com/aspx/cda/public1/news/articles/newsArticleDetails.aspx?CID=8248.

44. Klare, 'Energy Security': 488.

45. Williams, *The Tragedy of American Diplomacy*: 160. Williams adds that 'American leaders did not overlook the long-range importance of a new and as yet unrecognised [Middle Eastern] market of prodigious size' (Williams, *The Tragedy of American Diplomacy*: 159).

46. Kolko, *The Politics of War*: 624.

47. Bromley, *American Hegemony and World Oil*: 48, 53, 106, 123, 245. Bromley adds that control of oil is 'the centre of gravity' of US global economic preponderance (Simon Bromley, *American Power and the Prospects for International Order* (Cambridge: Polity, 2008): 105).

48. Svante Karlsson, *Oil and the World Order: American Foreign Oil Policy* (Warwickshire: Berg Publishers Limited, 1986): 279.

49. Jan H. Kalicki & David L. Goldwyn, 'Introduction', in Jan H. Kalicki & David L. Goldwyn (eds) *Energy and Security: Toward a New Foreign Policy Strategy* (Washington: Woodrow Wilson Centre Press, 2005): 11.

50. Bill Richardson, 'Foreword', in Jan H. Kalicki & David L. Goldwyn (eds) *Energy and Security: Toward a New Foreign Policy Strategy* (Washington: Woodrow Wilson Centre Press, 2005): xvii.

51. Fouskas & Gökay, *The New American Imperialism*: 24.

52. Noam Chomsky, *Imperial Ambitions: Conversations with Noam Chomsky on the Post-9/11 World* (London: Penguin Books, 2006): 6, 112; Noam Chomsky, *Interventions* (London: Penguin Books, 2007): 46, 47, 77, 85, 112, 135, 202, 208, and Noam Chomsky & Gilbert Achcar, *Perilous Power: The Middle East and US Foreign Policy* (London: Penguin Books, 2008): 55–57.

53. Rutledge, *Addicted to Oil*: xi.

54. Painter, *Oil and the American Century*: 1.

55. Kolko, *The Roots of American Foreign Policy*: 26.

56. Kolko, *The Politics of War*: 302.

57. Ibid.: 294–313; Kolko, *Main Currents in Modern American History*: 202–203 and Kolko & Kolko, *The Limits of Power*: 70–72; Painter, *Oil and the American Century*: 59, 199, 202.

58. Kolko & Kolko, *The Limits of Power*: 444, 447; Painter, *Oil and the American Century*: 155–156, and Karlsson, *Oil and the World Order*: 91. Oil was the largest single item on the dollar budget of most of the Marshall Plan countries. Due to the dominant position of the United States in petroleum equipment, technology and shipping, and due to the dominant position of the US oil companies and the US dollar, Europe needed dollars to pay for its energy needs with imported oil. Thus, oil became one of the key commodities in the European Recovery Program (ERP) (Painter, *Oil and the American Century*: 155–156).

59. Kolko & Kolko, *The Limits of Power*: 447.

60. This is the book which most other books on oil use as reference to the historical relation between the US Government and US oil corporations. It is thus an authority on the subject.

61. Anthony Sampson, *The Seven Sisters: The Great Oil Companies and The World They Shaped* (Kent: Hodder & Stoughton Limited: 1980): 82, 84, 105, 108, 117, 118.

62. Ibid: 107.

63. Ibid: 115.

64. Ibid: 243, 259.

65. Ibid: 287.

66. Ibid: 324.

67. Anthony Brown, *Oil, God and Gold: The Story of Aramco and the Saudi Kings* (New York: Houghton Mifflin Company, 1999): 299–300; William Clark, *Petrodollar Warfare: Oil, Iraq, and the Future of the Dollar* (Gabriola Island: New Society Publishers, 2005): 43; Karlsson, *Oil and the World Order*: 250, and Sampson, *The Seven Sisters*: 314.

68. John Morrissey, 'The Geoeconomic Pivot of the Global War on Terror: US Central Command and the War in Iraq', in David Ryan and Patrick Kiely (eds), *America and Iraq: Policy-Making, Intervention and Regional Politics* (Oxford: Routledge, 2009): 103–122. Also see Michael T. Klare, 'Energy Security', in Paul D. Williams (ed.), *Security Studies: An Introduction* (Oxford: Routledge, 2008): 487. It could be argued that the Carter Doctrine was modelled on the Truman Doctrine of 1947 and the Eisenhower Doctrine of 1957, as Brzezinksi told Carter that he 'had the opportunity to do what President Truman did on Greece and Turkey', and come up with a 'Carter Doctrine' (Karlsson, *Oil and the World Order*: 262 and Douglas Little, *American Orientalism: The United States and the Middle East since 1945* (London: I.B.Tauris, 2005): 152).

69. Yergin, *The Prize*: 14.

70. Michael Klare, *Resource Wars: The New Landscape of Global Conflict* (New York: Owl Books, 2002): xiii.

71. Ibid: 3.

72. Ibid: 27.

73. Fouskas & Gökay, *The New American Imperialism*: 152–153, 174; Dan Briody, *The Halliburton Agenda: The politics of oil and money* (New Jersey: Wiley, 2004): 199; Chalmers Johnson, *The Sorrows of Empire: Militarism, Secrecy and the End of the Republic* (London: Verso, 2006 edition): 145–146, 155; Michel Collon, *Media Lies and the Conquest of Kosovo: NATO's Prototype for the Next Wars of Globalization* (New York: Unwritten History: 2007): 6, 8, 29, 34, 35, 91, 93,

97, 99, 104, 105, 107, 108, 112–114, 118, 121, 123, 124, 127–130, 134–136, 145, 154, 156, 172, 210 and Peter Dale Scott, 'The Background of 9/11: Drugs, Oil and US Covert Operations', in David Ray Griffin and Peter Dale Scott (eds) *9/11 and American Empire: Intellectuals Speak Out* (Gloucestershire: Arris Books, 2007): 76.

74. Robert Bryce, *Cronies: Oil, the Bushes, and the Rise of Texas, America's Superstate* (New York: Public Affairs, 2004): 204; Rutledge, *Addicted to Oil*: 54; Craig Unger, *House of Bush, House of Saud* (London: Gibson Square, 2007): 39, 43 and Yergin, *The Prize*: 753.

75. Rutledge, *Addicted to Oil*: 7. Rutledge mentioned the drive for the 'American Imperium' in the Middle East in Rutledge, *Addicted to Oil*: xiii and 11.

76. Rutledge, *Addicted to Oil*: 93–94. For more information on increasing US dependence on Middle East oil, see Rutledge, *Addicted to Oil*: 9, 131, 197.

77. Juhasz, *The Bush Agenda*: 6.

78. Ibid: 4.

79. Ibid: 7, 102, 310.

80. Ibid: 5–6.

81. Michael Klare, *Blood and Oil* (London: Hamish Hamilton, 2004): 13, 56 and Rutledge, *Addicted to Oil*: 9.

82. Rutledge, *Addicted to Oil*: 130. The United States possesses only 4–5% of global oil reserves (Duncan Clarke, *Empires of Oil: Corporate Oil in Barbarian Worlds* (London: Profile Books, 2007): 292).

83. Fouskas & Gökay, *The New American Imperialism*: 71.

84. Kennedy, *The Rise and Fall of the Great Powers*: 316.

85. Ibid: 665–666.

86. Ibid: xv.

87. Bromley, *American Power*: 2.

88. Juhasz, *The Bush Agenda*: 8, 9.

89. Vassilis Fouskas & Bülent Gökay, *The New American Imperialism: Bush's War on Terror and Blood for Oil* (Connecticut: Praeger Security International, 2005): 71, 72, 136, 231.

90. Noam Chomsky, *The Essential Chomsky* (London: The Bodley Head, 2008.): 345 and Chalmers Johnson, *The Sorrows of Empire: Militarism, Secrecy and the End of the Republic* (London: Verso, 2006): 151–185.

91. Klare, *Blood and Oil*: 152–179.

92. Vassilis Fouskas & Bülent Gökay, *The New American Imperialism: Bush's War on Terror and Blood for Oil* (Connecticut: Praeger Security International, 2005): 150–160.

93. Lutz Kleveman, *The New Great Game: Blood and Oil in Central Asia* (London: Atlantic Books, 2004).

94. Ahmed Rashid, *Taliban: The Story of the Afghan Warlords* (London: Pan MacMillan, 2001): 128–217.

95. Jean-Charles Brisard & Guillaume Dasquié, *Forbidden Truth: US–Taliban Diplomacy and the Failed Hunt for Bin Laden* (New York: Nation Books, 2002).

96. Klare, *Blood and Oil*: 72–73; Michael Klare 'Resources' in John Feffer (ed.) *Power Trip: US Unilateralism and Global Strategy After September 11* (New York: Severn Stories Press, 2003): 50–51 and Michael T. Klare, 'The Bush/Cheney energy strategy: Implications for US foreign and military policy', *Information Clearing House*, 26–27 May 2003 [accessed 25 January 2010] http://www.informationclearinghouse.info/article4458.htm.

97. 'Quadrennial Defence Review Report', 30 September 2001, *US Department of Defence*, http://www.defenselink.mil/pubs/ qdr2001.pdf#search='quadrennial%20defence%20review%202001'.

98. Michael T. Klare, 'Resources': 50–51 and Klare, *Blood and Oil*: 72–73.

99. Lawrence Kaplan & William Kristol, *The War Over Iraq: Saddam's Tyranny and America's Mission* (California: Encounter Books: 2003): 73–74, 112; John Ikenberry, 'America's Imperial Ambition: The Lures of Preemption', *Foreign Affairs*, September/October 2002, 81(5), and Adrian Wooldridge, 'Can the Bush doctrine last?' from 'After Bush: A Special Report on America and the World', *The Economist*, 29 March 2008.

100. Bromley, *American Power*: 132, 134, 135–138.

101. Ibid: 142.

102. Rutledge, *Addicted to Oil*: xii, 11, 155–157, 198.

103. Ibid: xi.

104. Rutledge, *Addicted to Oil*: xiii, 11.

105. Klare, *Blood and Oil*: 59–73.

106. Rutledge, *Addicted to Oil*: 184–185.

107. Noam Chomsky, *Imperial Ambitions: Conversations with Noam Chomsky on the Post-9/11 World* (London: Penguin Books, 2006): 6, 112; Noam Chomsky, *Interventions* (London: Penguin Books, 2007): 46, 47, 77, 85, 112, 135, 202, 208, and Noam Chomsky and Gilbert Achcar, *Perilous Power: The Middle East and US Foreign Policy* (London: Penguin Books, 2008): 55–57; Chalmers Johnson, *The Sorrows of Empire: Militarism, Secrecy and the End of the Republic* (London: Verso, 2006): 151–185 and Stephen Pelletière, *America's Oil Wars* (Connecticut: Praeger, 2004): 133–134.

108. Roger Burbach & Jim Tarbell, *Imperial Overstretch: George W. Bush and the Hubris of Empire* (London: Zed Books, 2004): 51, 128–129, 199, 202.

109. Naomi Klein, *The Shock Doctrine: The Rise of Disaster Capitalism* (London: Penguin, 2008): 313.

110. Ibid: 322.

111. Ibid: 11, 56–57, 253.

112. Christian Euler, 'Resurrect Erhard', *The Shock Doctrine*, 7 April 2003 [accessed 5 January 2010] http://www.naomiklein.org/shock-doctrine/resources/milton-friedman-war-iraq. For the German original, please see Von Christian Euler, 'Lasst Erhard auferstehen', *Focus*, 7 April 2003 [accessed 5 January 2010] http://www.focus.de/finanzen/news/wirtschaft-lasst-auferstehen_aid_196501.html. Interestingly, in July 2006, Milton Friedman had an interview with the *Wall Street Journal*, where he said that he 'was opposed to going into war from the beginning' because it was an 'aggression' (Tunku Varadarajan, 'The romance of economics', *Wall Street Journal*, 22 July 2006 [accessed 5 January 2010] http://www.opinionjournal.com/editorial/feature.html?id=110008690).

113. Ron Suskind, *The Price of Loyalty* (London: Free Press, 2004): 72.

114. Ibid: 96.

115. Greg Palast, 'Secret US plans for Iraq's oil', *BBC Newsnight*, 17 March 2005 [accessed 29 May 2008] http://news.bbc.co.uk/1/hi/programmes/newsnight/4354269.stm.

116. Bromley, *American Hegemony and World Oil*: 48, 53, 239–241.

117. Clark, *Petrodollar Warfare*: 20, 21, 30 and Fouskas & Gökay, *The New American Imperialism*: 17–18. The 1974–75 agreement was necessary as there were fears that OPEC was discussing the pricing of oil in a basket of currencies, as the dollar was falling in value due to the financial strains of the Vietnam War (Clark, *Petrodollar Warfare*: 20, 30).

118. Fouskas & Gökay, *The New American Imperialism*: 18. Also see Bromley, *American Hegemony and World Oil*: 241.

119. Clark, *Petrodollar Warfare*: 22, 32–33. This is called 'petrodollar recycling,' and it works because every nation around the world needs to maximize its dollar holdings to import oil (Clark, *Petrodollar Warfare*: 32). As nations seek to maximize their dollar holdings, 'they do not just stack all these dollars in their vaults', they purchase dollar-assets from 'the issuer': the United States, which 'controls the dollar and prints it at fiat' (Clark, *Petrodollar Warfare*: 32).

120. Clark, *Petrodollar Warfare*: 22.

121. Ibid: 28. 70% of global trade is conducted in dollars (Clark, *Petrodollar Warfare*: 32).

122. Fouskas & Gökay, *The New American Imperialism*.

123. Clark, *Petrodollar Warfare* .

124. Ibid: 17, 27–28, 34, and Fouskas & Gökay, *The New American Imperialism*: 11–37.

125. Clark, *Petrodollar Warfare*: 28 and 33, and Fouskas & Gökay, *The New American Imperialism*: 11–37.

126. Clark, *Petrodollar Warfare*: 21–31 and Fouskas & Gökay, *The New American Imperialism*: 19, 30–31. Japan, for instance, would export goods to the US, receive payment in dollars, then use these dollars to buy oil from the Gulf States who would invest these dollars back in the US, and this is why it is called 'petrodollar recycling'. According to Clark, '70% of Saudi Arabia's entire wealth is held in one account; the Federal Reserve in New York'. One commentator says that as long as oil is traded in dollars, and the US can print them on fiat, then the US has its own 'money tree' (Simon Reeve, 'Conspiracy Theories: Iraq', *Sky One*, 10 January 2007, 22:00 GMT).

127. Clark, *Petrodollar Warfare* and Fouskas & Gökay, *The New American Imperialism*: 25–27.

128. Andrew J. Bacevich, 'Tragedy renewed: William Appleman Williams', *World Affairs*, Winter 2009 [accessed 22 April 2009] http://www.worldaffairs.org/2009%20-%20Winter/full-Bacevich.html, and Thomas McCormick 'What would William Appleman Williams say now?' *History News Network*, 24 September 2007 [accessed 22 April 2009] http://hnn.us/articles/42971.html.

129. Pelletière, *America's Oil Wars*: 137.

130. Roger Burbach & Jim Tarbell, *Imperial Overstretch: George W. Bush and the Hubris of Empire* (London: Zed Books, 2004): 51, 128–129, 199, 202.

Chapter 1

1. Daniel Yergin, *The Prize: The Epic Quest for Oil, Money and Power* (New York: Free Press, 1992): 753.

2. Michael Klare, *Resource Wars: The New Landscape of Global Conflict* (New York: Owl Books, 2002): 8–9.

3. George H.W. Bush, 'Address on administration goals before a joint session of Congress', *The American Presidency Project*, 9 February 1989 [accessed 19 October 2006] http://www.presidency.ucsb.edu/ws/index.php?pid=16660

4. Charles Krauthammer, 'The Unipolar Moment', *Foreign Affairs*, 70(1) 1991: 23, 24.

5. Krauthammer, 'The Unipolar Moment': 28–31.

6. Michael Cox, *US Foreign Policy After the Cold War: Superpower Without A Mission?* (London: Pinter, 1995): 4.

7. George H.W. Bush, 'Address Before a Joint Session of the Congress on the State of the Union', *The American Presidency Project*, 28 January 1992 [accessed 19 October 2006] http://www.presidency.uscb.edu/ws/index.php?pid=20544

8. James Boys, *An Evaluation of Engagement and Enlargement: The Clinton Doctrine (1993–1997)*, PhD thesis, University of Birmingham, July 2006: 1.

9. Andrew Bacevich, *American Empire: The Realities and Consequences of US Diplomacy* (Massachusetts: Harvard University Press, 2002): 36, 38–39, 88, 90.

10. Ibid: 71, 73, 85, 86.

11. Ibid: 126, 127, 128.

12. Maria Do Ceau Pinto, *Political Islam and the United States: A Study of US Policy Towards Islamist Movements in the Middle East* (Ithaca Press, 1999): x.

13. Noam Chomsky & Gilbert Achcar, *Perilous Power: The Middle East and US Foreign Policy* (London: Penguin Books, 2008): 42.

14. Bacevich, *American Empire*: 106.

15. Ian Rutledge, *Addicted to Oil: America's Relentless Drive for Energy Security* (London: I.B.Tauris, 2005): 80–101.

16. Jean-Charles Brisard & Guillaume Dasquié, *Forbidden Truth: US–Taliban Diplomacy and the Failed Hunt for Bin Laden* (New York: Nation Books, 2002):15.

17. Ahmed Rashid, *Taliban: The Story of the Afghan Warlords* (London: Pan MacMillan, 2001): 154, 163, 176.

18. Yergin, *The Prize*: 769.

19. Ibid: 768.

20. Michael Mandelbaum, 'The Bush Foreign Policy', *Foreign Affairs*, 70(1) (1991): 14–16.

21. Klare, *Resource Wars*: 6–7. Klare added that resources will not be the 'One Big Thing' that lies at the heart of international relations, but it will help explain such relations. Resource conflict will not be the only source of conflict in the 21st Century, but it will increasingly rise as a factor (Klare, *Resource Wars*: 14, 25).

22. Klare, *Resource Wars*: xi, 3, 27.

23. Robert S. Litwak, 'Iraq and Iran: From Dual to Differentiated Containment', in Robert J. Lieber (ed.) *Eagle Rules? Foreign policy and American Primacy in the Twenty-First Century* (New Jersey, Prentice Hall, 2002): 176.

24. Alan Friedman, *Spider's Web: The Secret History of How the White House Illegally Armed Iraq* (New York: Bantam Books: 1993): 133, 134.

25. George H.W. Bush, 'National Security Directive 26, US Policy Toward the Persian Gulf', *Foundation of American Scientists*, 2 October 1989 [accessed 13 October 2006] http://www.fas.org/irp/offdocs/nsd/nsd26.pdf

26. Ibid.

27. Ibid.

28. Peter Hahn, *Crisis and Crossfire: The United States and the Middle East since 1945* (Washington, DC: Potomac Books, 2005): 106.

29. Bush, 'National Security Directive 26'.

30. Friedman, *Spider's Web*: 163.

31. Robert Bryce, *Cronies: Oil, the Bushes, and the Rise of Texas, America's Superstate* (New York: Public Affairs, 2004): 204; Craig Unger, *House of Bush, House of Saud* (London: Gibson Square, 2007): 39, 43 and Yergin, *The Prize*: 753.

32. James Mann, *Rise of the Vulcans: The History of Bush's War Cabinet* (London: Penguin Books, 2004): 183. Another example of Bush's support for Iraq was that he voided a prohibition which the US Congress imposed on any new financing for Iraq from the Export-Import Bank, due to Saddam's human rights violations. On 17 January 1990, Bush signed a waiver which determined that the prohibition was 'not in the national interest of the United States', making Saddam once again eligible for US taxpayer-backed loan guarantees worth about $200 million (Friedman, *Spider's Web*: 157). The State Department said that lifting the ban was necessary to achieve the 'goal of increasing US exports and put us in a better position to deal with Iraq regarding its human rights record' (Noam Chomsky, *Hegemony or Survival: America's Quest for Global Dominance* (London: Penguin Books, 2004): 114)

33. Friedman, *Spider's Web*: 134.

34. Antonia Juhasz, *The Bush Agenda: Invading the World, One Economy at a Time* (London: Duckworth, 2006): 160, 161 and Joe Conason, 'The Iraq Lobby: Kissinger, the Business Forum & Co.', in Micah L. Sifry & Christopher Serf (eds) *The Gulf War Reader* (New York: Random House, 1991): 79–84.

35. This military cooperation was expanded to the extent that the CIA provided 'unusually detailed military intelligence to Iraq that Iraqi security agencies put to use in a genocide campaign against Iraq's Kurdish minority group' (Christopher Simpson, *National Security Directives of the Reagan and Bush Administrations: The Declassified History of US Political and Military Policy, 1981–1991* (Oxford: Westview Press, 1995): 897).

36. Reuven Hollo, *Oil and American Foreign policy in the Persian Gulf 1947–1991*, PhD, University of Texas, May 1995: 331. Even during his days as Vice President (under Ronald Reagan), Bush 'made it clear he wanted to help Iraq', said a former White House official. A gleaming example is the oil pipeline which Bechtel and Brown & Root planned to build from the Iraqi oilfields in Kirkuk to the Jordanian port of Aqaba, as Vice President Bush lobbied the Export-Import Bank so hard to extend financing to Iraq for this project. Also, Brown & Root have been working for the Baath Party since 1973 to build and develop the Mina al-Bakr and Khor al-Amaya oil terminals in Iraq (Bryce, *Cronies*: 124, 127–128, 129, 130; Friedman, *Spider's Web*: 117, 156–157 and Unger, *House of Bush, House of Saud*: 70). Reagan's Secretary of State George Shultz, who had business ties with

Bechtel, lobbied hard for the Aqaba pipeline. After he left government, he went back to work for Bechtel. Also, Reagan sent his special envoy, Donald Rumsfeld, to meet Saddam in November 1983 to discuss the Aqaba pipeline. After his initial approval, Saddam rejected the Aqaba pipeline deal in November 1985, citing concerns over the pipeline's security as it passed through Israel (Bryce, *Cronies*: 127–128; Juhasz, *The Bush Agenda*: 103, 165–167, and Dilip Hiro, *Secrets and Lies: The True Story of the Iraq War* (London: Politico's, 2005): 372).

37. Lawrence Freedman & Efraim Karsh, *The Gulf Conflict 1990–1991: Diplomacy and War in the New World Order* (London: Faber & Faber, 1994): 76; Pinto; *Political Islam and the United States*: 58, 117; Rutledge, *Addicted to Oil*: 51; Yergin, *The Prize*: 773; Maria Do Ceau Pinto, 'Persian Gulf Instability: A Threat to Western Interests'. From L.C. Montanheiro, R.H. Haigh & D.S. Morris (eds) *Essays on International Co-operation and Defence* (Sheffield: Sheffield Hallam University Press, 1998): 125.

38. George H.W. Bush, 'National Security Directive 45: US Policy in Response to the Iraqi Invasion of Kuwait', *Foundation of American Scientists*, 20 August 1990 [accessed 13 October 2006] http://www.fas.org/irp/offdocs/nsd/nsd_45.htm

39. Bush, 'National Security Directive 45'.

40. Ibid.

41. Yergin, *The Prize*: 774, and Faisal Islam, 'War: Who is it good for?', *Guardian*, 11 August 2002 [accessed 23 August 2002] http://www.guardian.co.uk/Iraq/story/0,2763,772666,00.html and Patrick E. Tyler & Richard W. Stevenson, 'Profound effect on US economy seen in a war on Iraq', *Truthout*, 30 July 2002 [accessed 5 August 2002] http://truthout.com/docs_02/07.31C.war.econo.htm.

42. 'Iraq overshadows world's biggest oil summit', CNN, 6 September 2002 [accessed 6 September 2002] http://www.cnn.com/2002/WORLD/meast/09/05/brazil.oil.congress.ap/index.html

43. F. Gregory Gause, 'Saudi Arabia over a barrel', *Foreign Affairs*, May/June 2000: 87. Saudi Arabia raised its production to make up for three-quarters of the lost supply. (Venezuela and the United Arab Emirates also increased their production of oil.) This Saudi action was not just a bow to US desires, but in the interest of the Saudis, since the Saudis have an interest in keeping oil prices within a range which is not too high or too low, since very low prices would decrease the oil exporter's revenues, and very high prices would lead to a decrease in demand for oil and the shift to other sources of energy and energy conservation measures (Yergin, *The Prize*: 774 and Gause, 'Saudi Arabia over a barrel': 87).

44. Klare, *Resource Wars*: 34 and Klare, *Blood and Oil* (London: Hamish Hamilton, 2004): 5 and 50.

45. Rutledge, *Addicted to Oil*: 52 and Yergin, *The Prize*: 773.

46. Klare, *Blood and Oil*: 50, and Klare, *Rising Powers, Shrinking Planet: How Scarce Energy is Creating a New World Order* (Oxford: Oneworld, 2008): 180–181.

47. George H.W. Bush, 'Address Before a Joint Session of the Congress on the State of the Union', *The American Presidency Project*, 29 January 1991 [Accessed 19 October 2006] http://www.presidency.ucsb.edu/ws/index. php?pid=19253.

48. Graham E. Fuller & Ian O. Lesser, 'Persian Gulf myths', *Foreign Affairs*, May/ June 1997: 43.

49. Ibid: 44.

50. Thomas J. McCormick, *America's Half-Century: United States Foreign Policy in the Cold War and After* (Baltimore: Johns Hopkins University Press: 1995): 248.

51. Ibid: 247–248.

52. Ibid: 246.

53. Ibid: 248.

54. Zbigniew Brzezinski, *Second Chance: Three Presidents and the Crisis of American Superpower* (New York: Basic Books, 2008): 69.

55. William G. Hyland, *Clinton's World: Remaking American Foreign Policy* (London: Praeger, 1999): 4–5. According to McCrisken, 'it is clear that almost from the outset the administration preferred the resort of force over the cooperative settlement' (Trevor B. McCrisken, *American Exceptionalism and the Legacy of Vietnam: US Foreign Policy Since 1974* (Hampshire: Palgrave Macmillan, 2003): 156). This is reiterated by John Morrissey, who argued that Washington had an 'emphasis on military force, rather than diplomacy, in dealing with the Middle East', especially that official documents say that Centcom's 'main mission' was 'guarding Gulf oil' as evident in the 1991 Gulf War (John Morrissey, 'The Geoeconomic Pivot of the Global War on Terror: US Central Command and the War in Iraq', in David Ryan and Patrick Kiely (eds), *America and Iraq: Policy-Making, Intervention and Regional Politics* (Oxford: Routledge, 2009): 108–111). Pelletière argued that Saddam indicated several times that he only wanted to use Kuwait as a bargaining chip to relieve his debts, and had Bush listened to him, Iraq would have withdrawn, but the United States wanted war (Stephen Pelletière, *America's Oil Wars* (Connecticut: Praeger, 2004): 105). In fact, the Americans showed Saudi King Fahd satellite images of the Kuwaiti–Saudi borders which, they alleged, contained images of Iraqi troops amassing near the Saudi border in attack formations, even though the pictures showed no evidence of that (Dilip Hiro, *Secrets and*

Lies: Operation Iraqi Freedom and After (New York: Nation Books, 2004): 48, and Unger, *House of Bush, House of Saud*: 139–140).

56. George H.W. Bush, 'National Security Directive 54: Responding to Iraqi Aggression in the Gulf', *Washington Post*, 15 January 1991 [accessed 13 October 2006] http://www.washingtonpost.com/wp-srv/inatl/longterm/fogofwar/docdirective.htm

57. Ibid.

58. Ibid

59. Adam Zagorin, 'All about the oil', *Time*, 17 February 2003: 25.

60. Kenneth Pollack, *The Threatening Storm: The Case for Invading Iraq* (New York: Random House, 2002): 46–47.

61. Norman H. Schwarzkopf, *It Doesn't Take a Hero* (London: Bantam Books, 1992): 578–579. Also see McCrisken, *American Exceptionalism and the Legacy of Vietnam*: 151.

62. Lawrence Kaplan & William Kristol, *The War Over Iraq: Saddam's Tyranny and America's Mission* (California: Encounter Books: 2003): 44, 45, 68 and McCrisken, *American Exceptionalism and the Legacy of Vietnam*: 151. Saddam was the only one strong enough to keep Iraq united. If Iraq was divided, it would be into a Kurdish north (which could threaten the security of Turkey, a NATO member and a vital ally to the United States), and a Shiite south (which could join Iran and increase Tehran's influence). The Shiites make up 60% of Iraq's population, 45% of Kuwait's, 25% of Saudi Arabia's, 70% of Bahrain's and 90% of Iran's. If Iraq was divided and the Shiites in the south of Iraq gained independence, then this would encourage the Shiites in other Arab Gulf States to do likewise. This might have an adverse effect on the Arab Gulf states, all of whom were pro-US allies. It would also de-stabilize the region, raising the price of oil or cutting the supplies from the region. Moreover, the Gulf's Arab Shiites might form alliances with Iran (Noam Chomsky, *Middle East Illusions* (Maryland: Rowman & Littlefield Publishing Group, 2003): 188, 189; Noam Chomsky, *Understanding Power: The Indispensable Chomsky* (London: Vintage Books, 2003): 168; Dilip Hiro, *Neighbours, Not Friends: Iraq and Iran After the Gulf War* (London: Routledge, 2001): 36–39; Henry Kissinger, *Does America need a foreign policy?* (New York: Simon & Schuster, 2001): 190; Nicholas J. Wheeler, *Saving Strangers: Humanitarian Intervention in International Society* (Oxford: Oxford University Press, 2000), 147–148, and 'Revolution Delayed,' *The Economist*, 17 September 2002: 30–31).

63. David Owen, *In Sickness and in Power: Illness in the Heads of Government During the Last 100 Years* (London: Meuthen, 2009): 273.

64. Pollack, *The Threatening Storm*: 47.

65. Joseph Stanislaw & Daniel Yergin, 'Oil: reopening the door,' *Foreign Affairs*, September/October 1993: 82.

66. Pinto, *Political Islam and the United States:* 117 and Pinto, 'Persian Gulf Instability': 123–124, 128.

67. Schwarzkopf, *It doesn't take a hero*: 504.

68. George H.W. Bush, 'Address Before a Joint Session of the Congress on the State of the Union', *The American Presidency Project*, 29 January 1991 [Accessed 19 October 2006]

69. http://www.presidency.ucsb.edu/ws/index.php?pid=19253 Hyland, *Clinton's World*: 5.

70. McCrisken, *American Exceptionalism and the Legacy of Vietnam*: 156–157.

71. Actually, Iraq had not had any real nuclear power since the bombing of Ozirak in 1981 (Pelletière, *America's Oil Wars*: 176).

72. Hollo, *Oil and American Foreign policy in the Persian Gulf 1947–1991*: 313, from George H.W. Bush 'President's address to a joint session of Congress', 6 March 1991.

73. George H.W. Bush, 'National Security Strategy of the United States', *Foundation of American Scientists*, August 1991 [accessed 19 October 2006] http://www.fas.org/man/docs/918015-nss.htm.

74. George H. W. Bush, 'Address Before a Joint Session of the Congress on the State of the Union', *The American Presidency Project*, 28 January 1992 [accessed 19 October 2006] http://www.presidency.uscb.edu/ws/index.php?pid=20544

75. Bush, 'Address Before a Joint Session of the Congress on the State of the Union', 28 January 1992.

76. Harvey Sicherman, 'A Cautionary Tale: The US and the Arab–Israeli Conflict', in Robert J. Lieber (ed.) *Eagle Rules? Foreign Policy and American Primacy in the Twenty-First Century* (New Jersey: Prentice Hall, 2002): 156.

77. Bush, 'National Security Strategy of the United States', August 1991.

78. Ibid.

79. Hollo, *Oil and American Foreign policy in the Persian Gulf 1947–1991*: 314 and 316 and Klare, 'Energy Security': 487.

80. David Gergen, 'America's missed opportunities', *Foreign Affairs*, 71(1) (1992): 12.

81. Robert E. Hunter, 'Starting at Zero: US Foreign Policy for the 1990s', in Brad Roberts (ed.) *US Foreign Policy After the Cold War* (Massachusetts: MIT Press, 1992): 7.

82. Michael Cox, *US Foreign Policy After the Cold War: Superpower Without A Mission?* (London: Pinter, 1995): 3.

83. Cormier, Pierre Raymond Joseph, *Understanding Bush's New World Order: Three Perspectives*, MA thesis submitted to the University of Manitoba, May 1995: 15, 16.

84. Cormier, *Understanding Bush's New World Order*: 17–18.

85. Pinto, *Political Islam and the United States*: x.

86. Hunter, 'Starting at Zero': 13–14.

87. Noam Chomsky, 'After the Cold War: US Middle East Policy', in Phyllis Bennis & Michel Moushabeck (eds) *Beyond the Storm: A Gulf Crisis Reader* (New York: Olive Branch Press, 1991): 79–80, 87. The 1991 Gulf War cost $60 billion, and the US allies paid $50 billion of this $60 billion ('Cost of war $200 bn, but that's nothing, says US advisor', *Guardian*, 16 September 2002 [accessed 17 September 2002] http://www.buzzle.com/editorials/9–16-2002–26488.asp)

88. Chomsky, 'After the Cold War': 79–80.

89. Klare, *Resource Wars*: 7.

90. Boys, *An Evaluation of Engagement and Enlargement*: 115.

91. William J. Clinton 'A new covenant for American security,' Georgetown University, 12 December 1991 [accessed 28 March 2008] http://www.ndol.org/ndol_ci.cfm?kaid=128&subid=174&contentid=250537

92. Boys, *An Evaluation of Engagement and Enlargement*: 21.

93. Ibid: 82, 106.

94. Ibid: 267.

95. William J. Clinton 'Remarks to the 48th session of the United Nations General Assembly, New York, 27 September 1993.' *The American Presidency Project*, 27 September 1993 [accessed 28 March 2008] http://www.presidency.ucsb.edu/ws/index.php?pid=47119&st=&st1=.

96. Anthony Lake, 'From containment to enlargement', Address at the School of Advanced International Studies, Johns Hopkins University, Washington, DC, 21 September 1993 [accessed 27 June 2008] http://www.globalsecurity.org/wmd/library/news/usa/1993/usa-930921.htm.

97. Rachel Bronson, *Thicker Than Oil: America's uneasy partnership with Saudi Arabia* (New York: Oxford University Press, 2006): 206, from Martin Indyk 'Back to the Bazaar', *Foreign Affairs*, January/February 81(1) (2002): 75–88.

98. Bronson, *Thicker Than Oil*, 206, from an interview with Martin Indyk, Washington, DC, 27 May 2005.

99. Indyk, 'Back to the Bazaar.'

100. Boys, *An Evaluation of Engagement and Enlargement*: 94.

101. Russel Watson and John Barry, 'A new kind of containment,' *Newsweek*, 12 July 1993 [accessed 28 March 2008] http://web.ebscohost.com/ehost/

detail?vid=4&hid=6&sid=b9af6e0e-b639–4582-b8fa-84ed9a11f13e%40sessionmgr7.

102. Anthony Lake, in a telephone interview with James Boys on 14 September 2004 (Boys, *An Evaluation of Engagement and Enlargement*: 95).

103. Martin Indyk, 'Transcript: Kuwait's Al-Qabas interviews Indyk, Ricciardone on Iraq: Changing regime requires partnership of Iraqi people, neighbours, US', *Embassy of the United States, Amman, Jordan*, 2 February 1999 [accessed 28 March 2008] http://www.usembassy-amman.org.jo/2Qabas.htm

104. Boys, *An Evaluation of Engagement and Enlargement*: 98.

105. Gause, 'The illogic of Dual Containment': 57.

106. Boys, *An Evaluation of Engagement and Enlargement*: 97, 100, 101.

107. Pelletière, *America's Oil Wars*: 119–120. A 1995 UN report stated that the sanctions on Iraq were not aimed at bringing down Saddam Hussein or deterring him from building 'some mythical nuclear bomb [sic]', but at preventing the market competition of Iraqi oil from forcing down the price of oil produced by Saudi Arabia (Pilger, *Hidden Agendas*: 2–3 and 56, and Phyllis Bennis, *Calling the Shots: How Washington dominates today's UN* (New York: Olive Branch Press, 2000): 166 and 169).

108. Bronson, *Thicker Than Oil*: 208–209.

109. Pelletière, *America's Oil Wars*: 125, 128.

110. Ibid: 125, 126.

111. Bronson, *Thicker Than Oil*: 216.

112. Ibid: 215.

113. Pelletière, *America's Oil Wars*: 127 and Rutledge, *Addicted to Oil*: 93–94. For more information on US dependence on Middle East oil, see Rutledge, *Addicted to Oil*: 9, 131, 197 and Thomas L. Friedman, 'Foreign Affairs: America's Oil Change', *New York Times*, 30 October 1997 [accessed 12 June 2006] http://www.nytimes.com/1997/10/30/opinion/foreign-affairs-america-s-oil-change.html.

114. The UN Oil-For-Food programme was established by the United Nations in 1996 as an exemption to sanctions against Iraq allowing Iraq to export $2 billion worth of oil every six months. Nevertheless, the monies from this programme were controlled by Saddam and his loyalists who distributed them as they wished (John Pilger, *Hidden Agendas* (London: Vintage, 1998): 54 and Ofra Bengio, 'How does Saddam hold on', *Foreign Affairs*, July/August 2000: 93–94).

115. Martin Indyk, 'Transcript: Kuwait's Al-Qabas interviews Indyk, Ricciardone on Iraq: Changing regime requires partnership of Iraqi people, neighbours, US', *US Embassy in Amman*, 2 February 1999 [accessed 28 March 2008] http://www.usembassy-amman.org.jo/2Qabas.htm

116. William J. Clinton, 'A statement by the President', *White House*, 31 October 1998 [accessed 4 July 2008] http://www.globalsecurity.org/wmd/library/news/iraq/1998/981031-wh-iraq.htm

117. Kenneth Pollack, *The Persian Puzzle: The Conflict Between Iran and America* (New York: Random House, 2005): 262–263.

118. R. Hrair Dekmejian & Hovann H. Simonian, *Troubled Waters: The Geopolitics of the Caspian Region* (London: I.B.Tauris, 2003): 37–38.

119. Ahmed Rashid, *Taliban: The Story of the Afghan Warlords* (London: Pan MacMillan, 2001): 173.

120. Dekmejian, *Troubled Waters*: 38.

121. Richard Clarke, *Against All Enemies: Inside America's War on Terror* (New York: Free Press, 2004): 102.

122. Pollack, *The Persian Puzzle*: 263, 270.

123. Clarke, *Against All Enemies*: 103.

124. William J. Clinton, 'Executive Order 12957: Prohibiting certain transactions with respect to the development of Iranian petroleum resources'. *The White House, Office of the Press Secretary, Iranian Trade*, 15 March 1995 [accessed 12 June 2005] http://www.iraniantrade.org/12957.htm

125. Clarke, *Against All Enemies*: 103 and Pollack, *The Persian Puzzle*: 271–272. Dick Cheney, then CEO of Halliburton, opposed these sanctions. Nevertheless, Clinton ordered Al Gore to coordinate efforts to build an oil pipeline to tap the resources of Central Asia and use routes that did not pass through Iranian territory (Clarke, *Against All Enemies*: 103).

126. William J. *Clinton, 'The Iran and Libya Sanctions Act of 1996'*, Foundation of American Scientists, 18 June 1996 [accessed 30 June 2005] http://www.fas.org/irp/congress/1996_cr/h960618b.htm.

127. Ibid.

128. Klare, *Resource Wars*: 100 and Klare, *Blood and Oil*: 107.

129. Juhasz, *The Bush Agenda*: 126–127. *CBS News* estimated that Halliburton sold about $40 million a year worth of oil-field services to Iran through its subsidiary, Halliburton Products and Services, which is registered in the Cayman Islands (Robert Bryce, *Gusher of Lies: The Dangerous Delusions of Energy Independence* (New York: PublicAffairs: 2008): 251, and Juhasz, *The Bush Agenda*: 126–127). Also, in May 1998, the United States and the European Union agreed that Washington would provide ILSA waivers to European corporations doing business in Iran, in return for the EU's commitments to increase its cooperation with the United States on non-proliferation and counter-terrorism. This waiver particularly concerned a consortium led by France's Total (Pollack, *The Persian Puzzle*: 288–289, 321).

130. Bryce, *Gusher of Lies*: 251.
131. Klare, *Blood and Oil*: 133 and Klare, *Resource Wars*: 4.
132. Rutledge, *Addicted to Oil*: 103 and George Monbiot, 'America's pipe dream', *Guardian*, 23 October 2001 [accessed 10 August 2009] http://www.guardian.co.uk/world/2001/oct/23/afghanistan.terrorism11
133. Klare, *Resource Wars*: 72–79.
134. Klare, *Resource Wars*: 3 and 83 and Rutledge, *Addicted to Oil*: 104.
135. Rutledge, *Addicted to Oil*: 112.
136. Rashid, *Taliban*: 145.
137. Klare, *Resource Wars*: 81–91.
138. Klare, *Blood and Oil*: 155, Klare, *Resource Wars*: 90 and Klare, *Rising Powers, Shrinking Planet*: 125.
139. Andrew Bacevich, *The Limits of Power: The End of American Exceptionalism* (New York: Metropolitan Books, 2008): 55
140. Klare, *Blood and Oil*: 13, 56 and Rutledge, *Addicted to Oil*: 9.
141. Klare, *Rising Powers, Shrinking Planet*: 123 and Rutledge, *Addicted to Oil*: 104.
142. Fiona Hill, 'A not-so-grand strategy: United States policy in the Caucasus and Central Asia since 1991', *Brookings*, February 2001 [accessed 16 February 2006] http://www.brook.edu/dybdocroot/views/articles/fhill/2001politique.htm.
143. Klare, *Resource Wars*: 3.
144. Klare, *Blood and Oil*: 133, Klare, *Resource Wars*: 4 and Klare, *Rising Powers, Shrinking Planet*: 124.
145. Klare, *Blood and Oil*: 154, Klare, *Resource Wars*: 89 and Rashid, *Taliban*: 174. In fact, Sheila Heslin was lobbying for an exclusive club known as the Foreign Oil Companies Group, a group of major petroleum companies doing business in the Caspian (Robert Baer, *See No Evil: The True Story of a Ground Soldier in the CIA's War on Terrorism* (London: Arrow Books, 2002): 363, 364).
146. Ahmed Rashid, *Jihad: The rise of militant Islam in Central Asia* (London: Yale University Press, 2002): 190.
147. Strobe Talbott, 'A Farewell to Flashman: American policy in the Caucasus and Central Asia', Address at the Johns Hopkins School of Advanced International Studies, Baltimore, Maryland, July 1997, *Tree Media Group* [accessed 14 May 2005] http://www.treemedia.com/cfrlibrary/library/policy/talbott.html.
148. Talbott , 'A Farewell to Flashman'.
149. Dekmejian, *Troubled Waters*: 135–136 and Klare, *Resource Wars*: 1.
150. 'Text: Pena announces Caspian Sea Initiative at Conference: Speech at Crossroads of the Worlds onference May 27', *US Embassy in Israel*, 1 June

1998 [accessed 23 July 2009] http://www.usembassy-israel.org.il/publish/press/energy/archive/1998/june/de1602.htm.

151. 'Text: Pena announces Caspian Sea Initiative at Conference'.

152. Richard L. Morningstar, 'Testimony by Ambassador Richard L. Morningstar, Special Advisor to the President and Secretary of State for Caspian Basin Energy Diplomacy, Before the Senate Subcommittee on International Economic Policy, Exports and Trade Promotion', Tree Media Group, 3 March 1999 [accessed 15 June 2005] http://www.treemedia.com/cfrlibrary/library/policy/morningstar.html.

153. Dekmejian, *Troubled Waters*: 29, and Rutledge, *Addicted to Oil*: 107. The agreement was signed with Azerbaijan's state-owned oil company, the Azerbaijan International Operating Company (AIOC) to exploit 4.3 billion barrels of estimated oil reserves in the Azeri-Chirag-Gunashli (AGC) fields, at a cost of $13 billion. Presiding over the assembled dignitaries from Azerbaijan and the world of multinational oil at the contract-signing ceremony was US Deputy Energy Secretary Bill White and UK Energy Minister Tim Eggar. The deal included the three US oil multinationals, Amoco, Pennzoil, Unocal, US energy services company McDermott, and the US–Saudi partnership Delta Hess, in addition to Britain's BP, Norway's Statoil, Russia's Lukoil and the Turkish State Oil Company. The United States intervened decisively to exclude Iran from the Contract of the Century (Dekemejian, *Troubled Waters*: 134; Klare, *Rising Powers, Shrinking Planet*: 120–121, 123 and Rutledge, *Addicted to Oil*: 107–108). Production Sharing Agreements will be discussed in more details in this book in the chapter on the American invasion of Iraq.

154. Klare, *Resource Wars*: 90 and Klare, *Blood and Oil*: 155.

155. Klare, *Resource Wars*: 102. Zbigniew Brzezinksi, former National Security Advisor to President Jimmy Carter, and consultant to US oil company Amoco, was a major supporter of the BTC pipeline, and was a key figure who played an influential role in directing the Clinton administration's attention towards the Caspian and Central Asia. He made extensive visits to the region in the early 1990s as a consultant to Amoco. He was asked by Clinton in October 1995 to deliver a personal letter from Clinton to President Heydar Aliyev of Azerbaijan to engage Aliyev in a dialogue regarding the prospects of the BTC pipeline, which, according to Brzezinski, was 'an initiative that subsequently has become an obstacle to a resurgence of Russian imperialism' (Brzezinksi, *Second Chance*: 121 and Rutledge, *Addicted to Oil*: 104–105). Israel, too, supported the BTC pipeline which would carry Caspian oil to the port of Haifa (Rashid, *Taliban*: 154).

156. Kleveman, *The New Great Game*: 160 and Rashid, *Taliban*: 160. Argentine's oil company Bridas was competing with Unocal over the construction of this pipeline (ibid). Actually, the Taliban and Pakistani Prime Minister Benazir Bhutto preferred to deal with Argentine's Bridas, not the United States' Unocal, since Bridas did not need loans from international financial institutions whose first requirement was international recognition of the Taliban regime. In March 1996, US ambassador to Pakistan, Tom Simmons, had a major row with Bhutto as he lobbied for Unocal. In 1996, the Pakistani President sacked Bhutto's government on charges of corruption, which many believe was a result of US pressure. The new government of Nawaz Sharif turned its back to Bridas and instead supported Unocal (Kleveman, *The New Great Game*: 243 and Rashid, *Taliban*: 165).

157. Rashid, *Taliban*: 154, 163, 176. The Saudis supported the Taliban in order to limit Iranian power, spread Wahhabism and help the American agenda in the region. Furthermore, Israel initially supported the Taliban, seeing that it was an anti-Iranian force (Rashid, *Taliban*: 154).

158. Brisard & Dasquié, *Forbidden Truth*: 21 and Rutledge, *Addicted to Oil*: 61.

159. Rashid, *Taliban*: 178. However, Raphel did tell the US Senate in May 1996 that 'Afghanistan has become a conduit for drugs, crime and terrorism' (Rashid, *Taliban*: 178).

160. Rashid, *Taliban*: 166.

161. John Maresca, 'US interests in the Central Asian Republics: Hearing before the Subcommittee on Asia and the Pacific of the Committee on International Relations, House of Representatives, One hundred fifth Congress, Second Session, 12 February 1998, Statement of John J. Maresca, Vice President of International Relations, Unocal Corporation', Hartford Web Publishing, 12 February 1998 [accessed 13 May 2006] http://www.hartford-hwp.com/archives/51/120.html

162. Steve Coll, *Ghost Wars: The Secret History of the CIA, Afghanistan and Bin Laden, From the Soviet Invasion to September 10, 2001* (London: Penguin Books, 2005): 330.

163. Kleveman, *The New Great Game*: 161, 162. Russia, India and Iran supported the Northern Alliance on this position, as all three countries had a reason to prevent the Unocal pipeline from going through Afghanistan. Moscow did not want the Turkmen to have alternative pipelines to the Russian pipelines, India did not want Pakistan's influence in the region to grow, and Iran is working on exporting gas to Pakistan itself (ibid).

164. Brisard & Dasquié, *Forbidden Truth*: 27.

165. Ibid: 9.

166. Ibid: 29.

167. Ibid: 6, 31.

168. Ibid: 8, 33. For instance, in January 2000, US Assistant Secretary of State Karl Inderfurth met Taliban officials in Pakistan (ibid). After the bombings, the United States urged Saudi Arabia to use its influence on the Taliban to help capture bin Laden. Riyadh suspended diplomatic relations with the Taliban and ceased all aid to them as a result of the insults to Prince Turki al Faisal during a visit to Mullah Omar in September 1998, where he tried to convince him to give up bin Laden, but they ended up exchanging insults. But the Saudis did not withdraw recognition of the Taliban government, despite this incident (Abdel Bari Atwan, *The Secret History of Al-Qaida* (London: Abacus, 2007): 48–49, Bronson, *Thicker Than Oil*: 227 and Rashid, *Taliban*: 138–139, 202).

169. Vassilis Fouskas & Bülent Gökay, *The New American Imperialism: Bush's War on Terror and Blood for Oil* (Connecticut: Praeger Security International, 2005): 152–153, 174; Dan Briody, *The Halliburton Agenda: The politics of oil and money* (New Jersey: Wiley, 2004): 199; Chalmers Johnson, *The Sorrows of Empire: Militarism, Secrecy and the End of the Republic* (London: Verso, 2006 edition): 145–146, 155; Michel Collon, *Media Lies and the Conquest of Kosovo: NATO's Prototype for the Next Wars of Globalization* (New York: Unwritten History: 2007): 6, 8, 29, 34, 35, 91, 93, 97, 99, 104, 105, 107, 108, 112–114, 118, 121, 123, 124, 127–130, 134–136, 145, 154, 156, 172, 210 and Peter Dale Scott, 'The Background of 9/11: Drugs, Oil and US Covert Operations', in David Ray Griffin & Peter Dale Scott (eds) *9/11 and American Empire: Intellectuals Speak Out* (Gloucestershire: Arris Books, 2007): 76.

170. Briody: *The Halliburton Agenda*: 199 and Johnson, *The Sorrows of Empire*: 145–146, 155.

171. Johnson, *The Sorrows of Empire*: 145, 215.

172. Bacevich, *American Empire*: 196–197.

173. Rutledge, *Addicted to Oil*: 104, 118, 119, 197.

174. Dekemjian, *Troubled Waters*: 29, Klare, *Resource Wars*: 85 and Rutledge, *Addicted to Oil*: 118, 119. Comparing the figures published by the US Energy Information Administration (EIA) on Caspian oil reserves for the four years of 1998, 2001, 2002 and 2003, we observe that the data for 'total' oil reserves have fluctuated as follows: 218 billion barrels in 1998, 262 billion in 2001, 243 in 2002, but down to 211 billion in 2003. And the figues for 'proven' reserves – those whose existence is known with a probability of more than 90% – have fluctuated even more: 32.5 billion barrels in 1998, but reduced to 25.8 billion in 2001, then down even further to 10 billion in 2002, then reviving to 25 billion in 2003 but still well below the 1998 figure (Rutledge, *Addicted to Oil*: 118, 119). Citing the over-optimistic initial estimates of Caspian energy reserves, energy expert Amy Myers Jaffe

wrote that the Caspian region had less energy sources 'than met the geological eye', forecasting that by 2015, it will produce only 3–4% of the world's oil supply (Amy Myers Jaffe & Robert A. Manning, 'Shocks of a world of cheap oil', *Foreign Affairs*, January/February 2000: 24).

175. Rashid, *Jihad*: 190; Rashid, *Taliban*: 176; Ahmed Rashid, 'The Taliban: exporting extremism', *Foreign Affairs*, November/December 1999: 35.

176. Rashid, *Jihad*: 190.

177. Hill, 'A not-so-grand strategy'.

178. Rutledge, *Addicted to Oil*: 82.

179. Ibid: 100. This was despite the fact that the National Security Council reported in 1998 that 'we are undergoing a fundamental shift away from reliance on Middle East oil ... Venezuela [has become] our number one foreign oil supplier and Africa supplies 15% of our imported oil' (Klare, *Resource Wars*: 46).

180. The Africa Growth and Opportunity Act (AGOA) was not as central to US foreign policy as the NAFTA. It also did not focus on energy trade as intensely as the NAFTA does. Generally speaking, AGOA was at best a promissory note, not exactly an agreement, especially considering that Africa accounts for only 1% of total US trade (Andrew Bacevich, *American Empire: The Realities and Consequences of US Diplomacy* (Massachusetts: Harvard University Press, 2002): 108).

181. For example, oil and uranium ores may have been a reason for the American military intervention in Somalia between 1992 and 1994, as American oil giants like Conoco held exploratory rights in large areas of land there, and were hoping that the American military intervention would put an end to the civil war and instability in Somalia, which threatened their highly expensive investments. Furthermore, the civil war in Somalia could have threatened the oil supply from the Middle East, since Somalia is located in the Horn of Africa, on the entrance of the Bab El Mandib Strait, which overlooks the Arabian Peninsula. Moreover, Brown & Root took part in the US military operation, building the base camps, supplying the troops with food, water and fuel, cleaning the latrines and even washing the clothes, making $109.7 million in Somalia, again showing that companies make a profit out of US military interventions (Bennis, *Calling the Shots*: 122; William Blum, *Rogue State: A Guide to the World's Only Superpower* (London: Zed Books, 2001): 158; Briody, *The Halliburton Agenda*: 186).

182. Klare, *Resource Wars*: 219–220.

183. 'US urges Africa to remove trade barriers', *BBC News*, 28 April 1999 [accessed 30 September 2005] http://news.bbc.co.uk/1/hi/world/africa/330137.stm

184. Klare, *Resource Wars*: 219.

185. Yergin, *The Prize*: 770.

Chapter 2

1. Bill Clinton, *My Life* (London: Arrow Books, 2005): 935. Also see Michael Gordon & Bernard Trainor, *Cobra II: The Inside Story of the Invasion and Occupation of Iraq* (London: Atlantic Books, 2006): 13–14.

2. Michael T. Klare, 'Resources', in John Feffer (ed.) *Power Trip: US Unilateralism and Global Strategy After September 11* (New York, Seven Stories Press, 2003): 50–51 and Klare, *Blood and Oil* (London: Hamish Hamilton, 2004): 72–73.

3. Klare, *Blood and Oil*: 13, 56 and Ian Rutledge, *Addicted to Oil: America's Relentless Drive for Energy Security* (London: I.B.Tauris, 2005): 9, 66. Domestic US oil production peaked in 1970. Geologist King Hubbert, who worked for Shell Oil in Huston, introduced Hubbert's Peak theory in 1956, which predicted that US oil production would peak between 1966 and 1972 (Robert Bryce, *Gusher of Lies: The Dangerous Delusions of Energy Independence* (New York: PublicAffairs: 2008): 34, 107 and William Clark, *Petrodollar Warfare: Oil, Iraq, and the Future of the Dollar* (Gabriola Island: New Society Publishers, 2005): 76). According to a study in May 2001 by the Department of Energy's Office of Transportation Technologies (OTT), the US will import 100% of its oil by 2050 (Rutledge, *Addicted to Oil*: 130). In 1973 the US was importing 35% of its oil (Bryce, *Gusher of Lies*: 104).

4. Rutledge, *Addicted to Oil*: 59 and Barton Gelman, *Angler: The Shadow Presidency of Dick Cheney* (London: Penguin Books, 2009): 94.

5. Craig Unger, *House of Bush, House of Saud* (London: Gibson Square, 2007):198.

6. Rutledge, *Addicted to Oil*: 59 and Unger, *House of Bush, House of Saud*: 191.

7. Joseph Stiglitz & Linda Blimes, *The Three Trillion Dollar War: The True Cost of the Iraq Conflict.* (London: Allen Lane, 2008): 15.

8. Unger, *House of Bush, House of Saud*: 218.

9. Bruce P. Montgomery, *The Bush-Cheney Administration's Assault on Open Government* (Connecticut: Praeger: 2008): 64 and Howard Fineman & Michael Isikoff. 'Big Energy at the Table', *Newsweek*, 14 May 2001: 48.

10. Montgomery, *The Bush-Cheney Administration's Assault on Open Government*: 64 and Fineman, 'Big Energy at the Table': 48–49.

11. Gelman, *Angler*: 81 and Fineman, 'Big Energy at the Table': 49–50.

12. Gelman, *Angler*: 81; Rutledge, *Addicted to Oil*: 58, 59 and Fineman, 'Big Energy at the Table': 50.

13. Rutledge, *Addicted to Oil*: 66.

14. Klare, *Blood and Oil*: 58.

15. Montgomery, *The Bush-Cheney Administration's Assault on Open Government*: 63.

16. Rutledge, *Addicted to Oil*: 54–56; Unger, *House of Bush, House of Saud*: 70, 115–122 and Jacob Weisberg, *The Bush Tragedy: The Unmaking of a President* (London: Bloomsbury, 2008): 49, 55.

17. Lawrence Freedman & Efraim Karsh, *The Gulf Conflict 1990–1991: Diplomacy and War in the New World Order* (London: Faber & Faber, 1994): 76, and Rutledge, *Addicted to Oil*: 51.

18. Dan Briody, *The Halliburton Agenda: The politics of oil and money* (New Jersey: Wiley, 2004): 196; Freedman & Karsh, *The Gulf Conflict 1990–1991*: 197; Roger Burbach & Jim Tarbell, *Imperial Overstretch: George W. Bush and the Hubris of Empire* (London: Zed Books, 2004): 78–84, 91–92, 207; Halper, Stefan & Jonathan Clarke, *America Alone: The Neo-conservatives and the Global Order* (Cambridge: Cambridge University Press, 2004): 48 and Rutledge, *Addicted to Oil*: 62.

19. Laura Flanders, *Bushwomen: How They Won the White House For Their Man.* (London: Verso, 2005): 56; Rutledge, *Addicted to Oil*: 62, 110, 116. Tengiz is the sixth largest oil bubble in the world, containing about 25 billion barrels of oil, discovered in 1979. When Chevron bought a drilling concession for Tengiz in 1993, forming the 50–50 joint venture Tengizchevroil with Kazakhstan's state-owned oil company, it became the first Western oil company to massively invest on post-Soviet territory (Klare, *Rising Powers, Shrinking Planet: How Scarce Energy is Creating a New World Order* (Oxford: Oneworld, 2008): 121, 123; Lutz Kleveman, *The New Great Game: Blood and Oil in Central Asia* (London: Atlantic Books, 2004): 80 and Rutledge, *Addicted to Oil*: 61).

20. Rutledge, *Addicted to Oil*: 62.

21. Briody, *The Halliburton Agenda*: 198; Montgomery, *The Bush-Cheney Administration's Assault on Open Government*: 62 and Rutledge, *Addicted to Oil*: 62.

22. Briody, *The Halliburton Agenda*: 197–198; Rutledge, *Addicted to Oil*: 62 and Unger, *House of Bush, House of Saud*: 225–226.

23. Rutledge, *Addicted to Oil*: 63 and Allan Sloan & Johnnie L. Roberts, 'Sticky business', *Newsweek*, 22 July 2002: 32–33.

24. Gellman, *Angler*: 94; T. Christian Miller, *Blood Money: Wasted Billions, Lost Lives and Corporate Greed in Iraq* (New York: Little, Brown & Company, 2006): 73, 303 and Unger, *House of Bush, House of Saud*: 225.

25. 'Memorandum for Commander US Army Corps of Engineers: Subject: Justification and Approval (J&A) for other than full and open competition for the execution of the Contingency Support Plan', 28 February 2003 (Declassified 22 April 2004) [accessed 11 January 2008] http://www.judicialwatch.org/archive/2004/kbr.pdf, pages 4–7.

26. Gellman, *Angler*: 90.
27. Ibid: 94.
28. Flanders, *Bushwomen*: 45,47, 51–53, 55–57 and Russell Baker, 'Condi and the boys', *The New York Review of Books*, 3 April 2008 [accessed 28 December 2010] http://www.nybooks.com/articles/archives/2008/apr/03/condi-and-the-boys/.
29. Unger, *House of Bush, House of Saud*: 157.
30. Ibid: 222, 284.
31. Ibid: 223.
32. Zbigniew Brzezinski, *Second Chance: Three Presidents and the Crisis of American Superpower* (New York: Basic Books, 2008): 137.
33. Ibid: 139.
34. Halper & Clarke, *America Alone*: 133.
35. John Ikenberry, 'America's Imperial Ambition: The Lures of Preemption', *Foreign Affairs*, 81(5) September/October 2002.
36. 'The First Gore-Bush Presidential Debate', *Commission on Presidential Debates*, 3 October 2000 [accessed 27 July 2008] http://www.debates.org/pages/trans2000a.html
37. 'The Second Gore-Bush Presidential Debate', *Commission on Presidential Debates*, 11 October 2000 [accessed 27 June 2008] http://www.debates/org/pages/trans2000b.html
38. Gordon & Trainor, *Cobra II*: 14, Joshua Hammer, 'On Patrol in Bosnia: US soldiers and the European allies agree Bush's threatened pullback could mean trouble', *Newsweek*, 5 February 2001: 24 and Ramesh Ratnesar, 'Present Danger', *Time*, 5 February 2001: 28.
39. Jacob Weisberg, *The Bush Tragedy: The Unmaking of a President* (London: Bloomsbury, 2008): 186–187. Also see G. John Ikenberry, *Liberal Order and Imperial Ambition* (Cambridge: Polity, 2006): 207.
40. Mark Thompson, 'The Secretary of Missile Defence', *Time*, 14 May 2001: 36.
41. G. John Ikenberry, *Liberal Order and Imperial Ambition* (Cambridge: Polity, 2006): 245–246.
42. Charles Krauthammer, 'The New Unilateralism', *Washington Post*, 8 June 2001 [accessed 28 December 2009] http://www.washingtonpost.com/ac2/wp-dyn/A38839–2001Jun7?language=printer
43. George W. Bush, 'A Distinctly American Internationalism', *Ronald Reagan Presidential Library, Simi Valley, California*, 19 November 1999 [accessed 10 October 2008] http://www.mtholyoke.edu/acad/intrel/bush/wspeech.htm and 'Bush backs NATO expansion', *BBC News*, 15 June 2001 [accessed 15 June 2001] http://news.bbc.co.uk/1/hi/world/europe/1389581.stm. According to Michael Meacher, the principal objective for the continued existence and expansion of NATO is the encirclement of Russia and the pre-emption of

China dominating access to oil and gas in the Caspian Sea and Middle East regions (Michael Meacher, 'The era of oil wars', *Guardian*, 29 June 2008 [accessed 20 June 2009] http://www.guardian.co.uk/commentisfree/2008/jun/29/oil.oilandgascompanies).

44. Bush, 'A Distinctly American Internationalism'. Nevertheless, Bush's main agenda as he entered office was not foreign policy, but his tax cuts program (James Mann, *Rise of the Vulcans: The History of Bush's War Cabinet* (London: Penguin Books, 2004): 290).

45. Halper & Clarke, *America Alone*: 14, 129, 130, and Ivo Daalder & James Lindsay, *America Unbound: The Bush Revolution in Foreign Policy* (1st edn) (Washington, DC: Brookings Institution Press, 2003): 15–16.

46. Halper & Clarke, *America Alone*: 131.

47. David Frum. *The Right Man: An Inside Account of the Surprise Presidency of George W. Bush* (London: Weidenfeld and Nicolson, 2003): 57.

48. Halper & Clarke, *America Alone*: 129.

49. Ron Suskind, *The Price of Loyalty* (London: Free Press, 2004): 72

50. Gordon & Trainor, *Cobra II*: 13, and George Tenet, *At the Centre of the Storm: My Years at the CIA* (London: Harper Press, 2007): 301.

51. Ivo Daalder & James Lindsay, *America Unbound: The Bush Revolution in Foreign Policy* (1st edn) (Washington, DC: Brookings Institution Press, 2003): 46.

52. Condoleezza Rice, 'Promoting the National Interest', *Foreign Affairs*, January/February 2000: 55.

53. Ibid: 55, 58–59.

54. Klare, *Rising Powers, Shrinking Planet*: 231–232.

55. Klare, *Resource Wars*: 9.

56. Rutledge, *Addicted to Oil*: 66 and 'Cheney, oil executives raise $8 million for GOP', *Quest for the Presidency*, 28 September 2000 [accessed 12 August 2002] http://quest.cjonline.com/stories/092800/gen_0928006149.shtml

57. 'Alaska or bust', *The Economist*, 10 February 2001: 21.

58. George W. Bush, 'Remarks by the President while touring Youth Entertainment Academy', *White House*, 14 March 2001 [accessed 6 February 2008] http://www.whitehouse.gov/news/releases/2001/03/20010314–2.html

59. George W. Bush, 'Remarks by the President in photo opportunity after meeting with National Energy Policy Development Group', *White House*, 19 March 2001 [accessed 6 February 2008] http://www.whitehouse.gov/news/releases/2001/03/20010320–1.html.

60. Rutledge, *Addicted to Oil*: 137, and 'Abraham Sees Nation Threatened by Energy Crisis', *USA Today*, 9 March 2001 [accessed 27 June 2008] http://www.usatoday.com/news/washington/2001–03-19-energy.htm.

61. Paul Sperry, 'White House energy task force papers reveal Iraqi oil maps: Judicial Watch lawsuit also uncovers list of foreign suitors for contracts' *World Net Daily*, 18 July 2003 [accessed 20 December 2007] http://www.worldnetdaily.com/news/article.asp?ARTICLE_ID=33642

62. Edward L. Morse & Amy Myers Jaffe, 'Strategic Energy Policy: Challenges for the 21st Century', Report of an independent task force cosponsored by the James A. Baker III Institute For Public Policy of Rice University and The Council on Foreign Relations, April 2001 [accessed 8 February 2008] http://www.cfr.org/content/publications/attachments/Energy%20TaskForce.pdf

63. Morse & Jaffe, 'Strategic Energy Policy: Challenges for the 21st Century': vi.

64. 'Alaska or bust': 21.

65. Rutledge, *Addicted to Oil*: 66.

66. George W. Bush, 'Remarks by the President to Capital City Partnership, River Centre Convention Centre St. Paul Minnesota', *White House*, 17 May 2001 [accessed 29 January 2008] http://www.whitehouse.gov/news/releases/2001/05/20010517-2.html

67. Ibid.

68. Ibid.

69. *National Energy Policy: Report of the National Energy Policy Development Group*, May 2001 [accessed 20 December 2005] http://www.whitehouse.gov/energy/National-Energy-Policy.pdf : 5–9.

70. National Energy Policy: 5–10.

71. Ibid: viii

72. Ibid: 1–10.

73. Ibid: 1–11.

74. Ibid: x.

75. Ibid: 1–11.

76. Ibid: x.

77. Ibid: x.

78. Ibid: 8–1

79. One barrel of oil is about 159 litres.

80. National Energy Policy: x.

81. Klare, *Blood and Oil*: 62.

82. Rutledge, *Addicted to Oil*: 130.

83. National Energy Policy: 8–4.

84. Klare, *Blood and Oil*: 62.

85. Ibid: 62–63.

86. Rutledge, *Addicted to Oil*: 66.

87. National Energy Policy: 8–6.

88. Ibid: 8–6.

89. Gellman, *Angler*: 81, 90.

90. Ibid: 94.

91. Ibid: 94–96 and Montgomery, *The Bush-Cheney Administration's Assault on Open Government*: 59–121.

92. 'Hot air rising', *The Economist*, 20 September 2003: 52. He invoked crisis once again to justify his policies in a speech after the blackout of August 2003, and said 'Lights went out … you know that'. Yet the lights did not go out because of any scarcity of power-generation capacity or of fossil-fuel supply (ibid).

93. Greg Palast, *The Best Democracy Money Can Buy* (London: Robinson, 2002): 86.

94. Paul Krugman, *The Great Unravelling* (London: Penguin Books, 2004): 330–331.

95. Gellman, *Angler*: 89 and Judy Pasternak, 'Going backwards: Bush's energy plan bares industry clout', *Los Angeles Times*, 26 August 2001 [accessed 6 February 2008] http://www.commondreams.org/headlines01/0826–02.htm.

96. George W. Bush, 'Remarks by the President in photo opportunity after meeting with National Energy Policy Development Group', *White House*, 19 March 2001 [accessed 6 February 2008]: http://www.whitehouse.gov/news/releases/2001/03/20010320–1.html. During the early noughties, OPEC was comfortable with the range of $22–28/barrel ('OPEC keeps oil output unchanged', *BBC News*, 19 September 2002 [accessed 19 September 2002] http://news.bbc.co.uk/1/hi/business/2267783.stm and Nelson D. Schwartz, 'Breaking OPEC's Grip', *Fortune*, 12 November 2001 [accessed 15 November 2001] http://www.fortune.com/indexw.jhtml?channel=artcol.jhtml&doc_id=2045854.)

97. Morse & Myers, 'Strategic Energy Policy: Challenges for the 21st Century': 43.

98. Unger, *House of Bush, House of Saud*: 234, 235, 241– 243 and Edward L. Morse & James Richard, 'The Battle for Energy Dominance', *Foreign Affairs*, March/April 2002: 21.

99. He is the son of Prince Sultan bin Abdulaziz, Saudi Defence Minister (Evan Thomas & Christopher Dickey, 'The Saudi Game', *Newsweek*, 19 November 2001: 29).

100. Thomas, 'The Saudi Game': 34.

101. Unger, *House of Bush, House of Saud*: 243 and Scott Macleod, 'With friends like these', *Time*, 12 November 2001: 42.

102. Unger, *House of Bush, House of Saud*: 243 and Scott Macleod, 'How to bring change to the Kingdom', *Time*, 4 March 2002: 33.

103. 'Regional: US energy review sidesteps sanctions issue', *EMAP Business International Ltd*, 1 June 2001 [accessed 17 September 2005] http://web.lexis-nexis.com/executive/form?_index=exec_en.html&_lang=en&ut=3311241244.

104. Morse & Myers, 'Strategic Energy Policy: Challenges for the 21st Century': 43.

105. Ibid: 44.

106. National Energy Policy: 8-18.

107. 'Regional: US energy review sidesteps sanctions issue'.

108. National Energy Policy: 8-4, 8-5.

109. Ibid: 8-5.

110. Morse & Myers, 'Strategic Energy Policy: Challenges for the 21st Century': 84. Upstream oil activities include exploration for oil and the extraction of oil, while downstream activities include transportation and refining.

111. Morse & Myers, 'Strategic Energy Policy: Challenges for the 21st Century': 22, 84.

112. Klare, *Blood and Oil*: 79.

113. Rutledge, *Addicted to Oil*: 143.

114. Klare, *Blood and Oil*: 79–80.

115. Rutledge, *Addicted to Oil*: 142, 143.

116. Klare, *Blood and Oil*: 80.

117. Gordon & Trainor, *Cobra II*: 13, and Tenet, *At the Centre of the Storm*: 301.

118. Suskind, *The Price of Loyalty*: 71 and 72, also see William Quandt, *Peace Process: American Diplomacy and the Arab–Israeli Conflict since 1967* (Washington, DC: Brookings Institution Press, 2005): 390 and Scott Lucas & Maria Ryan, 'Against Everyone and No-One: The Failure of the Unipolar in Iraq and Beyond', in David Ryan & Patrick Kiely (eds), *America and Iraq: Policy-Making, Intervention and Regional Politics* (Oxford: Routledge, 2009): 162.

119. Daalder & Lindsay, *America Unbound*: 65.

120. Ibid: 67.

121. Woodward, Bob, *Plan of Attack* (New York: Simon & Schuster, 2004):12.

122. Suskind, *The Price of Loyalty*: 72, 74.

123. Ibid: 72–75 and Lucas & Ryan, 'Against Everyone and No-One': 163.

124. Suskind, *The Price of Loyalty*: 77, 82, 85–86; Lucas & Ryan, 'Against Everyone and No-One': 163 and 'O'Neill: Bush planned Iraq invasion before 9/11', CNN, 14 January 2004 [accessed 2 March 2006] http://www.cnn.com/2004/ALLPOLITICS/01/10/oneill.bush/.

125. Gordon & Trainor, *Cobra II*: 15.

126. Woodward, *Bush at War*: 49. Also, according to head of MI6 Sir John Sawers (who was British Prime Minister Tony Blair's private secretary for

foreign affairs in the run-up to the Iraq war of 2003) discussions took place in January 2001 between Bush and Blair on how to change the regime in Iraq. 'There was no discussion of military invasion or anything like that', said Sawers, but preferred tactics were similar to those used to oust the former Serbian Leader Slobodan Milosevic (Michael Savage, 'UK and US talked of toppling Saddam in 2001', *Independent*, 11 December 2009: 25).

127. Greg Palast, 'Secret US plans for Iraq's oil', *BBC Newsnight*, 17 March 2005 [accessed 29 May 2008] http://news.bbc.co.uk/1/hi/programmes/ newsnight/4354269.stm. Insiders told Palast that the neo-conservatives favoured privatizing Iraq's oil sector and using Iraq's oil production to break OPEC, while the US oil corporations opposed privatization and breaking OPEC (ibid).

128. Suskind, *The Price of Loyalty*: 96.

129. Klare, *Blood and Oil*: 70; Jane Mayer, 'Contract Sport: What did the Vice-President do for Halliburton?', *The New Yorker*, 16 February 2004 [accessed 17 May 2007] http://web.lexis-nexis.com/executive/form?_ index=exec_en.html&_lang=en&ut=3361601930 and Jane Mayer, 'Jane Mayer on her article in The New Yorker about Dick Cheney's relation-ship with Halliburton', 19 February 2004, web.lexis-nexis.com/executive/ form?_index=exec_en.html&_lang=en&ut=3361601930.

130. Antonia Juhasz, *The Tyranny of Oil: The World's Most Powerful Industry – And What We Must Do To Stop It* (New York: Harper Collins, 2008): 341.

131. Johanna McGeary, 'Odd Man Out', *Time*, 10 September 2001: 54.

132. Gellman, *Angler*: 112; Thomas Kean *et al., The 9/11 Commission Report: Final Report of the National Commission on Terrorist Attacks Upon the United States* (New York: Norton, 2004): 204, and Frank Bruni, 'Bush Taps Cheney to Study Antiterrorism Steps', *New York Times*, 8 May 2001 [accessed 1 October 2005] http://web.lexis-nexis.com/executive/ form?_index=exec_en.html&_lang=en&ut=3310293205.

133. Cheney was 'uncharacteristically excited' during the meeting as CIA Director George Tenet was showing photos of what he said were Iraqi WMD factories (Suskind, *The Price of Loyalty*: 72).

134. Gordon & Trainor, *Cobra II*: 13, and Tenet, *At the Centre of the Storm*: 301.

135. Halper & Clarke, *America Alone*: 120 and Michael Hirsh & Daniel Klaidman, 'Condi's clout offensive', *Newsweek*, 14 March 2005: 31.

136. These lists and maps have been posted by the public interest law firm Judicial Watch on the following websites [all accessed 11 November 2007]: 'Foreign suitors for Iraqi oilfield contracts as of 5 March 2001 part 1', http:// www.judicialwatch.org/IraqOilFrgnSuitors.pdf, 'Foreign suitors for Iraqi oilfield contracts as of 5 March 2001 part 2'. http://www.judicialwatch.

org/IraqOilGasProj.pdf, 'Iraqi oilfields and exploration blocks', http://www.judicialwatch.org/IraqOilMap.pdf, 'Saudi Arabia major oil and natural gas development projects', http://www.judicialwatch.org/SAOilProj.pdf, 'Selected oil facilities in Saudi Arabia', http://www.judicialwatch.org/SAOilMap.pdf, 'Selected oil facilities of the United Arab Emirates', http://www.judicialwatch.org/UAEOilMap.pdf, 'United Arab Emirates major oil and gas development projects', http://www.judicialwatch.org/UAEOilProj.pdf

137. Suskind, *The Price of Loyalty*: 96.

138. Morse & Myers, 'Strategic Energy Policy: Challenges for the 21st Century': 46.

139. Ibid: 46.

140. Ibid: 47.

141. 'Regional: US energy review sidesteps sanctions issue'.

142. James Hoagland, 'Now an Iraqi war in Washington', *Washington Post*, 9 April 2001 [accessed 21 March 2008] http://www.globalpolicy.org/security/sanction/iraq1/turnpoint/2001/0409us.htm and 'US considers Iraqi coup', *Newsmax*, 28 April 2001 [accessed 11 March 2008] http://archive.newsmax.com/archives/articles/2001/4/27/211326.shtml.

143. Bacevich, *American Empire*: 208 and 'Analysis: A tougher line?' *BBC News*, 22 February 2001 [accessed 22 February 2001] http://news.bbc.co.uk/1/hi/world/middle_east/1174771.stm

144. 'Iraq sanctions in balance', CNN, 2 July 2001 [accessed 17 September 2005] http://archives.cnn.com/2001/WORLD/meast/07/02/iraq.sanctions/index.html

145. 'UN to discuss Iraq sanctions', CNN, 21 May 2001 [accessed 17 September 2005] http://archives.cnn.com/2001/WORLD/meast/05/21/iraq.sanctions/index.html

146. 'Bush seen backing Iraqi opposition against Saddam', *Gulf News*, 10 April 2001 [accessed 14 March 2008] http://www.gulf-news.com/Articles/print.asp?ArticleID=14219

147. 'Iraq sanctions plan submitted', CNN, 22 May 2001 [accessed 17 September 2005] http://archives.cnn.com/2001/WORLD/meast/05/22/iraq.sanctions/index.html

148. 'Iraq vows to stop oil exports', CNN, 2 June 2001, http://archives.cnn.com/2001/WORLD/meast/06/02/iraq.oil/index.html

149. 'Iraq sanctions in balance'.

150. 'Iraq rejects French sanctions proposal', CNN, 27 May 2001 [accessed 17 September 2005] http://archives.cnn.com/2001/WORLD/meast/05/26/iraq.sanctions/index.html

151. 'Iraq to resume oil exports', CNN, 6 July 2001 [accessed 17 September 2005] www.cnn.com/2001/WORLD/eurpoe/07/06/iraq.oilexports/index. html, 'Iraq vows to stop oil exports', 'Iraqi oil threat deadline looms', CNN, 3 June 2001 [accessed 17 September 2005] http://archives.cnn.com/2001/ WORLD/meast/06/03/oil.iraq/index.html, and 'OPEC bids to calm fears', CNN, 5 June 2001 [accessed 17 September 2005] edition.cnn.com/2001/ WORLD/meast/06/05/opec.iraq/index.html

152. 'Iraq to restart oil pumps', CNN, 3 July 2001 [accessed 17 September 2005] http://edition.cnn.com/2001/BUSINESS/07/03/iraq.opec/index.html and 'Iraq to resume oil exports', CNN, 6 July 2001 [accessed 17 September 2005] http:// www.cnn.com/2001/WORLD/eurpoe/07/06/iraq.oilexports/index.html.

153. 'Iraq to resume oil exports', 6 July 2001, www.cnn.com/2001/WORLD/ eurpoe/07/06/iraq.oilexports/index.html

154. Trudy Rubin, 'Powell's plan for Iraq is the best of bad options', *The Wichita Eagle*, 4 March 2001, http://cofax.wichitaeagle.com/content/ wichitaeagle/2001/03/04/editorial

155. However, during confirmation hearings, Wolfowitz was asked if there was a feasible way to remove Saddam from power, and he replied 'I haven't seen it yet' (Bacevich, *American Empire*, 207).

156. Woodward, *Plan of Attack*: 21.

157. Ibid: 23.

158. 'Bush seen backing Iraqi opposition against Saddam'.

159. Daalder & Lindsay, *America Unbound*: 46–47.

160. Gordon & Trainor, *Cobra II*: 13, and Tenet, *At the Centre of the Storm*: 301.

161. 'US considers Iraqi coup'.

162. Gordon & Trainor, *Cobra II*: 14 and Lewis D. Solomon, *Paul D. Wolfowitz: Visionary Intellectual, Policymaker, and Strategist* (London: Praeger Security International, 2007): 89.

163. Lucas & Ryan 'Against Everyone and No-One': 163 and Bryan Burrough, Evgenia Peretz, David Rose & David Wise. 'The Path to War', *Vanity Fair*, May 2004 [accessed 30 June 2008] http://www.accessmylibrary.com/coms2/ summary_0286–6240570_ITM

164. Lucas & Ryan, 'Against Everyone and No-One': 163.

165. Bush said that Saddam was 'still a menace, and we need to keep him in check, and will' (Christopher Dickey, 'Why not Saddam?' *Newsweek*, 20 August 2001: 28).

166. Clark, *Petrodollar Warfare*: 14–15, 95, and Michael Moore, *The Official Fahrenheit 9/11 Reader: The Must-Read Book of the Must-See Box Office Smash* (London: Penguin Books, 2004): 79.

167. Daalder & Lindsay, *America Unbound*: (1st edn): 129.

168. McGeary, 'Odd Man Out': 57.

169. Moore, *The Official Fahrenheit 9/11 Reader*: 79. Furthermore, in December 2000, the CIA and other intelligence agencies produced a National Intelligence Estimate carrying the headline 'Iraq: Steadily Pursuing WMD Capabilities' but the text was more cautious, stating that 'Iraq did not appear to have taken significant steps toward the reconstruction' of WMDs (Gordon & Trainor, *Cobra II*: 125 and Lucas & Ryan, 'Against Everyone and No-One': 161).

170. Tenet, *At the Centre of the Storm*: 303.

171. Daalder & Lindsay, *America Unbound* (1st edn): 66.

172. Jonny Dymond, 'Powell turns new Mid-East page', *BBC News*, 24 February 2001 [accessed 3 September 2005] http://news.bbcoc.uk/1/hi/world/middle_east/1187238.stm; 'Iraq to dominate Powell trip', *BBC News*, 24 February 2001 [accessed 3 September 2005] http://news.bbc.co.uk/1/hi/world/middle_east/1187406.stm, 'Powell ends Mid-East tour', *BBC News*, 26 February 2001 [accessed 3 September 2005] http://news.bbc.co.uk/1/hi/world/middle_east/1191228.stm, 'Powell presses anti-Saddam message', *BBC News*, 24 February 2001 [accessed 3 September 2005] http://news.bbc.co.uk/1/hi/world/middle_east/1188048.stm

173. William Hartung, 'Military' in John Feffer, *Power Trip: US Unilateralism and Global Strategy After September 11* (New York: Seven Stories Press, 2003): 68.

174. Hartung, 'Military': 67.

175. Bacevich, *American Empire*: 223.

176. Litwak, 'Iraq and Iran: From Dual to Differentiated Containment': 189, 193 and Michael Nacht, 'Weapons Proliferation and Missile Defence: New Patterns, Tough Choices', In Robert Lieber (ed.) *Eagle Rules? Foreign Policy and American Primacy in the Twenty-First Century* (New Jersey, Prentice Hall, 2002): 284.

177. Frances Fitzgerald, 'The black comedy of missile defence', *Foreign Policy in Focus*, 21 December 2001 [accessed 30 June 2008] http://www.fpif.org/presentations/wmd01/fitzgerald.html

178. 'US Senate extends sanctions law', *BBC News*, 26 July 2001 [accessed 26 July 2001] http://news.bbc.co.uk/1/hi/world/middle_east/1457639.stm.

179. Pollack, *The Persian Puzzle*: 343 and Litwak, 'Iraq and Iran: From Dual to Differentiated Containment': 193.

180. Litwak, 'Iraq and Iran: From Dual to Differentiated Containment': 193.

181. Pollack, *The Persian Puzzle*: 343–344.

182. Ibid: 344.

183. Paul Reynolds. 'President Bush's first foreign test', *BBC News*, 1 February 2001 [accessed 1 February 2001] http://news.bbc.co.uk/1/hi/world/americas/1149017.stm

184. 'When will sanctions be lifted?' *BBC News*, 1 February 2001 [accessed 1 February 2001] http://news.bbc.co.uk/1/hi/world/1148057.stm

185. Eli J. Lake, 'Analysis: Energy report hedges on Iran', 17 May 2001 [accessed 30 June 2008] http://web.lexis-nexis.com/executive/form?_index=exec_en.html&_lang=en&ut=3311241244. The oil lobby would be disappointed to see that the NEP of May 2001 did not lift the sanctions against Iran and Libya.

186. Lake, 'Analysis: Energy report hedges on Iran'

187. National Energy Policy: 8–6.

188. Lake, 'Analysis: Energy report hedges on Iran'

189. Rob Watson. 'US may extend Iran Libya sanctions', *BBC News*, 24 May 2001 http://news.bbc.co.uk/1/hi/world/americas/1348330.stm

190. 'US Senate extends sanctions law', *BBC News*, 26 July 2001 [accessed 26 July 2001] http://news.bbc.co.uk/1/hi/world/middle_east/1457639.stm.

191. 'Libya and Iran hit by new sanctions', *BBC News*, 4 August 2001 [accessed 4 August 2001] http://news.bbc.co.uk/1/hi/world/americas/1473176.stm.

192. National Energy Policy: 8-5.

193. Tim Weiner, 'Bush due to visit Mexico to discuss obtaining energy', *New York Times*, 13 February 2001 [accessed 6 February 2008], http://query.nytimes.com/gst/fullpage.html?res=9C01E4D81E31F930A25751C0A9679C8B63. The NEP said that 'In 2000, nearly 55% of US gross oil imports came from four countries: 15% from Canada, 14% each from Saudi Arabia and Venezuela, and 12% from Mexico' (National Energy Policy: 8-4).

194. National Energy Policy: 8-9.

195. Klare, *Blood and Oil*: 118 and 'Bush and Fox forge links', *BBC News*, 17 February 2001 [accessed 17 February 2001] http://news.bbc.co.uk/1/hi/world/americas/1173045.stm

196. National Energy Policy: 8-10.

197. Ibid: 8-11.

198. Ibid: 8-10 and 8-19.

199. Ibid: 8-11.

200. Barnaby Mason, 'Analysis: Powell's interest in Africa', *BBC News*, 23 May 2001 http://news.bbc.co.uk/1/hi/world/africa/1347282.stm

201. Donald Rothchild. 'The United States and Africa: Power and Limited Influence', in Robert J. Lieber (ed.) *Eagle Rules? Foreign Policy and American Primacy in the Twenty-First Century*. (New Jersey, Prentice Hall, 2002): 238, from Ian Fisher 'Africans ask if Washington's sun will shine on them', *New York Times*, 8 February 2001, p. A3 http://query.nytimes.com/gst/fullpage.html?res=9A0E7DB1731F93BA335751C0A9679C8B63.

202. Klare, *Blood and Oil*: 144.

203. Ibid: 74.

204. Ibid: 79 from the US Department of Energy, Energy Information Administration (DOE/EIA) International Energy Outlook 2001, table D6 page 240. However, in the period between 1999 and 2009, total oil production in the Persian Gulf states still averaged around 20 mbpd ('US Energy Information Administration/Monthly Energy Review May 2010', *Energy Information Administration*, May 2010 [accessed 28 May 2010] http://www.eia.gov/emeu/mer/pdf/pages/sec11_2.pdf)

205. Bryce, *Gusher of Lies*: 97; Klare, *Blood and Oil*: 82; Klare, *Rising Powers, Shrinking Planet*: 180; Klare, 'Energy Security': 487 and Morrissey, 'The Geoeconomic Pivot of the Global War on Terror': 103.

206. Richard Seymour, 'The Real Cost of the Iraq War', *The Middle East*, May 2009: 49.

207. Klare, *Blood and Oil*: 82.

208. Rutledge, *Addicted to Oil*: 11.

209. Ibid: 104, 118, 119.

210. Klare, *Rising Powers, Shrinking Planet*: 125.

211. Doug Blum,'America's Caspian Policy Under the Bush Administration', *Centre for Strategic and International Studies*, March 2001, http://www.csis.org/ruseura/ponars/policymemos/pm_0190.pdf.

212. Blum, 'America's Caspian Policy Under the Bush Administration'. The companies thought that the BTC pipeline was too long and too expensive (Kleveman, *The New Great Game*: 27). Even though Cheney was previously sceptical about the economics of the BTC, he became an advocate of the pipeline once he became Vice President. Another former sceptic, BP's CEO John Browne, became a supporter. It is not clear what made Browne change his mind. One reason might have been that BP would receive a US Government subsidy if it agreed to the pipeline. Another reason might have been that one of the conditions for the US Government's approval of BP's 1999 merger with Amoco was that BP 'came on board' for the BTC project (Rutledge, *Addicted to Oil*: 116–117).

213. National Energy Policy: 8–7. Also see Klare, *Rising Powers, Shrinking Planet*: 125–126.

214. National Energy Policy: 8–13

215. Ibid: 8–13.

216. Richard Boucher, 'State Department Regular Briefing', 29 August 2001 [accessed 12 November 2005] http://web.lexis-nexis.com/executive/form?_index=exec_en.html&_lang=en&ut=3311242783

217. Brisard & Dasquié, *Forbidden Truth*: 5–6 and Moore, *The Official Fahrenheit 9/11 Reader*: 48–49.

218. Paul Findley, *Silent No More: Confronting America's False Images of Islam* (Maryland: Amana Publications, 2001): 106.

219. 'Rockets hit Kabul at Taliban parade', CNN, 20 August 2001 [accessed 2 March 2008] http://archives.cnn.com/2001/WORLD/asiapcf/central/08/19/afghan.indep/

220. Brisard & Dasquié, *Forbidden Truth*: 41–42.

221. Michael Meacher, 'This war on terrorism is bogus', *Guardian*, 6 September 2003 [accessed 1 November 2009] http://www.guardian.co.uk/politics/2003/sep/06/september11.iraq.

222. Brisard & Dasquié, *Forbidden Truth*: 42, 236.

223. Brisard & Dasquié, *Forbidden Truth*: 42, and Jonathan Steele, Ewen MacAskill, Richard Norton-Taylor & Ed Harriman 'Threat of US strike passed to Taliban weeks before NY attack', *Guardian*, 22 September 2001 [accessed 30 January 2008] http://www.guardian.co.uk/international/story/0,3604,556254,00.html.

224. Brisard & Dasquié, *Forbidden Truth*: 42.

225. Kean *et al.*, *The 9/11 Commission Report*: 203.

226. Ibid: 208.

227. Woodward, *Bush at War*: 35–36.

228. Kean et al. *The 9/11 Commission Report*: 206.

229. Woodward, *Bush at War*: 83.

230. Woodward, *Plan of Attack*: 23.

231. Steve Coll, *Ghost Wars: The Secret History of the CIA, Afghanistan and Bin Laden, From the Soviet Invasion to September 10, 2001* (London: Penguin Books, 2005): 551.

Chapter 3

1. Zbigniew Brzezinski, *Second Chance: Three Presidents and the Crisis of American Superpower* (New York: Basic Books, 2008): 136.

2. Noam Chomsky & Gilbert Achcar, *Perilous Power: The Middle East and US Foreign Policy* (London: Penguin Books, 2008): 59.

3. Roger Burbach & Jim Tarbell, *Imperial Overstretch: George W. Bush and the Hubris of Empire* (London: Zed Books, 2004): 128–129.

4. Andrew Bacevich, *American Empire: The Realities and Consequences of US Diplomacy* (Massachusetts: Harvard University Press, 2002): 239–240.

5. Eichengreen, Barry & Douglas A. Irwin, 'A Shackled Hegemon', in Melvin P. Leffler & Jeffrey W. Legro (eds) *To Lead the World: American Strategy After the Bush Doctrine* (New York: Oxford University Press, 2008): 184.

6. Bacevich, *American Empire:* 232–233 and G. John Ikenberry, 'America's Imperial Ambition', *Foreign Affairs*, September/October 2002.

7. George W. Bush, 'The National Security Strategy of the United States of America', *White House*, September 2002 [accessed 3 April 2006] http://www. whitehouse.gov/nsc/nss/html

8. Klare, *Blood and Oil* (London: Hamish Hamilton, 2004):72–73 and Klare, 'Resources', in John Feffer (ed.) *Power Trip: US Unilateralism and Global Strategy After September 11* (New York, Seven Stories Press, 2003): 50–51.

9. 'Quadrennial Defence Review Report', 30 September 2001, US Department of Defence, http://www.defenselink.mil/pubs/qdr2001.pdf#search='quad rennial%20defence%20review%202001'.

10. Klare, *Blood and Oil*: 71. Bromley mentioned this 'arc of instability' and its link to oil resources in *American Power*, pp. 132, 134, 135–138.

11. 'Quadrennial Defence Review': 4.

12. Klare, *Blood and Oil*: 71. Klare argues that 'the principal thrust of the [Bush] administration's military policy' was not the NMD system, but the enhancement of America's 'power projection' forces, meaning that US forces had to be able to be transported to distant combat zones, especially in energy-vital regions. This was evident in Bush's Citadel speech where he said that 'our forces in the next century must be agile, lethal, readily deployable'. Bush gave top priority to US power projection while at the same time endorsing a National Energy Policy that entailed increased US dependence on oil from areas of crisis and conflict. This shows that oil and military power were 'merged' into a single, integrated design for American world dominance (Michael T. Klare, 'The Bush/Cheney energy strategy: Implications for US foreign and military policy', *Information Clearing House*, 26–27 May 2003 [accessed 25 January 2010] http://www.informationclearinghouse.info/article4458.htm).

13. Klare, *Blood and Oil*: 72–73 and Klare, 'Resources': 50–51.

14. Hartung, 'Military': 64, from page 18 of the Quadrennial Defence Review 2001.

15. George W. Bush, 'President Delivers State of the Union Address', *White House*, 29 January 2002 [accessed 30 June 2004] http://www.whitehouse. gov/news/releases/2002/01/20020129–22.html.

16. George W. Bush, 'President Bush Delivers Graduation Speech at West Point, United States Military Academy, West Point, New York', *White House*, 1 June 2002 [accessed 30 June 2007] http://www.whitehouse.gov/ news/releases/2002/06/20020601–3.html, also quoted in Daalder & Lindsay, *America Unbound* (1st edn): 121–122. Bush hinted in the West Point Speech that a display of military capabilities would prompt the Middle East to follow America's political and economic agendas (Lucas & Ryan, 'Against

Everyone and No-One': 169), as he said that 'the peoples of the Islamic nations want and deserve the same freedoms and opportunities as people in every nation ... An advancing nation will pursue economic reform' (Bush, 'President Bush Delivers Graduation Speech at West Point'). Again, this shows the military–economic link in US foreign policy.

17. Robert J. Lieber, *The American Era: Power and Strategy for the 21st Century* (New York: Cambridge University Press, 2005): 43–44.

18. See, for example, Zbigniew Brzezinksi, *The Choice: Global Domination or Global Leadership.* (New York: Basic Books, 2005): 42, 59, 60; Richard Nixon, *Beyond Peace* (New York: Random House, 1994): 141; Anthony Sampson, 'West's greed for oil fuels Saddam fever', *Guardian*, 11 August 2002 [accessed 3 August 2009] http://www.guardian.co.uk/world/2002/aug/11/iraq.oil and Michael Meacher, 'The era of oil wars', *Guardian*, 29 June 2008 [accessed 20 June 2009] http://www.guardian.co.uk/commentisfree/2008/jun/29/oil. oilandgascompanies. Brzezinski said that the US 'could slide into a collision with the world of Islam' in the unstable region which he called the 'Global Balkans', which he defined as 'the crucial swathe of Eurasia between Europe and the Far East ... heavily dominated by Muslims'. Citing that the 'Global Balkans' contains 68% of the world's proven oil reserves and 41% of the world's natural gas reserves, he said that 'the combination of oil and volatility gives the United States no choice' but to be involved in this region (Brzezinksi, *The Choice*: 42, 59, 60).

19. George W. Bush, 'The National Security Strategy of the United States of America', *White House*, September 2002 [accessed 3 April 2006] http://www. whitehouse.gov/nsc/nss/html (emphasis added). Interestingly, the focus on Middle East oil was absent from the list of geographic energy regions in the NSS 2002. As seen above, the Western Hemisphere, Africa, Central Asia and the Caspian region were mentioned, but not the Middle East. It seems that Bush did not want to mention Middle East oil, due to the talk of the upcoming war on Iraq in the late summer of 2002 (Juhasz, *The Bush Agenda*: 47–48).

20. George W. Bush, 'President Delivers State of the Union Address', *White House*, 29 January 2002 [accessed 30 June 2004] http://www.whitehouse. gov/news/releases/2002/01/20020129-22.html.

21. Duncan Clarke, *Empires of Oil: Corporate Oil in Barbarian Worlds* (London: Profile Books, 2007): 118.

22. Spencer Abraham, 'Secretary of Energy Spencer Abraham's keynote address to the Hoover Institution's Conference on California's electricity problem', US Department of Energy, Office of Policy and International Affairs, 18 October 2001 [accessed 12 September 2002] http://www.pi.energy.gov/ CA011018.html.

23. Ibid.
24. Ibid.
25. John Morrissey, 'The Geoeconomic Pivot of the Global War on Terror: US Central Command and the War in Iraq', in David Ryan & Patrick Kiely (eds), *America and Iraq: Policy-Making, Intervention and Regional Politics* (Oxford: Routledge, 2009): 114.
26. Clarke, *Against All Enemies*: 283.
27. Trevor B. McCrisken, 'George W. Bush, American Exceptionalism and the Iraq War', in David Ryan & Patrick Kiely (eds), *America and Iraq*: 185.
28. Halper, Stefan & Jonathan Clarke, *America Alone: The Neo-conservatives and the Global Order* (Cambridge: Cambridge University Press, 2004): 308.
29. 'Does He Know Where It's Leading?' *The Economist*, 30 July 2005: 23–25.
30. George Monbiot, 'America's pipe dream', *Guardian*, 23 October 2001 [accessed 10 August 2009] http://www.guardian.co.uk/world/2001/oct/23/afghanistan.terrorism11, also see Lutz Kleveman, *The New Great Game: Blood and Oil in Central Asia* (London: Atlantic Books, 2004): 226–227.
31. Vassilis Fouskas & Bülent Gökay, *The New American Imperialism: Bush's War on Terror and Blood for Oil* (Connecticut: Praeger Security International, 2005): 156.
32. Nicholas Lemann, 'The Next World Order', *The New Yorker*, 1 April 2002 [accessed 4 June 2007] http://www.globalpolicy.org/wtc/analysis/2002/0401newyorker.com
33. Lawrence Kaplan & William Kristol, *The War Over Iraq: Saddam's Tyranny and America's Mission* (California: Encounter Books: 2003): 113.
34. Bob Woodward, *State of Denial: Bush at War*, Part III (London: Pocket Books, 2006): 32.
35. Ibid: 67.
36. David Ray Griffin, *The 9/11 Commission Report: Omissions and Distortions* (Gloucestershire: Arris Books, 2005): 116, from 'Secretary Rumsfeld interview with the New York Times', *New York Times*, 12 October 2001. Also cited in Bacevich, *American Empire*: 227. Also cited in David Ray Griffin, '9/11, The American Empire, and Common Moral Norms' in David Ray Griffin & Peter Dale Scott (eds.) *9/11 and American Empire; Intellectuals Speak Out* (Gloucestershire: Arris Books: 2007):13.
37. Griffin, *The 9/11 Commission Report: Omissions and Distortions*: 121.
38. Bush, 'The National Security Strategy of the United States of America'.
39. Woodward, *Bush at War*: 194.
40. Richard Seymour, 'The Real Cost of the Iraq War', *The Middle East*, May 2009: 49.

41. Brzezinksi, *Second Chance*: 121 and Ian Rutledge, *Addicted to Oil: America's Relentless Drive for Energy Security* (London: I.B.Tauris, 2005): 104–105.

42. Zbigniew Brzezinski, 'A geostrategy for Eurasia', *Foreign Affairs*, September/October 1997: 52, also quoted in Fouskas & Gökay, *The New American Imperialism*: 21, 150–151.

43. Brzezinksi, 'A geostrategy for Eurasia': 50–51.

44. Zbigniew Brzezinksi, *The Grand Chessboard: American Primacy and its Geostrategic Imperatives* (New York: Basic Books, 1997): 40.

45. Fouskas & Gökay, *The New American Imperialism*: 21.

46. Zalmay Khalilzad, former US ambassador to Afghanistan, was an ex-UNOCAL official (Fouskas & Gökay, *The New American Imperialism*: 157 and Rutledge, *Addicted to Oil*: 61). So was Hamid Karzai, the Afghan interim president. This illustrates Kolko's points about corporatism, and the business-government career cycle. Also, Karzai's personal security detail is from Dyncorp, and Guantanamo Bay prison was built by Brown & Root, again showing the corporate link with the US military (Burbach & Tarbell, *Imperial Overstretch*: 200, 201). This is 'a convergence of political and economic interests under the rubric of Operation Enduring Freedom' (Fouskas & Gökay, *The New American Imperialism*: 157).

47. Fouskas & Gökay, *The New American Imperialism*: 23, 151.

48. Brzezinksi, 'A geostrategy for Eurasia': 57.

49. Rutledge, *Addicted to Oil*: 115

50. Kleveman, *The New Great Game*: 92.

51. Fouskas & Gökay, *The New American Imperialism*: 157. The *New York Times* mentioned that the region has more than 6% of the world's proven oil reserves and almost 40% of its gas reserves. These figures were based on initial over-optimistic estimates of the energy reserves in Central Asia, which turned out to be exaggerated (ibid).

52. Brzezinksi, *The Grand Chessboard*: 40 and Rutledge, *Addicted to Oil*: 117.

53. Ahmed Rashid, 'Central Asia', in John Feffer *Power Trip: US Unilateralism and Global Strategy After September 11* (New York: Seven Stories Press, 2003): 123.

54. Klare, *Blood and Oil*: 137, from the US Department of State (DoS), Congressional Budget Justification: Foreign Operations, Fiscal Year 2005 (Washington, DC: DoS 2004), pp. 345, 363.

55. Chalmers Johnson, *The Sorrows of Empire: Militarism, Secrecy and the End of the Republic* (London: Verso, 2006 edition): 146. Burbach & Tarbell mention a 'petro-military complex' which flourished 'particularly under George W. Bush' (Burbach & Tarbell, *Imperial Overstretch*: 51, 199).

56. Klare, *Blood and Oil*: 137.

57. Ibid: 7, 72, cited in William Clark, *Petrodollar Warfare: Oil, Iraq, and the Future of the Dollar* (Gabriola Island: New Society Publishers, 2005): 69, 71.

58. Anthony Sampson, 'West's greed for oil fuels Saddam fever', *Guardian*, 11 August 2002 [accessed 3 August 2009] http://www.guardian.co.uk/world/2002/aug/11/iraq.oil, also cited in Johnson, *The Sorrows of Empire*: 234.

59. Noam Chomsky, *The Essential Chomsky* (London: The Bodley Head, 2008): 345.

60. Rashid, 'Central Asia': 122.

61. Fiona Hill, 'Areas for Future Cooperation or Conflict in Central Asia and the Caucasus', *Yale University Conference, 'The Silk Road in the 21st Century'*, Brookings, 19 September 2002 [accessed 16 February 2006] http://www.brook.edu/views/speeches/hillf/20020919.htm

62. Klare, *Blood and Oil*: 156, 157.

63. Jean-Charles Brisard & Guillaume Dasquié, *Forbidden Truth: US–Taliban Diplomacy and the Failed Hunt for Bin Laden* (New York: Nation Books, 2002): 145.

64. Rashid, 'Central Asia': 118, 127.

65. Kean et al. *The 9/11 Commission Report*: 171.

66. Michael C. Ruppert, 'Terror in America, Asian financial markets plummet', *Centre for Research on Globalisation*, 11 September 2001 [accessed 14 September 2001] http://globalresearch.ca/articles/RUP109A.html.

67. 'OPEC fails to halt slide', CNN, 15 November 2001 [accessed 15 November 2001] http://europe.cnn.com/2001/business/11/15/opec/index.html. The demand for oil decreased because the collapse of the World Trade Center of New York, one of the nerve centres of the global economy where a large share of world trade took place, had the effect of decreasing global investment, so the demand for oil fell and the price decreased. Actually, the world economy was already suffering from a recession before 11 September, but the attacks exacerbated the global economic slowdown (ibid).

68. Christopher Dickey, 'The Once and Future Petro Kings', *Newsweek*, 8 April 2002: 37.

69. Griffin, *The 9/11 Commission Report: Omissions and Distortions*: 85; Craig Unger, *House of Bush, House of Saud* (London: Gibson Square, 2007):11 and Dickey, 'The Once and Future Petro Kings': 37.

70. For example, there were allegations that when the US Congress released the report of the bipartisan inquiry into the 11 September attacks in July 2003 (coincidentally, a year before the 9/11 Commission Report of July 2004), Bush censored 27 pages that discussed the Saudi role in the attacks (Unger, *House of Bush, House of Saud*: 277, 281 and Griffin, *The 9/11 Commission Report: Omissions and Distortions*: 67–68, from Bob Graham *Intelligence Matters: The*

CIA, the FBI, Saudi Arabia and the failure of America's war on terror (New York: Random House, 2004): 12–13, 167, 169, 215, 223, 224, 225, and Dana Priest 'White House, CIA kept key portions of report classified', *Washington Post*, 25 July 2003).

71. George W. Bush, 'President to send Secretary Powell to Middle East', *White House*, 4 April 2002 [accessed 2 March 2008] http://www.whitehouse.gov/news/releases/2002/04/20020404–1.html

72. Rachel Bronson, *Thicker Than Oil: America's uneasy partnership with Saudi Arabia* (New York: Oxford University Press, 2006): 239.

73. Ibid: 238.

74. Oil exploration in Iraq slowed down significantly from the 1980s, due to the war against Iran, the 1991 Gulf War, and the sanctions imposed on Iraq. When analysts say that Iraq possesses 12 or 13% of global oil reserves, this figure does not include the oilfields that are probably not explored yet. Analysts estimate that the amount of unknown, unexplored oil underneath Iraqi soil is probably double the known amount explored today, as Iraq's oil reserves were officially estimated at 112 or 115 billion barrels, but more optimistic estimates put the figure at possibly 350 billion barrels ('Iraq forecasts oil reserves would reach 350 billion barrels', Al Jazeera, 29 April 2008 [accessed 29 April 2008] http://www.aljazeera.net/NR/exeres/ED98BDFF-9C9B-4848-9011-059EFFACF7C4.htm).

75. 'Saudi Arabia's blinking yellow light', *The Economist*, 9 November 2002: 41

76. Leon T. Hadar, 'What Green Peril?' *Foreign Affairs*, 72(2), 1993: 39.

77. For information on Iraq's Sunni, Shiite and Kurdish populations, see Kissinger, *Does America need a foreign policy*: 190 and Wheeler, *Saving Strangers*:147–148.

78. 'Friend or foe', *The Economist*, 10 August 2002: 43. Similarly, Irwin Stelzer of the *Weekly Standard* said that the United States must make clear that 'in the event of an upheaval in Saudi Arabia, we will take control of, protect, and run the kingdom's oilfields.' (Rutlege, *Addicted to Oil*: 173).

79. *The Economist*, 10 August 2002: 6 and 'Friend or foe': 43.

80. On 7 August 2002, the Saudi Foreign Minister, Prince Saud Al Faisal, publicly confirmed that his country would not allow the Americans to use Saudi military bases to attack Iraq ('Friend or foe': 43).

81. 'Who backs war?' *BBC News*, 19 September 2002 [accessed 19 September 2002], http://news.bbc.co.uk/1/hi/world/Americas/2267767.stm.

82. 'Your moves', *The Economist*, 21 September 2002: 49. Despite Riyadh's pullback, however, Washington was not completely satisfied with Saudi cooperation. In November 2002, Press Secretary Ari Fleischer stated that 'Saudi Arabia is a good partner in the war against terrorism, but it can do more'

(Ibrahim Warde, *The Price of Fear: Al-Qaeda and the Truth Behind the Financial War on Terror* (London: I.B.Tauris, 2007): 67, from David E. Sanger, 'Bush officials praise Saudis for aiding terror fight,' *New York Times*, 27 November 2002).

83. Rutledge, *Addicted to Oil*: 174. See Woodward, Bob, *Plan of Attack* (New York: Simon & Schuster, 2004), for discreet communication between the Saudis and Washington over the invasion of Iraq.

84. Kenneth Pollack, 'Next Stop Baghdad?', *Foreign Affairs*, March/April 2002: 43.

85. Rutledge, *Addicted to Oil*: 102, 119.

86. Williams, *The Tragedy of American Diplomacy*: 129.

87. Klare, *Blood and Oil*: 175.

88. Klare, *Blood and Oil*: 5 and Klare, 'Energy Security': 487.

89. Paul Wolfowitz, 'Deputy Secretary Wolfowitz interview with Sam Tannenhaus, Vanity Fair', US Department of Defence, 9 May 2003 [accessed 10 August 2009] http://www.defenselink.mil/transcripts/transcript.aspx?transcriptid=2594

90. Halper & Clarke, *America Alone*: 319; Dilip Hiro, *Secrets and Lies: Operation Iraqi Freedom and After* (New York: Nation Books, 2004): 47 and T. Christian Miller, *Blood Money: Wasted Billions, Lost Lives and Corporate Greed in Iraq* (New York: Little, Brown & Company, 2006): 25, 26.

91. Walter Pincus, 'Ex-CIA official faults use of data on Iraq', *Washington Post*, 10 February 2006 [accessed 19 December 2010] http://www.washingtonpost.com/wp-dyn/content/article/2006/02/09/AR2006020902418.html

92. Pilger, John, 'Iraq: The Lying Game', *The Daily Mirror*, 27 August 2002: 8.

93. Robert Draper, *Dead Certain: The Presidency of George W. Bush* (New York: Free Press, 2008): 309 and George Tenet, *At the Centre of the Storm: My Years at the CIA* (London: Harper Press, 2007): 453, 454, 455, 462. In fact, Iraq could not build an atomic weapon since Israel bombed Ozirak in 1981 (Stephen Pelletière, *America's Oil Wars* (Connecticut: Praeger, 2004): 176).

94. Hiro, *Secrets and Lies* (2004): 36.

95. Pelletière, *America's Oil Wars*: 142.

96. Woodward, *Plan of Attack*: 354–356. Former CIA analyst Stephen Pelletière said that the whole propaganda over Iraq's WMD was 'founded on lies' (Pelletière, *America's Oil Wars*: 141).

97. Phyllis Bennis, *Calling the Shots: How Washington dominates today's UN* (New York: Olive Branch Press, 2000): 267–271 and Chalmers Johnson, *The Sorrows of Empire: Militarism, Secrecy and the End of the Republic* (London: Verso, 2006 edition): 230.

NOTES   239

98. 'WMD intelligence wrong, says Kay', BBC News, 28 January 2004 [accessed 5 August 2009] http://news.bbc.co.uk/1/hi/world/middle_east/3439673.stm

99. 'Bush pledge over US intelligence', BBC News, 31 March 2005 [accessed 1 April 2005] http://news.bbc.co.uk/1/hi/world/americas/4396457.stm and 'Report: Iraq intelligence dead wrong', CNN, 1 April 2005 [accessed 1 April 2005] http://www.cnn.com/2005/POLITICS/03/31/intel.report/. In July 2002, UN weapons inspector Scott Ritter wrote that 'I bear personal witness through seven years as a chief weapons inspector in Iraq for the United Nations to both the scope of Iraq's weapons of mass destruction programs and the effectiveness of the UN weapons inspectors in ultimately eliminating them' (Johnson, *Sorrows of Empire*: 230). In December 2002 when Iraq handed in a 12,000-page report detailing its WMD history to the United Nations, members of the Bush administration hurried to New York to obtain the report before anyone else would see it, in order to censor 8,000 pages which named US and Western companies which sold weapons and dual-use technologies to Iraq, including Honeywell, Hewlett-Packard, Eastman Kodak and others (Hiro, *Secrets and Lies* (2004): 106–107 and Johnson, *Sorrows of Empire*: 224).

100. Tenet, *At the Centre of the Storm*: 321.

101. George W. Bush, 'President Delivers "State of the Union"', *White House*, 28 January 2003 [accessed 26 June 2006] www.whitehouse.gov/news/releases/2003/01/20030128-19.html

102. 'Bush administration on Iraq 9/11 links', *BBC News*, 18 September 2003 [accessed 18 September 2003] http://news.bbc.co.uk/1/hi/world/americas/3119676.stm

103. Clarke, *Against All Enemies*: 32–33 and Kean *et al.*, *The 9/11 Commission Report*: 334.

104. Hiro, *Secrets and Lies* (2004): 28 and Kean *et al.*, *The 9/11 Commission Report*: 228. For more details on the fallacy of Saddam–al Qaeda links, see Hiro, *Secrets and Lies* (2004): 28, 29, 121.

105. Johnson, *The Sorrows of Empire*: 305.

106. 'Bush rejects Saddam 9/11 link', *BBC News*, 18 September 2003 [accessed 18 September 2003] http://news.bbc.co.uk/1/hi/world/americas/3118262.stm. 70% of Americans thought that Saddam was involved in the attacks, due, at best, to a lack of clarity by the Bush administration and, at worst, deliberate obfuscation (ibid).

107. 'Saddam "had no link to al Qaeda"', *BBC News*, 9 September 2006 [accessed 9 September 2006] http://news.bbc.co.uk/1/hi/world/americas/5328592.stm. For more details on the fallacy of Saddam–al Qaeda links, see Hiro, *Secrets and Lies* (2004): 28, 29, 121.

108. Michael Scheuer, *Imperial Hubris: Why the West is Losing the War on Terror* (Washington, DC: Potomac Books: 2005): xvi–xvii.

109. John Esterbrook, 'Rumsfeld: It would be a short war', CBS, 15 November 2002 [accessed 5 August 2009] http://www.cbsnews.com/ stories/2002/11/015/main529569.shtml

110. Donald Rumsfeld, 'Secretary Rumsfeld interview with Al Jazeera TV', US Department of Defence, 25 February 2003 [accessed 10 August 2009] http://www.defenselink.mil/transcrpits/transcript.aspx?transcriptid=1946

111. Alan Greenspan, *The Age of Turbulence*. (London: Penguin Books, 2008): 463.

112. Antonia Juhasz, *The Tyranny of Oil: The World's Most Powerful Industry – And What We Must Do To Stop It* (New York: Harper Collins, 2008): 319.

113. Joseph Stiglitz & Linda Blimes, *The Three Trillion Dollar War: The True Cost of the Iraq Conflict* (London: Allen Lane, 2008): 9 and Donald Lambro, 'Economic effect of war seen as small', *Washington Times*, 16 September 2002 [accessed 16 September 2002] http://www.washtimes.com/ national/20020916–8081695.htm

114. Robert Dreyfuss, 'The Thirty-Year Itch', *Mother Jones*, March/April 2003 [accessed 18 December 2010] http://motherjones.com/politics/2003/03/ thirty-year-itch.

115. Rutledge, *Addicted to Oil*: xi.

116. Noam Chomsky, *Imperial Ambitions: Conversations with Noam Chomsky on the Post-9/11 World* (London: Penguin Books, 2006): 6, 112; Noam Chomsky, *Interventions* (London: Penguin Books, 2007): 46, 47, 77, 85, 112, 135, 202, 208, and Chomsky, *Perilous Power*: 55–57.

117. Chomsky, *Perilous Power*: 56.

118. Chomsky, *Interventions*: 88, 162 and Chomsky, *Hegemony or Survival*: 125.

119. Chomsky, *Perilous Power*: 58.

120. Clark, *Petrodollar Warfare*: 39.

121. Brzezinski, *The Choice*: 63.

122. Chomsky, *Interventions*: 85–86, 112 and Chomsky, *Perilous Power*: 54–55.

123. Brzezinski, *The Choice*: 71–72.

124. Stephen J. Sniegoski, *The Transparent Cabal: The Neoconservative Agenda, War in the Middle East, and the National Interest of Israel* (Virginia: Enigma Editions, 2008): 344–345.

125. David Harvey, *The New Imperialism* (Oxford: Oxford University Press, 2005): 85.

126. Dreyfuss, 'The Thirty-Year Itch'.

127. Johnson, *The Sorrows of Empire*: 151–185.

128. Ibid: 215, 226, 234.

129. Pelletière, *America's Oil Wars*: 137.

130. Ibid: 151.
131. Ibid: 133–134.
132. Ibid: 144.
133. Klare, *Blood and Oil*: 152.
134. Bromley, *American Power*: 105.
135. Clark, *Petrodollar Warfare*: 68.
136. Bromley, *American Power*: 105.
137. Morrissey, 'The Geoeconomic Pivot of the Global War on Terror': 103, 104, 115, 116.
138. Clark, *Petrodollar Warfare*: 92.
139. Clark, *Petrodollar Warfare*: 98; Klare, *Rising Powers, Shrinking Planet: How Scarce Energy is Creating a New World Order* (Oxford: Oneworld, 2008): 193 and Rutledge, *Addicted to Oil*: 179.
140. Bromley, *American Power*: 102.
141. Ibid: 103.
142. Michael Klare, 'Bush-Cheney energy strategy: Procuring the rest of the world's oil', *Foreign Policy in Focus*, January 2004 [accessed 10 June 2004] http://www.fpif.org/papers/03petropol/politics_body.html.
143. Clark, *Petrodollar Warfare*: 119.
144. Naomi Klein, *The Shock Doctrine: The Rise of Disaster Capitalism* (London: Penguin, 2008): 313.
145. Bromley, *American Power*: 140.
146. John Chapman, 'The Real Reasons Bush Went to War', *Guardian*, 28 July 2004 [accessed 9 August 2009] http://www.guardian.co.uk/world/2004/jul/28/iraq.usa.
147. Richard Benson, 'Oil, the Dollar and US Prosperity', *Information Clearing House*, 8 August 2003 [accessed 10 August 2009] http://www.informationclearing-house.info/article4404.htm, also cited in Clark, *Petrodollar Warfare*: 115.
148. Simon Reeve, 'Conspiracy Theories: Iraq', *Sky One*, 10 January 2007, 22:00 GMT.
149. Clark, *Petrodollar Warfare*: 28, 31 and Fouskas & Gökay, *The New American Imperialism*: 25; Ed Blanche, 'Iran takes on US but at what cost?' *The Middle East*, March 2006: 21.
150. Charles Recknagel, 'Iraq: Baghdad moves to euro', *RFERL*, 1 November 2000 [accessed 9 August 2009] http://www.rferl.org/content/article/1095057.html.
151. Clark, *Petrodollar Warfare*: 117–118.
152. Robert Block, 'Some Muslim nations advocate dumping the dollar for the euro', *Wall Street Journal*, 15 April 2003 [accessed 10 August 2009] http://online.wsj.com/article/SB105032550519373700.html or http://www.

ipdirect.home.pl/dinar/USDvsEU/dumping_usd_wsj.htm   and   Faisal
Islam, 'Iraq nets handsome profit by dumping dollar for euro', *Guardian*,
16 February 2003 [accessed 10 August 2009] http://www.guardian.co.uk/
business/2003/feb/16/iraq.theeuro.

153. F. William Engdahl, 'A New American Century? Iraq and the hidden euro-
dollar wars', *Current Concerns*, April 2003 [accessed 10 August 2009] http://
www.currentconcerns.ch/archive/2003/04/20030409.php.

154. Patrick Brethour, 'OPEC mulls move to euro for pricing crude oil', *Energy
Bulletin*, 11 January 2004 [accessed 10 August 2009] http://www.ener-
gybulletin.net/node/123.

155. Clark, *Petrodollar Warfare*: 38.

156. Clark, *Petrodollar Warfare*: 139, Gabriel Kolko, *The Age of War: The United
States Confronts the World* (London: Lynne Rienner Publishers, 2006):
164 and Steve Johnson & Javier Blas, 'OPEC sharply reduces dollar ex-
posure', *Financial Times*, 6 December 2004 [accessed 10 August 2009]
http://www.ft.com/cms/s/0/67f88f7c-47cb-11d9-a0fd-00000e2511c8.
html?nclick_check=1.

157. Clark, *Petrodollar Warfare*: 145.

158. Ibid: 159.

159. Ibid: 4, 68, 119, 122.

160. Bromley, *American Power*: 104; Pelletière, *America's Oil Wars*: 134. For figures and statistics on OPEC countries' annual oil production, see the table
on world oil production in 2005 (accessed 30 October 2010) from BP's
website: http://www.bp.com/liveassets/bp_internet/globalbp/globalbp_uk_
english/publications/energy_reviews_2006/STAGING/local_assets/down-
loads/pdf/table_of_world_oil_production_2006.pdf.

161. Peter Beaumont & Faisal Islam, 'Carve-up of oil riches begins', *The Observer*,
3 November 2002 [accessed 3 November 2002] http://observer.guardian.
co.uk/Iraq/story/0,12239,825105,00.html.

162. Claudia Rosett, 'One purely evil cartel', *The Wall Street Journal*, 30 July
2003 [accessed 9 August 2009] http://www.opinionjournal.com/columnists/
cRosett/?id=110003813

163. Clarke, *Empires of Oil*: 42, 43.

164. Bryce, *Gusher of Lies*: 56–57.

165. Greg Palast, 'Secret US plans for Iraq's oil', *BBC Newsnight*, 17 March
2005 [accessed 29 May 2008] http://news.bbc.co.uk/1/hi/programmes/
newsnight/4354269.stm. Also see Clark, *Petrodollar Warfare*: 122.

166. Andrew Walker, 'US playing with fire, warns Yamani', *BBC News*,
14 March 2003 [accessed 14 March 2003] http://news.bbc.co.uk/1/hi/
business/2851723.stm

167. Pelletière, *America's Oil Wars*: 86 and Rutledge, *Addicted to Oil*: 186–187. Rutledge argues that the US depends on OPEC to keep prices higher than free-market level (ibid).

168. Pelletière, *America's Oil Wars*: 134; Rutledge, *Addicted to Oil*: 187 and Greg Palast, 'Secret US plans for Iraq's oil'. Rutledge argued that the US needs OPEC to put an upper limit on prices, so that they will not get too high (ibid).

169. Daniel Yergin, 'The fight over Iraq's oil', *BBC News*, 14 March 2003 [accessed 14 March 2003] http://news.bbc.co.uk/1/hi/business/2847905. stm. Yergin added that Iraq would not have the capacity to flood the market even if it wanted to (ibid). But I disagree with him, because some forecasts say that Iraq has unexplored oil reserves that, if explored, would make it equal to Saudi Arabia in terms of total oil reserves (although, since the Iraqi oil sector is seriously underdeveloped due to years of sanctions, it would take several years and tens of billions of dollars to reconstruct the Iraqi oil sector).

170. In October 2008, OPEC's preferred price was about $80/barrel (Pamela Ann Smith, 'OPEC faces a difficult year', *The Middle East*, January 2009: 57; Jad Mouawad, 'Oil prices slip below $70 a barrel', *New York Times*, 16 October 2008 [accessed 16 October 2008] http://www.nytimes. com/2008/10/17/business/worldbusiness/17oil.html?th&emc=th, Nelson D. Schwartz and Jad Mouawad, 'OPEC says it will cut oil output', *New York Times*, 24 October 2008 [accessed 24 October 2008] http://www.nytimes. com/2008/10/25/business/worldbusiness/25oil.html?th&emc=th, 'OPEC president: The crisis cost member countries $700 *billion*', *Al Jazeera, 19* November 2008 [accessed 19 November 2008] http://www.aljazeera.net/ NR/exeres/B164049D-36C1-41E0-8666-00903E6183F5.htm, and 'Saudi Arabia demands $75 per barrel', Al Jazeera, 29 November 2008 [accessed 29 November 2008] http://www.aljazeera.net/NR/exeres/B2C4EB00-4DD9-49A7-A767-407EA7E3E036.htm)

171. Ariel Cohen & William Schirano, 'Why should OPEC be immune from *lawsuits*', *The Heritage Foundation*, 18 August 2005 [accessed 24 February 2009] http://www.heritage.org/Press/Commentary/ed081805c.cfm, and Thomas W. Evans, 'Sue OPEC', *New York Times*, 19 June 2008 [accessed 24 February 2009] http://www.nytimes.com/2008/06/19/opinion/19evans. html?ex=1371528000&en=a65de029c28fbd1c&ei=5124&partner=permali nk&exprod=permalink

172. Chris Baltimore, 'Bush admin opposes Democratic push to sue OPEC', *Reuters*, 22 May 2007 [accessed 24 February 2009] http://www.reuters.com/ article/politicsNews/idUSN2242484120070522?feedType=RSS

173. Tom Doggett, 'House passes bill to sue OPEC over oil prices', *Reuters*, 20 May 2008 [accessed 24 February 2009] http://www.reuters.com/article/top-News/idUSWAT00953020080520?feedType=RSS&feedName=topNews& rpc=22&sp=true

174. Edmund L. Andrews, 'Panel supports tax breaks for coal and non-oil fuel', *New York Times*, 20 June 2007 [accessed 24 February 2009] http://www.nytimes.com/2007/06/20/business/20energy.html?_r=1&scp=3&sq=nopec&st=cse; Chris Baltimore, 'House okays bill to sue OPEC over prices', *Reuters*, 22 May 2007 [accessed 24 February 2009] http://www.reuters.com/article/politics-News/idUSN2246922520070522?feedType=RSS&rpc=22, and 'Oil price still near record $142', *BBC News*, June 27, 2008 [accessed 24 February 2009] http://news.bbc.co.uk/1/hi/business/7476910.stm. Also see 'Bush Administration opposes Democrat attempt to sue OPEC', Al Jazeera, 23 May 2007 [accessed 21 June 2008] http://www.aljazeera.net/News/archive/archive?ArchiveId=1058769, and 'House of Representatives approves draft law to sue OPEC', Al Jazeera, 23 May 2007 [accessed 21 June 2008] http://www.aljazeera.net/News/archive/archive?ArchiveId=1058798)

175. Andrew Walker, 'US playing with fire, warns Yamani', *BBC News*, 14 March 2003 [accessed 14 March 2003] http://news.bbc.co.uk/1/hi/business/2851723.stm

176. Kaplan & Kristol, *The War Over Iraq*: 99.

177. William Kristol, 'Memorandum to: Opinion leaders', *Project for the New American Century*, 25 April 2002 [accessed 9 August 2009] http://www.newamericancentury.org/saudi-042502.htm

178. William Kristol, 'A new approach to the Middle East', *The Weekly Standard*, 22 May 2002 [accessed 9 August 2009] http://www.weeklystandard.com/content/public/articles/000/000/001/273awdje.asp.

179. Schwartz, 'Breaking OPEC's grip'.

180. Clarke, *Against All Enemies*: 283.

181. Bromley, *American Power*: 138.

182. Johnson, *The Sorrows of Empire*: 236–242 and Klare, *Blood and Oil*: 90.

183. Paul Wolfowitz, 'Deputy Secretary Wolfowitz interview with Sam Tannenhaus, Vanity Fair', US Department of Defence, 9 May 2003 [accessed 10 August 2009] http://www.defenselink.mil/transcripts/transcript.aspx?transcriptid=2594

184. Cary Fraser, 'The Middle East and the Persian Gulf as the gateway to imperial crisis', in David Ryan & Patrick Kiely (eds), *America and Iraq: Policy-Making, Intervention and Regional Politics* (Oxford: Routledge, 2009): 208–211. Also see Johnson, *The Sorrows of Empire*: 236–242 and Klare, *Blood and Oil*: 90.

185. Rutledge, *Addicted to Oil*: 156, 157 and Klare, *Blood and Oil*: 89–90.

186. See Bob Woodward, *Plan of Attack*: passim, for the discreet communications between the Saudis and Washington over the invasion of Iraq.

187. Halper & Clarke, *America Alone*: 148.

188. Ibid: 155–156.

189. Bronson, *Thicker Than Oil*: 240.

190. Pinto, *Political Islam and the United States*: x.

191. Charles Krauthammer, 'Coming ashore', *Time*, 17 February 2003: 27.

192. Halper & Clarke, *America Alone*: 308.

193. Ibid: 309.

194. Ibid: *America Alone*: 156.

195. Kaplan & Kristol, *The War Over Iraq*: 101. They dismiss the 'war for oil theory', insisting that the real reason is imposing democracy to fight terrorism: 'Characterizing the US effort to oust Saddam as a war for oil, critics of the Bush Doctrine at home and abroad try to portray the United States as a bully state intoxicated by the arrogance of power' (ibid: 104).

196. 'Rep Ellen Tauscher decries logic used by DOD Deputy Wolfowitz', *House of Representatives*, 22 June 2004 [accessed 4 July 2008] http://www.house.gov/tauscher/press/06-22-04.htm

197. As described in Bob Woodward, *State of Denial: Bush at War*, Part III (London: Pocket Books, 2006): 440.

198. Greg Miller, 'Democracy Domino Theory "not credible"', *Los Angeles Times*, 14 March 2003 [accessed 12 April 2010] http://www.commondreams.org/headlines03/0314–06.htm and Sam Tanenhaus, 'The rise and fall of the Domino Theory', *New York Times*, 23 March 2003 [accessed 12 April 2010] http://www.nytimes.com/2003/03/23/weekinreview/23TANE.html?pagewanted=1

199. Pincus, 'Ex-CIA official faults use of data on Iraq'.

200. Brzezinski, *Second Chance*: 143.

201. Kaplan & Kristol: *The War Over Iraq*: 100.

202. 'Does he know where it's leading?' *The Economist*, 30 July 2005: 23–25.

203. Johnson, *The Sorrows of Empire*: 235 and Rutledge, *Addicted to Oil*: 177.

204. Kolko, *The Age of War*: 141.

205. Hoagland, 'Now an Iraqi war in Washington', and 'US considers Iraqi coup'.

206. Clarke, *Against All Enemies*: 31. After the meeting, Bush asked Clarke to 'see if Iraq was involved'. Clarke replied that his department has 'checked and there were no state sponsors'. But Bush insisted upon the request, so Clarke repeated the investigation and confirmed that Iraq was not linked to the 11 September attacks (Clarke, *Against All Enemies*: 32–33 and Kean *et al.*, *The*

9/11 Commission Report: 334). Furthermore, on 15 September Bush spoke privately with Henry Shelton, who was in his final month as JCS Chairman, asking him if he was making a mistake by focusing on al Qaeda instead of Iraq. Shelton reassured Bush that they were making the right choice, and that attacking Iraq without a clear reason would destabilize the Middle East and hamper the coalition-building efforts, especially seeing that there was no reason to link Iraq to the 11 September attacks. 'That's what I think', Bush told Shelton, adding that 'we will get [Saddam] but at a time and place of our choosing' (Michael Gordon & Bernard Trainor. *Cobra II: The Inside Story of the Invasion and Occupation of Iraq* (London: Atlantic Books, 2006): 17).

207. Griffin, *The 9/11 Commission Report: Omissions and Distortions*: 129.

208. Kean *et al.*, *The 9/11 Commissions Report*: 335.

209. Daalder & Lindsay, *America Unbound* (1st edn): 130, and Woodward, *Bush at War*: 99.

210. Kean *et al.*, *The 9/11 Commission Report*: 335.

211. Daalder & Lindsay, *America Unbound* (1st edn): 105.

212. Ibid: 105 and Kean *et al.*, *The 9/11 Commission Report*: 336. Wesley Clark, the commander of Operation Allied Force in Serbia in 1999, said that two weeks after the 11 September attacks, a 'senior general' at the Joint Staff told him that 'We're going to attack Iraq. The decision has basically been made.'(Wesley Clark, *A Time to Lead: For Duty, Honor and Country* (Hampshire: Palgrave Macmillan, 2007): 230–231.

213. Daalder & Lindsay, *America Unbound* (1st edn): 131.

214. Ibid: 117, 131.

215. Ibid: 117–118.

216. Ibid: 118.

217. Clark, *Petrodollar Warfare*: 102.

218. Bob Woodward, *State of Denial: Bush at War*, Part III (London: Pocket Books, 2006): 81.

219. George W. Bush, 'President Delivers State of the Union Address', *White House*, 29 January 2002 [accessed 30 June 2004] www.whitehouse.gov/news/releases/2002/01/20020129-11.html

220. According to administration officials, the speechwriters who wrote that speech did not have any policy guidance to put Iran in the speech. Iran was just a 'prop' needed to make a point. North Korea was added because a third country was needed in order to form an 'axis', and to ensure that the list was not entirely Islamic (Pollack, *The Persian Puzzle*: 352 and Julian Borger, 'How I created the axis of evil', *Guardian*, 28 January 2003 [accessed 28 May 2007] http://foi.Missouri.edu/polinfoprop/howicreated.htm)

221. Daalder & Lindsay, *America Unbound* (1st edn): 132; Daalder & Lindsay, *America Unbound* (2nd edn):129–130 and James Carney, 'Frenemies', *Time*, 28 July 2008: 29.

222. *Time*, 25 March 2002: 17.

223. Hiro, *Secrets and Lies* (2004): 31.

224. Daalder & Lindsay, *America Unbound* (1st edn): 135, from Nicholas Lemann 'How it came to war' *New Yorker*, 31 March 2003 [accessed 3 March 2008] http://www.newyorker.com/archive/2003/03/31/030331fa_fact?currentPage=2

225. Daalder & Lindsay, *America Unbound* (1st edn): 138 and Klare, *Rising Powers, Shrinking Planet*: 181. Cheney also said that inspections of Iraqi facilities 'may not be as crucial if you've got other measures in place' (Lucas & Ryan, 'Against Everyone and No-One': 163).

226. Hiro, *Secrets and Lies* (2004): 44.

227. Ibid: 5.

228. 'Mr. Bush lays out his case', *The Economist*, 21 September 2002: 49.

229. 'Saddam plays his ace', *The Economist*, 21 September 2002: 61.

230. Johanna McGeary, '6 reasons why so many allies want Bush to slow down', *Time*, February 2003: 20.

231. 'US convinced Iraq has banned weapons', *BBC News*, 10 January 2003 [accessed 23 October 2006] http://news.bbc.co.uk/1/hi/world/middle_east/2466783.stm

232. Bromley, *American Power*: 105.

233. See Andrew J. Bacevich, 'Tragedy renewed: William Appleman Williams', *World Affairs*, Winter 2009 [accessed 22 April 2009] http://www.worldaffairs.org/2009%20-%20Winter/full-Bacevich.html, and Thomas McCormick 'What would William Appleman Williams say now?' *History News Network*, 24 September 2007 [accessed 22 April 2009] http://hnn.us/articles/42971.html.

234. Rashid, 'Central Asia': 118.

235. Phase I of a military conflict is the preparation for combat. Phase II consists of initial operations. Phase III is the main combat stage, and Phase IV is concerned with post-combat operations (Thomas E. Ricks, 'Army Historian Cites Lack of Postwar Plan', *Washington Post*, 25 December 2004, page A01, www.washingtonpost.com/wp-dyn/articles/A24891-2004Dec24.html)

Chapter 4

1. William Clark, Petrodollar Warfare: Oil, Iraq, and the Future of the Dollar (Gabriola Island: New Society Publishers, 2005): 98; Klare, Rising Powers, Shrinking Planet: How Scarce Energy is Creating a New World Order

(Oxford: Oneworld, 2008): 193 and Ian Rutledge, Addicted to Oil: America's Relentless Drive for Energy Security (London: I.B.Tauris, 2005): 179.

2. Donald Rumsfeld & Rik Kirkland, 'Don Rumsfeld talks guns and butter', Fortune, 18 November 2002 [accessed 4 July 2008] http://money.cnn.com/ magazines/fortune/fortune_archive/2002/11/18/332271/ and http://web. ebscohost.com/ehost/detail?vid=5&hid=15&sid=f1a96962–8b4e-4c6e-86ff-5e1ae5fc7590%40sessionmgr2

3. 'Past comments about how much Iraq would cost' [accessed 4 July 2008] http://fairuse.100webcustomers.com/sf/iraqwarcost.htm, from House of Representatives [accessed 20 July 2006] http://www.house.gov/schakowsky/ iraqquotes_web.htm.

4. 'Rep Ellen Tauscher decries logic used by DOD Deputy Wolfowitz', House of Representatives, 22 June 2004 [accessed 4 July 2008] http://www.house. gov/tauscher/press/06–22-04.htm

5. 'Past comments about how much Iraq would cost' [accessed 4 July 2008] http://fairuse.100webcustomers.com/sf/iraqwarcost.htm, from House of Representatives [accessed 20 July 2006] http://www.house.gov/schakowsky/ iraqquotes_web.htm.

6. Ibid. Furthermore, Press Secretary Ari Fleischer said at a White House press briefing on 18 February 2003: 'The reconstruction costs remain ... an issue for the future. And Iraq, unlike Afghanistan, is a rather wealthy country. Iraq has tremendous resources that belong to the Iraqi people. And so there are a variety of means that Iraq has to be able to shoulder much of the burden for their own reconstruction' (Ibid.)

7. Halper, Stefan & Jonathan Clarke, America Alone: The Neo-conservatives and the Global Order (Cambridge: Cambridge University Press, 2004): 223; Joseph Stiglitz & Linda Blimes, The Three Trillion Dollar War: The True Cost of the Iraq Conflict (London: Allen Lane, 2008): 7–8 and 'Nightline: Project Iraq', Foundation of American Scientists, 23 April 2003 [accessed 22 February 2008] http://www.fas.org/sgp/temp/natsios042303.html.

8. Stiglitz & Blimes, The Three Trillion Dollar War: 7–8 and 'Nightline: Project Iraq'. Interestingly, Lawrence Lindsey was fired from the Bush administration for saying in September 2002 that the cost of the war might reach $200 billion (Stiglitz & Blimes, The Three Trillion Dollar War: 7; Linda Blimes & Joseph Stiglitz, 'Is this any way to rebuild Iraq?', Los Angeles Times, 15 August 2008 [accessed 10 May 2010] http://www.latimes.com/news/opinion/la-oe-bilm-es15-2008aug15,0,4432303.story, Linda Blimes & Joseph Stiglitz, 'The Iraq war will cost us $3 trillion, and much more', Washington Post, 9 March 2008 [accessed 10 May 2010] http://www.washingtonpost.com/wp-dyn/content/ article/2008/03/07/AR2008030702846.html; Blimes & Stiglitz, 'The US in

Iraq: An economics lesson', Los Angeles Times, 2 July 2009 [accessed 10 May 2010] http://articles.latimes.com/2009/jul/02/opinion/oe-bilmes2; Blimes & Stiglitz, 'War's price tag', Los Angeles Times, 16 March 2008 [accessed 10 May 2010] http://articles.latimes.com/2008/mar/16/opinion/op-bilmes16; Stiglitz & Blimes, '$3 trillion may be too low', Guardian, 6 April 2008 [accessed 10 May 2010] http://www.guardian.co.uk/commentisfree/2008/apr/06/3trillionmaybetoolow; Stiglitz & Blimes, 'Ask a question: The Three Trillion Dollar War', McClatchy, 15 April 2008 [accessed 10 May 2010] http://www.mcclatchydc.com/qna/forum/three_trillion_dollar_war/index.html; Joseph Stiglitz, 'It's not the economy, stupid', Financial Times, 28 September 2004 [accessed 10 May 2010] http://www2.gsb.columbia.edu/faculty/jstiglitz/download/opeds/Economy_Stupid.pdf;Joseph Stiglitz, 'The Three Trillion Dollar War', Project Syndicate, 7 March 2008 [accessed 10 May 2010] http://www.project-syndicate.org/commentary/stiglitz97/English; Stiglitz & Blimes, 'The Three Trillion Dollar War', Times Online, 23 February 2008 [accessed 10 May 2010] http://www.timesonline.co.uk/tol/comment/columnists/guest_contributors/article3419840.ece; Stiglitz & Blimes, 'The $3 Trillion War', Vanity Fair, April 2008 [accessed 10 May 2010] http://www.vanityfair.com/politics/features/2008/04/stiglitz200804; Joseph Stiglitz, 'War at any cost? The total economic costs of the war beyond the federal budget: Testimony before the Joint Economic Committee', Columbia University, 28 February 2008 [accessed 10 May 2010] http://www2.gsb.columbia.edu/faculty/jstiglitz/download/papers/Stiglitz_testimony.pdf; Peter Wilson, 'Stiglitz: Iraq war "caused slowdown in the US" ', The Australian, 28 February 2008 [accessed 10 May 2010] http://www.theaustralian.com.au/news/iraq-war-caused-slowdown-in-the-us/story-e6frg6tf-1111115661208).

9. Halper & Clarke, America Alone: 155 and 'Corrections and clarifications', The Guardian, 6 June 2003 [accessed 28 December 2010] http://www.guardian.co.uk/theguardian/2003/jun/06/correctionsandclarifications.

10. Dilip Hiro, Secrets and Lies: Operation Iraqi Freedom and After: (2005 edition): 274; Chalmers Johnson, The Sorrows of Empire: Militarism, Secrecy and the End of the Republic (London: Verso, 2006 edition): 234 and Rutledge, Addicted to Oil: 181. Also, the oilfields around Kirkuk, which produced a third of Iraq's oil output, were secured by the 173rd Airborne Divisoin (Hiro, Secrets and Lies (2005 edition): 234, 281). After the capture of Qaim, US troops cut off an oil pipeline to Syria (Hiro, Secrets and Lies (2005 edition): 292).

11. Raymond W. Baker, Shereen T. Ismael & Tareq Y. Ismael, 'Preface', in Raymond W. Baker, Shereen T. Ismael & Tareq Y. Ismael (eds), Cultural Cleansing in Iraq: Why Museums Were Looted, Libraries Burned and Academics Murdered (London: Pluto Press, 2010): xi and Raymond W.

Baker, Shereen T. Ismael & Tareq Y. Ismael, 'Ending the Iraqi State', in Baker et al., Cultural Cleansing in Iraq: 26.

12. Barton Gelman, Angler: The Shadow Presidency of Dick Cheney (London: Penguin Books, 2009):106; Juhasz, Antonia, The Bush Agenda: Invading the World, One Economy at a Time (London: Duckworth, 2006): 180; Jane Mayer, 'Contract Sport: What did the Vice-President do for Halliburton?', The New Yorker, 16 February 2004 [accessed 17 May 2007] http://web.lexis-nexis.com/executive/form?_index=exec_en.html&_lang=en&ut=3361601930; Mayer, 'Jane Mayer on her article in The New Yorker about Dick Cheney's relationship with Halliburton' and 'Cheney energy task force documents feature map of Iraqi oilfields' Judicial Watch, 17 July 2003 [accessed 11 January 2008] http://www.judicialwatch.org/printer_iraqi-oilfield-pr.shtml. The revelation of the documents was a result of a Freedom of Information Act (FOIA) lawsuit by Judicial Watch, a public-interest group that investigates government corruption and abuses.

13. Paul Sperry, 'White House energy task force papers reveal Iraqi oil maps: Judicial Watch lawsuit also uncovers list of foreign suitors for contracts' World Net Daily, 18 July 2003 [accessed 20 December 2007] http://www.worldnetdaily.com/news/article.asp?ARTICLE_ID=33642. Also see Gellman, Angler: 106.

14. 'Cheney energy task force documents feature map of Iraqi oilfields' Judicial Watch, 17 July 2003 [accessed 11 January 2008] http://www.judicial-watch.org/printer_iraqi-oilfield-pr.shtml. The maps and lists are found on the following websites: 'Foreign suitors for Iraqi oilfield contracts as of 5 March 2001 part 1', http://www.judicialwatch.org/IraqOilFrgnSuitors.pdf, 'Foreign suitors for Iraqi oilfield contracts as of 5 March 2001 part 2', http://www.judicialwatch.org/IraqOilGasProj.pdf; 'Iraqi oilfields and exploration blocks', http://www.judicialwatch.org/IraqOilMap.pdf; 'Saudi Arabia major oil and natural gas development projects', http://www.judicialwatch.org/SAOilProj.pdf, 'Selected oil facilities in Saudi Arabia', http://www.judicialwatch.org/SAOilMap.pdf; 'Selected oil facilities of the United Arab Emirates', http://www.judicialwatch.org/UAEOilMap.pdf; 'United Arab Emirates major oil and gas development projects', http://www.judicialwatch.org/UAEOilProj.pdf

15. Juhasz, The Bush Agenda: 320–321; T. Christian Miller, Blood Money: Wasted Billions, Lost Lives and Corporate Greed in Iraq (New York: Little, Brown & Company, 2006): 186–201; Rutledge, Addicted to Oil: 189; Joel Bainerman, 'How much will the war on Iraq cost the US taxpayer?' The Middle East, February 2006: 17. It took until April 2004 to appoint an auditor to the DFI (Bainerman, 'How much will the war on Iraq cost the US taxpayer?': 17).

16. Hiro, Secrets and Lies: (2005): 310.

17. 'Fact sheet: UN Security Council high level meeting on Iraq', White House, 15 December 2010 [accessed 29 December 2010] http://www.whitehouse. gov/the-press-office/2010/12/15/fact-sheet-un-security-council-high-level-meeting-iraq and 'UN ends Iraq's Saddam era sanctions', Iraq Business News, 21 December 2010 [accessed 29 December 2010] http://www.iraq-businessnews.com/tag/united-nations/

18. Pratap Chaterjee, Iraq, Inc.: A Profitable Occupation (New York: Seven Stories Press, 2004): 179 and Rutledge, Addicted to Oil: 189.

19. Rutledge, Addicted to Oil: 190.

20. Chaterjee, Iraq, Inc.: 179, and Rutledge, Addicted to Oil, 189–190.

21. Juhasz, The Bush Agenda: 211–221 and Richard Seymour, 'The Real Cost of the Iraq War', The Middle East, May 2009: 50. The transformation of the laws of an occupied country is illegal under the Geneva Conventions and US Army rules. So the contracts signed on the basis on these changes in Iraq's law in favour of US companies are also illegal (Juhasz, The Bush Agenda: 318–320). Thomas C. Foley, an investment banker, a major Republican donor and Bush's former classmate at Harvard Business School, was the CPA official in charge of Iraq's privatization. When told that international law prevents the sale of assets by an occupation government, he said 'I don't give a sh*t about international law. I made a commitment to the president that I'd privatize Iraq's business.' However, Foley's efforts to privatize the Iraqi economy were widely regarded as a failure (Miller, Blood Money: 38 and Rajiv Chandrasekaran, Imperial Life in the Emerald City: Inside Baghdad's Green Zone (London: Bloomsbury Publishing, 2008): 106, 140).

22. Gabriel Kolko, The Age of War: The United States Confronts the World (London: Lynne Rienner Publishers, 2006): 149 and Naomi Klein, The Shock Doctrine: The Rise of Disaster Capitalism (London: Penguin, 2008): 11, 56–57, 253.

23. Juhasz, The Bush Agenda: 212.

24. Ibid: 197. In fact, before the invasion, Bremer used to work for Kissinger Associates, and was chairman and CEO of insurance company Marsh and McLennan's Crisis Consulting Practice, specializing in risk assessment and services for multinational corporations. In a November 2001 report which he wrote, he said that privatization would be hazardous to locals, but beneficial to the multinational corporations (Juhasz, The Bush Agenda: 191, 192).

25. David L. Phillips, Losing Iraq: Inside the postwar reconstruction fiasco (New York: Basic Books, 2005): 147 and Max Fuller & Dirk Adriaensens, 'Wiping the Slate Clean', in Raymond W. Baker, Shereen T. Ismael & Tareq Y. Ismael (eds), Cultural Cleansing in Iraq: Why Museums Were Looted, Libraries

Burned and Academics Murdered (London: Pluto Press, 2010): 173–174. See also Paul Bremer, My Year in Iraq: The struggle to build a future of hope (New York: Simon and Schuster, 2006): 40 and 44, and Juhasz, The Bush Agenda: 201.

26. Dilip Hiro, Secrets and Lies (2005): 419.

27. Juhasz, The Bush Agenda: 253, 254; Michael T. Klare, Blood and Oil (London: Hamish Hamilton, 2004): 103, and Rutledge, Addicted to Oil: 182–184, 187–188, 192.

28. Klein, The Shock Doctrine: 345.

29. Rutledge, Addicted to Oil: 188.

30. Dilip Hiro, Blood of the Earth: The global battle for vanishing oil resources (London: Politico's, 2007): 145, from Greg Palast, 'Secret US plans for Iraq's oil', BBC Newsnight, 17 March 2005 [accessed 29 May 2008] http://news.bbc.co.uk/1/hi/programmes/newsnight/4354269.stm.

31. Hiro, Blood of the Earth: 146.

32. Rutledge, Addicted to Oil: 184.

33. Ibid: 184.

34. Ibid: 184. PSAs are highly unusual in the Middle East, where the oil industry in Saudi Arabia and Iran is state controlled (Danny Forston, Andrew Murray-Watson & Tim Webb, 'Future of Iraq: The spoils of war', Independent, 7 January 2007 [accessed 11 January 2007] http://news.independent.co.uk/world/middle_east/article2132569.ece). Nevertheless, PSAs are not new to Iraq, as PSAs, or Production and Developments Contracts which are very similar to PSAs, have been used in Iraq in 1997 with Russian, Chinese and Brazilian companies, although for reasons which were more political than commercial, to try to convince these countries to break the embargo on Iraq. However, these contracts failed to break the embargo on Iraq (Fouad Al-Amir, 'Discussion on the Iraq law' [accessed 1 March 2007] http://www.al-ghad.org/2007/02/20/discussion-on-the-iraq-oil-law/ and 'Effect of production sharing agreements on Iraq revenues', Al Jazeera, 8 October 2007 [accessed 13 January 2008] http://www.aljazeera.net/NR/exeres/80E8DE78-A0E4-4268-9547-3D4D89D0E930.htm and 'Future of Iraq: Oil and Energy Working Group: Considerations relevant to an oil policy for a liberated Iraq', 27 January 2003: 6).

35. Klare, Blood and Oil: 103–104.

36. Upstream oil activities include the exploration for oil and the extraction of oil, while downstream activities include transportation and refining.

37. 'Effect of production sharing agreements on Iraq revenues'.

38. Al-Amir, 'Discussion on the Iraq law' and 'Effect of production sharing agreements on Iraq revenues'.

39. 'Evaluation of new Iraqi oil law', Al Jazeera, 17 July 2007 [accessed 13 January 2008] http://www.aljazeera.net/NR/exeres/311D9F61-D2DD-4004 -80CC-AD6A9A5AB6F0.htm?wbc_purpose=Basic

40. Al-Amir, 'Discussion on the Iraq law' and Sean Michael Wilson & Lee O'Connor, Iraq: Operation Corporate Takeover (London: War on Want/ Boychild Productions Books, 2007): 37.

41. Juhasz, The Bush Agenda: 325, from Greg Muttitt, 'Crude Designs'. Juhasz argued that it is not obvious that Iraq needed foreign investments, since the end of the American occupation in itself would bring the stability necessary for Iraq investment, and Iraqis had the expertise and skills necessary to invest on their own (Juhasz, The Bush Agenda: 325).

42. Al-Amir, 'Discussion on the Iraq law' and 'Effect of production sharing agreements on Iraq revenues' and 'Evaluation of new Iraqi oil law'.

43. 'Effect of production sharing agreements on Iraq revenues' and Greg Muttitt, 'Crude designs: The rip off of Iraq's oil wealth', Global Policy Forum, November 2005 [accessed 28 May 2008] http://www.globalpolicy. org/security/oil/2005/crudedesigns.htm

44. Rutledge, Addicted to Oil: 178.

45. Klare, Blood and Oil: 99, from David Rieff 'Blueprint for a mess', New York Times Magazine, 2 November 2003

46. Rutledge, Addicted to Oil: 178.

47. Ibid: 179, from US Department of State, Press Release, 19 December 2002.

48. Rutledge, Addicted to Oil: 179–180, from US Department of State, Press Release, 19 December 2002.

49. Juhasz, The Bush Agenda: 253.

50. 'Future of Iraq: Oil and Energy Working Group: Considerations relevant to an oil policy for a liberated Iraq', 27 January 2003 [accessed 11 January 2008] http://www.thememoryhole.org/state/future_of_iraq/future_oil.pdf: 2.

51. Ibid: 1.

52. Ibid: 2.

53. Ibid: 2.

54. Ibid: 2 Al-Amir, 3, and 5.

55. Ibid: 3

56. Ibid: 5

57. Ibid: 9

58. Rutledge, Addicted to Oil: 179.

59. Juhasz, The Bush Agenda: 255 Al-Amir, 256.

60. 'Iraq tenders only for US allies,' BBC News, 10 December 2003 [accessed 10 December 2003] http://news.bbc.co.uk/1/hi/business/3305505.stm.

Moreover, Paul Bremer made it clear from the outset that London's reconstruction remit would not include the Iraqi oil industry; that would be run by the Americans. The exclusivity of American influence caused deep unease back in the boardrooms of British and European oil companies. Fearing that they would lose out to their American competitors, British oil companies held talks with 10 Downing Street before the invasion about the post-war distribution of contracts, insisting on a level playing field (Julian Borger, Terry Macalister & Martin Chulov, 'Iraq: the legacy – Basra's failed oil bonanza', Guardian, 15 April 2009 [accessed 20 June 2009] http://www.guardian.co.uk/world/2009/apr/15/iraq-oil-legacy).

61. 'US lets Canada bid for Iraq work', BBC News, 14 January 2004 [accessed 14 January 2004] http://news.bbc.co.uk/1/hi/world/Americas/3393113.stm

62. Klare, Blood and Oil: 102 and Rutledge, Addicted to Oil: 188.

63. Klare, Blood and Oil: 102.

64. Rutledge, Addicted to Oil: 192.

65. Klare, Blood and Oil: 103.

66. 'Iraq plans to double liquid petroleum gas output through Restore Iraqi Oil program', Portaliraq, 25 November 2004 [accessed 26 June 2006] www.portaliraq.com/news/Iraq+plans+to+double+liquid+petroleum+gas+output+through+Restore+Iraqi+Oil+program__653.html

67. Dan Briody, The Halliburton Agenda: The politics of oil and money (New Jersey: Wiley, 2004):184–187; Miller, Blood Money: 77–79; Rutledge, Addicted to Oil: 181–182.

68. Miller, Blood Money: 25–27 and 83–84.

69. Briody, The Halliburton Agenda: 221.

70. Juhasz, The Bush Agenda: 121–122.

71. Ibid. 129 and Wilson & O'Connor, Iraq: Operation Corporate Takeover: 30.

72. Juhasz, The Bush Agenda: 181. Shultz wrote an article in the Washington Post in September 2002, saying that Saddam had weapons of mass destruction, and that, by removing Saddam, 'a model can emerge that other Arab societies may look to and emulate for their own transformation and that of the entire region' (Juhasz, The Bush Agenda: 183).

73. Juhasz, The Bush Agenda: 183.

74. Briody, The Halliburton Agenda: 231.

75. Rutledge, Addicted to Oil: 181–182.

76. KBR had a role in devising the LOGCAP program in 1992, and has worked under its umbrella in several locations around the world. See Briody, The Halliburton Agenda: 177–189; Juhasz, The Bush Agenda: 226 and Kolko, The Age of War: 156.

77. 'Memorandum for Commander US Army Corps of Engineers: Subject: Justification and Approval (J&A) for other than full and open competition for the execution of the Contingency Support Plan', 28 February 2003 (Declassified 22 April 2004) [accessed 11 January 2008] http://www.judicialwatch.org/archive/2004/kbr.pdf, pages 4–7. New York Times columnist Maureen Dowd has spoken of 'a secret 500-page document prepared by Halliburton on what to do with Iraq's oil industry – a plan it wrote several months before the invasion of Iraq, and before it got a no-bid contract to implement the plan' (Maureen Dowd, 'Vice axes that 70's show', New York Times, 28 December 2005 [accessed 12 March 2008] http://query.nytimes.com/gst/fullpage.html?res=9F00E4D61230F93BA15751C1A9639C8B63&scp=2&sq=maureen%20dowd%20vice%20axes%20that%2070's%20show&st=cse). In the end, however, Saddam did not order the torching of the oil wells (Hiro, Secrets and Lies (2nd edn): 393). He never actually planned to torch the Iraqi oil wells, as his miscalculations were so great that he never expected to lose the war (Michael Gordon & Bernard Trainor. Cobra II: The Inside Story of the Invasion and Occupation of Iraq. (London: Atlantic Books, 2006): 505). This showed that there were great flaws in planning on the American side, as well as on the Iraqi side (Gordon & Trainor, Cobra II: 504–506). However, US troops claim that Iraqi troops did set some oil wells in Rumeila on fire, and tried to set other oil facilities on fire but did not have the time (Hiro, Secrets and Lies (2005 edition): 186, 195).

78. 'Wolfowitz declares non-competitive bids in Iraq to be in the public interest', US Foreign Aid Watch, 9 December 2003 [accessed 18 February 2009] http://www.foreignaidwatch.org/modules.php?op=modload&name=News&file=article&sid=542

79. Miller, Blood Money: 85–86.

80. Ibid: 43, 87.

81. Ibod: 43–44.

82. Stiglitz & Blimes, The Three Trillion Dollar War: 15.

83. Miller, Blood Money: 73.

84. 'Iraq contracts won by Bush donors', BBC News, 31 October 2003 [accessed 1 November 2003] http://news.bbc.co.uk/1/hi/business/3231345.stm.

85. Juan Cole, 'Critique of US policy in Iraq', Informed Comment, 24 May 2006 [accessed 20 June 2009] http://www.juancole.com/2006/05/critique-of-us-policy-in-iraq-bush.html.

86. David Teather, 'Halliburton units file for bankruptcy', Guardian, 17 December 2003 [accessed 20 June 2009] http://www.guardian.co.uk/world/2003/dec/17/iraq.dickcheney.

87. Gellman, Angler: 95 and Helen Dewar & Dana Milbank, 'Cheney dismisses critic with obscenity', Washington Post, 25 June 2004 [accessed 23 March 2008] http://www.washingtonpost.com/wp-dyn/articles/A3699–2004Jun24.html.

88. Klein, The Shock Doctrine: 298.

89. For more information on how private US corporations are profiting from US domestic security contracts, read Klein, The Shock Doctrine, chapters 14 and 15.

90. Ibid: 7.

91. Ibid: 6.

92. Ibid: 9.

93. Ibid: 309.

94. Hiro, Secrets and Lies: The True Story of the Iraq War: 419.

95. Ibid: 419, from New York Times, 24 April 2004. There was a lot of speculation over the replacement of Garner by Paul Bremer. One reason might have been the bureaucratic competition between the State Department and the Department of Defence, as Paul Bremer was working for the State Department while Jay Garner was a retired army lieutenant general. According to Jay Garner, however, the reason for the change had more to do with differences over how far the US should dominate the political and economic process in Iraq (Juhasz, The Bush Agenda: 189).

96. Juhasz, The Bush Agenda: 189 from Greg Palast's BBC television interview with Jay Garner, 14 March 2004.

97. Juhasz, The Bush Agenda: 190 and David Leigh, 'General sacked by Bush says he wanted early elections', Guardian, 18 March 2004 [accessed 2 July 2009] http://www.guardian.co.uk/world/2004/mar/18/iraq.usa.

98. Phillips, Losing Iraq: 179.

99. Rutledge, Addicted to Oil: 195.

100. Chandrasekaran, Imperial Life in the Emerald City: 207.

101. Phillips, Losing Iraq: 178–179.

102. Toby Dodge, 'The ideological roots of failure: The application of kinetic neo-liberalism to Iraq', International Affairs, 86(6), November 2010: 1274, 1279.

103. Klein, The Shock Doctrine: 364.

104. Chandrasekaran, Imperial Life in the Emerald City: 206–207 and 'Iraq cleric condemns US plans', BBC News, 1July 2003 [accessed 22 February 2008] news.bbc.co.uk/1/hi/world/middle_east/3032988.stm

105. Phillips, Losing Iraq: 179.

106. Bremer, My Year in Iraq: 19.

107. Phillips, Losing Iraq: 170, from Partap Chaterjee 'Iraq: selections not elections' Corpwatch, 1 July 2004 [accessed 22 February 2008] http://www.corpwatch.org/article.php?id=11403

108. Chandrasekaran, Imperial Life in the Emerald City: 214, and 'Draft version of Ambassador Bremer's statement to the Senate Armed Service committee', Global Security, 25 September 2003 [accessed 11 March 2008] http://www.globalsecurity.org/military/library/congress/2003_hr/bremer.pdf.

109. George Packer, The Assassins' Gate: America in Iraq (London: Faber & Faber Limited, 2007): 316.

110. Chandrasekaran, Imperial Life in the Emerald City: 209.

111. Ibid: 213. Bremer's seven-step plan to Iraqi sovereignty was: 1 the formation of a Governing Council; 2 the establishment of a committee to determine how to write a constitution; 3 the Governing Council's assumption of more day-to-day tasks; 4 the writing of a constitution, and this is where the plan stalled as al-Sistani called for writing the constitution by elected representatives (ibid: 206, 207).

112. Chandrasekaran, Imperial Life in the Emerald City: 213–215.

113. Ibid: 219–220 and Packer, The Assassins' Gate: 316.

114. Woodward, State of Denial: 269.

115. Packer, The Assassins' Gate: 316.

116. Thomas E. Ricks, 'Army Historian Cites Lack of Postwar Plan', Washington Post, 25 December 2004 [accessed 26 May 2006] http://www.washingtonpost.com/wp-dyn/articles/A24891-2004Dec24.html. Phase I is the preparation for combat. Phase II consists of initial operations. Phase III is the main combat stage, and Phase IV is concerned with post-combat operations (ibid).

117. Gordon & Trainor, Cobra II: 160.

118. Ibid: 495–496.

119. Daalder & Lindsay, America Unbound (1st edn): 151 and Daalder & Lindsay, America Unbound (2nd edn): 166.

120. Gordon & Trainor, Cobra II: 158.

121. Daalder & Lindsay, America Unbound (2nd edn): 166 and Gordon & Trainor, Cobra II: 495, 501.

122. 'Rumsfeld 'wanted cheap war'', BBC News, 30 March 2003 [accessed 26 June 2008] http://news.bbc.co.uk/1/hi/world/americas/2899823.stm.

123. 'Iraq war in figures', BBC News, 29 August 2010 [accessed 30 August 2010] http://www.bbc.co.uk/news/world-middle-east-11107739

124. Daalder & Lindsay, America Unbound (2nd edn): 164.

125. Ibid: 163–164.

126. Tenet, At the Centre of the Storm: 308–309.

127. Gordon & Trainor, Cobra II: 502.

128. Larry Diamond, 'What Went Wrong in Iraq', Foreign Affairs, September/October 2004 [accessed 16 June 2008] http://www.foreignaffairs.org/20040901faessay83505/larry-diamond/what-went-wrong-in-iraq.html.

129. 'Led by donkeys', The Economist, 8 April 2006: 92–93.

130. Gordon & Trainor, Cobra II: 159 and Halper & Clarke, America Alone: 225. Jay Garner was so impressed by Tom Warrick, the head of the Future of Iraq Project, that he asked Warrick to join the Office of Reconstruction and Humanitarian Assistance (ORHA) and go to Iraq. However, Warrick was dismissed due to the feud between the State Department and the Department of Defence, as Rumsfeld asked Garner to take Warrick off the team. Rumsfeld later told Powell that the decision to block Warrick came from Dick Cheney, who despised Warrick (Gordon & Trainor, Cobra II: 159 and Packer, The Assassins' Gate: 124.)

131. Gordon & Trainor, Cobra II: 502. Eric Shinseki, the US Army Chief of Staff, told the Senate Armed Service Committee on 25 February 2003, that several hundred thousand troops would be needed to sustain security in Iraq in the period after the war. But Rumsfeld and Wolfowitz criticized Shinseki, saying that he was 'widely off the mark', as Rumsfeld and the neo-conservatives wanted to demonstrate that the invasion could be done with minimum cost and with high levels of technology (Halper & Clarke, America Alone: 222). The Pentagon replaced Shinseki a year before his term expired (Diamond, 'What went wrong in Iraq').

132. Phillips, Losing Iraq: 11. The Future of Iraq Project did offer a useful way of bringing Iraqi exiles together to discuss the problems of Iraq, and proposed some good ideas, but it was far short of a viable plan. David Kay, a senior member of the Garner team (who would later lead the CIA effort to investigate Iraq's missing WMDs) has read the Future of Iraq Study, and summed it up: 'It was un-implementable. It was a series of essays to describe what the future could be. It was not a plan to hand to a task force and say "go implement". If it had been carried out it would not have made a difference.' Similarly, Colonel Paul Hughes who was also a senior member of Jay Garner's staff, said that there was 'a real lack of planning capacity at the Department of State, hence, just about any study gets labelled a plan … While [the Future of Iraq Project] produced some useful background information, it had no chance of really influencing the post-Saddam phase of the war' (Gordon & Trainor, Cobra II: 150, 158, 159.)

133. Gordon & Trainor, Cobra II: 497.

134. Bremer, My Year in Iraq: 109–112. US troops were so stretched that they were unable to guard the pipelines or power grids (Miller, Blood Money: 266–267).

135. Miller, Blood Money: 96.

136. Halper & Clarke, America Alone: 223.

137. Miller, Blood Money: 46, 111, 117, 118, 124, 196, 292.

138. James Glanz & T. Christian Miller, 'Official history spotlights Iraqi rebuilding blunders', New York Times, 13 December 2008 [accessed 14 December 2008] http://www.nytimes.com/2008/12/14/world/middleeast/14reconstruct.html?th&emc=th and 'Report shows wasted Iraq spending', BBC News, 2 February 2009 [accessed 2 February 2009] http://news.bbc.co.uk/1/hi/world/americas/7865840.stm.

139. Bremer, My Year in Iraq: 112.

140. Diamond, 'What Went Wrong in Iraq'.

141. The AEI, which had a leading role in calling for the invasion of Iraq, was among the conservative think tanks financed by the American corporate right to promote free enterprise, free market principles, limited government intervention and a strong US military to protect America's business interests around the world (Burbach & Tarbell, Imperial Overstretch: 78–84, 91–92, 207 and Halper & Clarke, America Alone: 48). For more on the funding network of neo-conservative think tanks, see Halper & Clarke, America Alone: 108–109.

142. George W. Bush, 'President discusses the future of Iraq', White House, 26 February 2003 [accessed 22 February 2008] http://www.whitehouse.gov/news/releases/2003/02/20030226-11.html

143. George W. Bush, 'President Bush discusses freedom in Iraq and Middle East', White House, 6 November 2003 [accessed 22 February 2008] http://www.whitehouse.gov/news/releases/2003/11/20031106-2.html

144. Burbach & Tarbell, Imperial Overstretch: 188.

145. 'Report: AT&T contract under fire', CNN Money, 7 March 2005 [accessed 27 May 2010] http://money.cnn.com/2005/03/07/news/international/iraq_att/ and Gene Retske, 'American legion acts on AT&T/Iraq issue', The Prepaid Press, 14 February 2006 [accessed 27 May 2010] http://www.prepaid-press.com/news_detail.php?t=paper&id=1132.

146. Miller, Blood Money: 251–255 and T. Christian Miller, 'Army and insurer at odds', Los Angeles Times, 13 June 2005 [accessed 20 May 2010] http://www.corpwatch.org/article.php?id=12377.

147. Burbach & Tarbell, Imperial Overstretch: 177.

148. Matt Kennard, 'An interview: Joseph Stiglitz on Latin America and Iraq', The Comment Factory, 18 November 2008 [accessed 12 December 2010] http://www.thecommentfactory.com/an-interview-joseph-stiglitz-on-latin-america-and-iraq-642/

149. Dodge, 'The ideological roots of failure': 1269–1286.

150. Morrissey, 'The Geoeconomic Pivot of the Global War on Terror': 115.

151. Clark, Petrodollar Warfare: 121; Fouskas & Gökay, The New American Imperialism: 27 and Simon Reeve, 'Conspiracy Theories: Iraq', Sky One, 10 January 2007, 22:00 GMT.

152. Rutledge, Addicted to Oil: 190.

153. Maya Schenwar, 'In final days, Bush pushes for Iraq's oil', Truthout, 11 November 2008 [accessed 11 August 2009] http://www.truthout. org/111108A. Also see Dana Milbank, 'Democrats play house to rally against the war', Washington Post, 17 June 2005 [accessed 2 October 2009]http://www.washingtonpost.com/wp-dyn/content/article/2005/06/16/ AR2005061601570.htm and http://warnewstoday.blogspot.com/2008/10/ war-news-for-tuesday-october-14–2008.html.

154. Juhasz, The Bush Agenda: 259.

155. Ibid: 262, 263 and Klein, The Shock Doctrine: 329.

156. Juhasz, The Bush Agenda: 269.

157. Bryce, Gusher of Lies: 118.

158. Juhasz, The Bush Agenda: 274–275. The US also signed Free Trade Agreements with key Gulf States, like Bahrain in 2004 and Oman in 2005, with negotiations over similar FTAs with UAE, Kuwait and Qatar underway. These agreements have already opened up the market for a host of foreign oil and gas companies, including ExxonMobil, Total and Royal Dutch/Shell (Morrissey, 'The Geoeconomic Pivot of the Global War on Terror': 114).

159. Robert Z. Lawrence, A US–Middle East Trade Agreement: A Circle of Opportunity? (Washington, DC: Peterson Institute for International Economics: 2006): 5–6 and Richard Seymour, 'The Real Cost of the Iraq War', The Middle East, May 2009: 50.

160. Lawrence, A US–Middle East Trade Agreement: 7.

161. See Juhasz, The Bush Agenda: 277–290.

162. Ibid: 267.

163. Ibid: 268.

164. Ibid: 276–277.

165. George W. Bush, 'President Sworn-In to Second Term', White House, 20 January 2005 [accessed 30 June 2007] http://www.whitehouse.gov/news/ releases/2005/01/20050120–1.html

166. Woodward, State of Denial: 371.

167. George W. Bush: 'State of the Union Address', White House, 2 February 2005 [accessed 30 June 2007] http://www.whitehouse.gov/news/releases/ 2005/02/20050202–11.html

168. 'Don't be fooled: Middle East democracy has only the most tenuous link with war in Iraq', Independent, 8 March 2005: 28.

169. In January 2005 Abbas won presidential elections in Palestine, and the Bush administration boasted, falsely, that this democracy was due to Bush's policies in the region. What made the elections possible was the simple fact

that Arafat, the biggest political heavyweight in the Palestinian arena, was dead, and thus the elections became necessary to elect a successor. Arafat had the habit of surrounding himself with lightweights like Abu Mazen, and it was the death of the heavyweight that allowed the United States to put the lightweight in power.

170. Fareed Zakaria, 'Don't blame the Saudis', Newsweek, 6 September 2004: 17.

171. George W. Bush, 'President Bush discusses global war on terror,' White House, 19 March 2008 [accessed 19 March 2008] http://www.whitehouse. gov/news/releases/2008/03/20080319-2.html.

172. George W. Bush, 'Full text: Bush address on Iraq', BBC News, 11 January 2007 [accessed 12 January 2007] http://news.bbc.co.uk/1/hi/world/americas/ 6250687.stm

173. Bush, 'Full text: Bush address on Iraq'; George W. Bush, 'President Bush delivers State of the Union address,' White House, 23 January 2007 [accessed 9 January 2008] http://www.whitehouse.gov/news/ releases/2007/01/20070123-2.html; 'Bush counts on allies in Iraq, Asia to stop violence, nuclear weapons', Fox News, 16 October 2006 [accessed 16 October 2006] http://www.foxnews.com/story/0,2933,221418,00. html?sPage=fnc.world/iraq.

174. 'Rice meets Iraqi Kurdish leaders', BBC News, 6 October 2006 [accessed 6 October 2006] http://news.bbc.co.uk/1/hi/world/middle_east/5412178.stm

175. Vivienne Walt, 'Energy: pump it up', Time, 7 December 2009: 51 and Steven Mufson, 'Iraq struggles to finish oil law', Washington Post, 24 January 2007 [accessed 9 January 2008] http://www.washingtonpost.com/ wp-dyn/content/article/2007/01/23/AR2007012301534.stm.

176. Michael Howard, 'The struggle for Iraq's oil flares up as Kurds open doors to foreign investors', Guardian, 7 August 2007: 17; Walt, 'Energy: pump it up': 51, 'Breakthrough in Iraq oil standoff,' BBC News, 27 February 2007 [accessed 27 February 2007] http://news.bbc.co.uk/1/hi/world/middle_east/6399257.stm; Robert Dreyfuss, 'Nationalists stirring in Iraq' The Nation, 16 January 2008 [accessed 26 January 2008] http://www.thenation.com/doc/20080128/dreyfuss; 'Iraqi cabinet backs draft oil law,' BBC News, 3 July 2007 [accessed 3 July 2007] http://news.bbc.co.uk/1/hi/world/middle_east/6264184.stm; 'Iraqi government approves draft oil law', Al Jazeera, 3 July 2007 [accessed 4 July 2007] http://www.aljazeera.net/NR/exeres/302B5BC3-1821-40F6-BDBA-9271E6506328.htm; 'US soldier and 12 Iraqis killed and oil law crisis deepens', Al Jazeera, 6 July 2007 [accessed 6 July 2007] www.aljazeera.net/NR/exeres/F1D386A-79F8-4A23-99AF-348C12045325.htm.

177. 'Iraq oil law might take months to pass – minister', Iraq Updates, 17 November 2007 [accessed 4 July 2008] http://www.iraqupdates.com/p_articles.php/article/24038, 'Iraqi oil minister dismisses near passing of oil law', Al Jazeera, 16 November 2007 [accessed 17 November 2007] http://www.aljazeera.net/NR/exeres/25A301A-AF08–4B59–9770-73451E586.html and Thomas E Ricks & Karen De Young. 'For US, the goal is now Iraqi solutions: Approach acknowledges benchmarks aren't met', Washington Post, 10 January 2008 [accessed 19 January 2008] www.washingtonpost.com/wp-dyn/content/article/2008/01/09/AR2008010903701.html?wpisrc=newsletter

178. Ed Crooks & Roula Khalaf, 'Shell in Iraq gas deal worth up to $4bn', Financial Times, 8 September 2008 [accessed 9 September 2008] http://www.ft.com/cms/s/0/e72e708a-7dec-11dd-bdbd-000077b07658.html, Sam Dagher, 'Shell opens an office in Baghdad after a 36-year absence', New York Times, 22 September 2008 [accessed 23 September 2008] http://www.nytimes.com/2008/09/23/world/middleeast/23iraq.html?_r=1&oref=slogin; 'Iraq strikes gas deal with Shell', BBC News, 23 September 2008 [accessed 23 September 2008] http://news.bbc.co.uk/1/hi/business/7631108.stm and 'Gas deal between Iraq and Shell delayed', Al Jazeera, 6 September 2009 [accessed 6 September 2009] http://www.aljazeera.net/NR/exeres/906798D6-C5D4-43A7-8268-D5BAFCA1A734.htm.

179. Williams, 'Warily moving ahead on oil contracts'; Sarah Arnott, 'The battle for Iraq's oil', Independent, 15 April 2009 [accessed 20 June 2009] http://www.independent.co.uk/news/business/analysis-and-features/the-battle-for-iraqs-oil-1668882.html and Mark Gregory, 'Firms challenge Iraqi oil auction', BBC News, 8 July 2009 [accessed 8 July 2009] http://news.bbc.co.uk/1/hi/business/8130791.stm

180. 'Iraq: Key facts and figures', BBC News, 29 January 2009 [accessed 29 January 2009] http://news.bbc.co.uk/1/hi/world/middle_east/7856618.stm

181. 'Maliki: Our over-dependence on oil exposes us to the global crisis', Al Jazeera, 25 February 2009 [accessed 25 February 2009] http://www.aljazeera.net/NR/exeres/DD743A92-FF43-4F86-991D-32D1D4F58CEF.htm.

182. 'Leading article: Iraq wrestles back to control its own destiny', Independent, 1 July 2009 [accessed 1 July 2009] http://www.independent.co.uk/opinion/leading-articles/leading-article-iraq-wrestles-back-control-of-its-own-destiny-1726048.html; Patrick Cockburn, 'Collapse in Iraqi oil price shatters hope of recovery', Independent, 20 March 2009 [accessed 20 March 2009] http://www.independent.co.uk/news/world/middle-east/collapse-in-iraqi-oil-price-shatters-hope-of-recovery-1649553.html. For more information, see Alissa J. Rubin, 'Oil revenues dropping, Iraq's parliament cuts the

budget', New York Times, 5 March 2009 [accessed 6 March 2009] http://
www.nytimes.com/2009/03/06/world/middleeast/06iraq.html?ref=world.

183. 'Al Qaeda in Iraq rises again', Al Jazeera, 23 September 2010 [accessed
14 September 2010] http://www.aljazeera.net/NR/exeres/3B855310–7711-
4152–9CAF-4654E980310A.htm?GoogleStatID=9

184. Timothy Williams & Duraid Adnan, 'Sunnis in Iraq allied with US quitting
to rejoin rebels', New York Times, 16 October 2010 [accessed 17 October
2010] http://www.nytimes.com/2010/10/17/world/middleeast/17awakening.
html?th&emc=th

185. Al Jazeera English TV, Thursday 10 December 2009.

186. Cockburn, 'Collapse in Iraqi oil price shatters hope of recovery', 'Leading
article: Iraq wrestles back to control its own destiny', 'No victory in Iraq
says Petraeus', BBC News, 11 September 2008 [accessed 11 September
2008] http://news.bbc.co.uk/1/hi/world/middle_east/7610405.stm, 'Odierno
warning on Iraq security', BBC News, 16 September 2008 [accessed 16
September 2008] http://news.bbc.co.uk/1/hi/world/middle_east/7618553.
stm; Alissa J. Rubin, 'Oil revenues dropping, Iraq's parliament cuts the
budget', New York Times, 5 March 2009 [accessed 6 March 2009] http://
www.nytimes.com/2009/03/06/world/middleeast/06iraq.html?ref=world;
Milan Vesley, 'Iraq: A success or just a political ball game?', The Middle
East, November 2008: 33.

187. Arnott, 'The battle for Iraq's oil'.

188. Ibid; Cockburn, 'Collapse in Iraqi oil price shatters hope of recovery'

189. Pamela Ann Smith, 'Kurdistan's $21-billion oil bonanza', The Middle
East, August/September 2009: 35; Cockburn, 'Collapse in Iraqi oil price
shatters hope of recovery'; 'Iraq estimates cost of developing its oilfields is
$52 billion', Al Jazeera, 1 May 2009 [accessed 1 May 2009] http://www.
aljazeera.net/NR/exeres/17356807-E540-49E0-8D2A-1345A25B00DF.
htm; Williams, 'Warily moving ahead on oil contracts'.

190. Ed Blanche, 'Bleeding Iraq Dry', The Middle East, April 2006: 18–23;
Cockburn, 'Collapse in Iraqi oil price shatters hope of recovery' and Rod
Nordland & Jad Moawwad, 'Iraq considers giving foreign oil investors better
terms', New York Times, 18 March 2009 [accessed 20 June 2009] http://www.
nytimes.com/2009/03/19/world/middleeast/19iraq.html?scp=6&sq=iraq%20
oil&st=cse

191. Walt, 'Energy: pump it up': 51; Cockburn, 'Collapse in Iraqi oil price shatters
hope of recovery', and Williams, 'Warily moving ahead on oil contracts'.

192. Richard A. Oppel, 'Iraq's insurgency run on stolen oil profits', New York
Times, 16 March 2008 [accessed 16 March 2008] http://www.nytimes.
com/2008/03/16/world/middleeast/16insurgent.html?th&emc=th

193. Daalder & Lindsay, America Unbound (2nd edn): 169.

194. Smith, 'Kurdistan's $21-billion oil bonanza': 35; Cockburn, 'Collapse in Iraqi oil price shatters hope of recovery'; Williams, 'Warily moving ahead on oil contracts'.

195. Miller, Blood Money: 101–106.

196. 'Iraq ready to review old Russian oil contracts', Al Jazeera, 12 February 2008 [accessed 9 April 2009] http://www.aljazeera.net/NR/exeres/2B0905C9-F287-4525-BB15-C4A97E2EC036.htm and 'Russia and Iraq agree to revive large oil contract', Al Jazeera, 9 April 2009 [accessed 9 April 2009] http://www.aljazeera.net/NR/exeres/78E9927E-3B13-468C-9510-00D4D19AA2B1.htm

197. Crooks & Khalaf, 'Shell in Iraq gas deal worth up to $4bn'; Dagher, 'Shell opens an office in Baghdad after a 36-year absence', 'Iraq strikes gas deal with Shell', and 'Gas deal between Iraq and Shell delayed'.

198. 'Economic agreements between Iraq and France', Al Jazeera, 3 July 2009 [accessed 3 July 2009] http://www.aljazeera.net/NR/exeres/9480872C-58ED-4F79-A4A5-C690FE361702.htm; 'Iraq invites France to invest and build a nuclear reactor', Al Jazeera, 22 February 2009 [accessed 22 February 2009] http://www.aljazeera.net/NR/exeres/09BDF5B5-AA21-43DA-A554-47F2C075F9F4.htm; 'Iraq invites France to participate in infrastructure', Al Jazeera, 2 July 2009 [accessed 3 July 2009] http://www.aljazeera.net/NR/exeres/038EC5DD-4226-47EB-B2F1-C4671E6C3EE8.htm; Williams, 'Warily moving ahead on oil contracts'.

199. Cockburn, 'Collapse in Iraqi oil price shatters hope of recovery'; Arnott, 'The battle for Iraq's oil', 'Iraq opens 11 oil and gas fields to foreign companies'; Al Jazeera, 1 January 2009 [accessed 3 January 2009] http://www.aljazeera.net/NR/exeres/7A506DFE-BF36-42D2-BE92-B0ADC81F9CC2.htm; 'Iraq sets conditions for oil contracts', Al Jazeera, 25 August 2009 [accessed 25 August 2009] http://www.aljazeera.net/NR/exeres/DC21F342-E143-45F9-92C8-6CCE43E3209C.htm and 'Oil companies reject Iraq's terms', BBC News, 30 June 2009 [accessed 30 June 2009] http://news.bbc.co.uk/1/hi/business/8125731.stm.

200. Walt, 'Energy: pump it up': 50; Patrick Cockburn, 'Bidding war for Iraq's huge oil contracts sputters into life', Independent, 1 July 2009 [accessed 1 July 2009] http://www.independent.co.uk/news/world/middle-east/bidding-war-for-iraqs-huge-oil-contracts-sputters-into-life-1726205.html; David Prosser, 'No need for Iraq to sell its future cheap', Independent, 1 July 2009 [accessed 1 July 2009] http://www.independent.co.uk/news/business/comment/david-prosser-no-need-for-iraq-to-sell-its-future-cheap-1726221.html; Matthew Weaver, 'Iraq invites oil investment but the

majors stay away', Guardian, 29 June 2008 [accessed 29 June 2008] http://www.guardian.co.uk/business/2008/jun/29/oil.oilandgascompanies1 and Williams, 'Warily moving ahead on oil contracts'.

201. Smith, 'Kurdistan's $21-billion oil bonanza': 34, 'Leading article: Iraq wrestles back to control its own destiny'; Mark Gregory, 'Firms challenge Iraqi oil auction', BBC News, 8 July 2009 [accessed 8 July 2009] http://news.bbc.co.uk/1/hi/business/8130791.stm; 'Oil companies reject Iraq's terms', and Cockburn, 'Bidding war for Iraq's huge oil contracts sputters into life'.

202. Prosser, 'No need for Iraq to sell its future cheap'.

203. Gregory, 'Firms challenge Iraqi oil auction'.

204. Smith, 'Kurdistan's $21-billion oil bonanza': 34.

205. Walt, 'Energy: pump it up': 50 and 'Exxon develops an Iraqi oilfield', Al Jazeera, 5 November 2009 [accessed 5 November 2009] www.aljazeera.net/NR/exeres/1C61E2E0–0F97-4D83-94AD-2842CFDAA2E1.htm.

206. Martin Chulov, 'Shell wins rights over vast Iraqi oilfield as foreign firms get access', Guardian, 12 December 2009: 38; Patrick Cockburn, 'Rush for Iraq's oil in defiance of bombers', Independent, 12 December 2009: 28; Carola Hoyos, 'Shell wins 'gold rush' Iraqi oilfields auction', Financial Times, 12 December 2009: 8; 'Iraq oil development rights contacts awarded', BBC News, 11 December 2009 [accessed 11 December 2009] http://news.bbc.co.uk/1/hi/business/8407274.stm; 'Lukoil wins Iraq oilfield contract', Al Jazeera English, 12 December 2009 [accessed 13 December 2009] http://english.aljazeera.net/news/middleeast/2009/12/20091212944 47518191.html, 'Oil firms awarded Iraq contracts', Al Jazeera English, 11 December 2009 [accessed 11 December 2009] http://english.aljazeera.net/news/middleeast/2009/12/200912117243440687.html.

207. Cockburn, 'Collapse in Iraqi oil price shatters hope of recovery'.

208. Arnott, 'The battle for Iraq's oil'.

209. Williams, 'Warily moving ahead on oil contracts'.

210. Smith, 'Kurdistan's $21-billion oil bonanza': 33; 'Baghdad accuses Kurds of wasting Iraq's oil', Al Jazeera, 17 May 2009 [accessed 17 May 2009] http://www.aljazeera.net/NR/exeres/69E90EBB-DF90-4EA1-B98F-3-ECCEC89DB67.htm; 'Baghdad: Kurdish oil contracts illegal', Al Jazeera, 10 June 2009 [accessed 10 June 2009] http://www.aljazeera.net/NR/exeres/A8E328D3-20CD-4574-9FFF-967D5A71792F.htm and 'Disagreement between Baghdad and Irbil over oil dossier', Al Jazeera, 12 May 2009 [accessed 12 May 2009] http://www.aljazeera.net/NR/exeres/C7EA9186-92C6-495F-8193-2155DFA26051.htm

211. 'Kurdish parliament approves regional oil and gas law', Al Jazeera, 7 August 2007 [accessed 21 June 2008] http://www.aljazeera.net/news/archive/

archive?ArchiveId=1065230www.aljazeera.net, and Howard. 'The struggle for Iraq's oil flares up as Kurds open doors to foreign investors': 17.

212. 'KRG demands Shahrestani resignation', Al Jazeera, 14 September 2007 [accessed 21 June 2008] http://www.aljazeera.net/news/archive/archive?ArchiveId=1068852

213. Alissa J. Rubin & Andrew Kramer 'Official calls Kurd oil deal at odds with Baghdad', New York Times, 28 September 2007 [accessed 29 September 2007] http://www.nytimes.com/2007/09/28/world/middleeast/28iraq.html1?th&emc=th.

214. Richard Wolfee & Gretel C. Kovach, 'Let's make an oil deal: A Bush family friend may be undermining Iraqi peace', Newsweek, 1 October 2007: 26; James Glanz, 'Panel questions State Dept. role in Iraq oil deal', New York Times, 3 July 2008 [accessed 13 July 2008] http://www.nytimes.com/2008/07/03/world/middleeast/03kurdistan.html?_r=1&scp=3&sq=U.S.%20Advised%20Iraqi%20Ministry%20on%20Oil%20Deals&st=cse&oref=login, and Steven Mufson, 'Bush officials condoned regional Iraqi oil deal: Contract contradicted State Dept.'s public stance', Washington Post, 3 July 2008 [accessed 10 July 2008] http://www.washingtonpost.com/wp-dyn/content/article/2008/07/02/AR2008070203322.html.

215. Arnott, 'The battle for Iraq's oil'; Cockburn, 'Collapse in Iraqi oil price shatters hope of recovery'; Patrick Cockburn, 'Iraq faces a new war as tensions rise in north', Independent, 23 February 2009 [accessed 23 February 2009] http://www.independent.co.uk/news/world/middle-east/iraq-faces-a-new-war-as-tensions-rise-in-north-1629343.html; Patrick Cockburn, 'Kurdish faultline threatens to spark new war', Independent, 10 August 2009 [accessed 10 August 2009] http://www.independent.co.uk/news/world/middle-east/kurdish-faultline-threatens-to-spark-new-war-1769954.html; Sam Dagher, 'Kurds defy Baghdad, laying claim to land and oil', New York Times, 9 July 2009 [accessed 9 July 2009] http://www.nytimes.com/2009/07/10/world/middleeast/10kurds.html?th&emc=th and Williams, 'Warily moving ahead on oil contracts'.

216. Ed Blanche, 'Iran tightens its grip on the 'new' Iraq', The Middle East, May 2010: 12–17 and 'Leading article: Iraq wrestles back to control its own destiny'.

217. Natalia Antelava, 'Sunnis fearful of Iraq future', BBC News, 13 May 2009 [accessed 8 December 2009] http://news.bbc.co.uk/1/hi/world/middle_east/8045696.stm; Martin Chulov, 'Iraq bombs linked to Sunni militias who fought against al-Qaida', Guardian, 7 April 2009 [accessed 8 December 2009] http://www.guardian.co.uk/world/2009/apr/07/iraq-bombs-baghdad;

Martin Chulov, 'Iraq disbands Sunni militia that helped defeat insurgents', Guardian, 2 April 2009 [accessed 8 December 2009] http://www.guardian.co.uk/world/2009/apr/02/iraq-sunni-militia-disbanded; Patrick Cockburn, 'Iraq militia fear reprisals after US exit', Independent, 13 April 2009 [accessed 8 December 2009] http://www.independent.co.uk/news/world/middle-east/iraq-militia-fear-reprisals-after-us-exit-1668029.html; 'Iraq's Awakening Councils', BBC News, 1 October 2008 [accessed 8 December 2009] http://news.bbc.co.uk/1/hi/world/middle_east/7644448.stm; Aamer Madhani, 'US withdrawal date approaches in Iraq', USA Today, 22 June 2009 [accessed 8 December 2009] http://www.usatoday.com/news/military/2009–06-21-iraqwithdraw_N.htm; Jim Muir, 'Is Iraq sliding back into chaos?', BBC News, 25 April 2009 [accessed 8 December 2009] http://news.bbc.co.uk/1/hi/world/middle_east/8018866.stm.

218. James Glanz & Steven Lee Myers, 'Assault by Iraq on Shiite forces stalls in Basra,' New York Times, 28 March 2008 [accessed 28 March 2008] http://www.nytimes.com/2008/03/28/world/middleeast/28iraq.html?th&emc=th and Michael Kamber & James Glanz, 'Iraqi crackdown on Shiite forces sets off fighting', New York Times, 26 March 2008 [accessed 26 March 2008] http://www.nytimes.com/2008/03/26/world/middleeast/26iraq.html?th&emc=th.

219. Cordula Meyer, 'A lot of blood for little oil', Der Spiegel, 9 December 2010 [accessed 9 December 2010] http://www.spiegel.de/international/world/0,1518,732984,00.html.

220. James Glanz & T. Christian Miller, 'Official history spotlights Iraqi rebuilding blunders', New York Times, 13 December 2008 [accessed 14 December 2008] http://www.nytimes.com/2008/12/14/world/middleeast/14reconstruct.html?th&emc=th and 'Report shows wasted Iraq spending', BBC News, 2 February 2009 [accessed 2 February 2009] http://news.bbc.co.uk/1/hi/world/americas/7865840.stm.

221. Meyer, 'A lot of blood for little oil'.

222. Miller, Blood Money: 162–164, 184–185.

223. 'US Bechtel wraps up Iraq projects', BBC News, 2 November 2006 [accessed 3 November 2006] http://news.bbc.co.uk/2/hi/americas/6112164.stm

224. Anthony Cordesman, 'Playing the Course: A strategy for reshaping US policy in Iraq and the Middle East', CSIS, November 2004, www.csis.org/media/csis/pubs/iraq-playingcourse.pdf and 'Iraq expert predicts uncertainty', BBC News, 14 November 2003 [accessed 3 August 2006] http://news.bbc.co.uk/1/hi/world/americas/32772333.stm

225. 'US–Iraqi contract in disarray', BBC News, 23 October 2007 [accessed 23 October 2007] http://news.bbc.co.uk/1/hi/world/Americas/7057629.stm.

226. 'US army contracting alarms panel', BBC News, 2 November 2007, http://news.bbc.co.uk/1/hi/world/Americas/7074285.stm.

227. Cordesman, 'Playing the course.'

228. As described in Woodward, State of Denial: 440.

229. Fareed Zakaria, 'Elections are not democracy', Newsweek, 7 February 2005: 13.

230. George W. Bush, 'President Bush discusses importance of freedom in the Middle East', 13 January 2008 [accessed 19 January 2008] http://www.whitehouse.gov/news/releases/2008/01/20080113–1.html

231. Paul Adams, 'Iran imperative spurs US aid move,' BBC News, 31 July 2007 [accessed 31 July 2007] http://news.bbc.co.uk/1/hi/world/middle_east/6923347.stm.

232. Klare, Blood and Oil: 2–3, 71–72, 107 and Klare, Resource Wars: 72–73.

233. Ed Blanche, 'Iran-Israel covert war', The Middle East, July 2009: 28–31; David E. Sanger, 'US rejected aid for Israeli raid on Iranian nuclear site', New York Times, 10 January 2009 [accessed 10 January 2009] http://www.nytimes.com/2009/01/11/washington/11iran.html?ref=todayspaper, and Jonathan Steele, 'Israel asked US for green light to bomb nuclear sites in Iran', Guardian, 25 September 2008 [accessed 25 September 2008] http://www.guardian.co.uk/world/2008/sep/25/iran.israelandthepalestinians1.

234. 'Fund and find your opposition' ,The Economist, 25 February 2006: 61.

235. Samantha Power, 'Rethinking Iran', Time, 28 January 2008: 27 and 'Leading article: Iraq wrestles back to control its own destiny', Independent, 1 July 2009 [accessed 1 July 2009] http://www.independent.co.uk/opinion/leading-articles/leading-article-iraq-wrestles-back-control-of-its-own-destiny-1726048.html.

236. 'American disappointment over oil deal between China and Iran', Al Jazeera, 11 December 2007 [accessed 21 June 2008] http://aljazeera.net/News/archive/archive?ArchiveId=1077712 ; 'China defends Iran gas deal talks', BBC News, 11 January 2007 [accessed 11 January 2007] http://news.bbc.co.uk/1/hi/business/6251365.stm.; 'China develops Yadavaran oilfield for $2 billion', Al Jazeera, 10 December 2007 [accessed 21 June 2008] http://aljazeera.net/News/archive/archive?ArchiveId=1077620;' China increases dependence on Iran's oil despite pressure', Al Jazeera, 14 December 2007 [accessed 21 June 2008] http://aljazeera.net/news/archive/archive?ArchiveId=1078018; 'Gas deal worth billions of dollars between Iran and Chinese consortium', Al Jazeera, 14 March 2009 [accessed 14 March 2009] http://www.aljazeera.net/NR/exeres/D85C46BD-5760–42B9-B5F2-BDD18128307D.htm; 'Iran is largest oil exporter to China',

Al Jazeera, 22 June 2009 [accessed 22 June 2009] http://www.aljazeera. net/NR/exeres/339A39B9–7396-4BD2-A67A-4914E5F8E0B4.htm

237. Ed Blanche, 'Pipeline politics', The Middle East, October 2008: 23 , Somini Sengupta & Heather Timmons, 'Iranian President's visit a test for India,' New York Times, 30 April 2008 [accessed 30 April 2008] http://www. nytimes.com/2008/04/30/world/asia/30india.html?ref=world

238. 'America calls on Turkey to seek alternatives to Iranian gas', Al Jazeera, 23 September 2007 [accessed 21 June 2008] http://aljazeera.net/news/archive/ archive?ArchiveId=1069731.) and 'Nabucco agreement signed in Turkey', Al Jazeera, 13 July 2009 [accessed 13 July 2009] http://www.aljazeera.net/ NR/exeres/0272AEF4–2E51–4030-8AD5-BB3BAECB6DF9.htm.

239. 'Malaysia stands by Iran gas deal', BBC News, 2 February 2007 [accessed 22 June 2008] http://news.bbc.co.uk/2/hi/business/6323401.stm.

240. Ed Crooks, 'Shell Iran plan will come under US scrutiny', MSNBC, 30 January 2007 [accessed 30 January 2007] http://www.msnbc.man.com/ id/16876969/wid/11915829)/ and 'Signed by Shell and Repsol for $10 bil- lion: Washington considers sanctions on gas agreements in Iran', Al Jazeera, 30 January 2007 [accessed 21 June 2008] http://www.aljazeera.net/news/ archive/archive?ArchiveId=1031268, and 'Iran reveals secret agreements to face sanctions', Al Jazeera, 24 January 2007 [accessed 21 June 2008] http:// www.aljazeera.net/news/archive/archive?ArchiveId=1030761; 'Washington opposes agreement by Norwegian company to develop gas in Iran', Al Jazeera, 24 April 2007 [accessed 21 June 2008] http://www.aljazeera.net/ news/archive/archive?ArchiveId=1042102.

241. Klare, Blood and Oil: 171; 'Glittering towers in a war zone', The Economist, 9 December 2006: 23; 'No strings', A Ravenous Dragon: A special report on China's quest for resources, The Economist, 15 March 2008: 12; 'Chinese leader boosts Sudan ties', BBC News, 2 February 2007 [accessed 2 February 2007] http://news.bbc.co.uk/1/hi/world/africa/6323017.stm; 'Jintao in Sudan to strengthen economic and oil cooperation', Al Jazeera, 2 February 2007 [accessed 21 June 2008] http://www.aljazeera.net/news/ archive/archive?ArchiveId=1031508; 'Last chance for Darfur troops', BBC News, 13 April 2007 [accessed 13 April 2007] http://news.bbc.co.uk/1/hi/ world/africa/6551405.stm; 'US to toughen sanctions on Sudan', BBC News, 29 May 2007 [accessed 29 May 2007], http://news.bbc.co.uk/1/hi/world/ africa/6699479.stm; Joseph Winter, 'Khartoum booms as Darfur burns', BBC News, 24 April 2007 [accessed 24 April 2007] http://news.bbc. co.uk/1/hi/world/africa/6573527.stm.

242. Fareed Zakaria, 'Don't blame the Saudis', Newsweek, 6 September 2004: 17.

243. Jad Mouawwad, 'China's growth shifts the geopolitics of oil', New York Times, 19 March 2010 [accessed 19 March 2010] http://www.nytimes.com/2010/03/20/business/energy-environment/20saudi.html?th&emc=th.

244. 'Decline in American oil imports', Al Jazeera, 9 April 2010 [accessed 9 April 2010] http://www.aljazeera.net/NR/exeres/33D177BA-C531-4A09-B162-DA92986F309C.htm.

245. Ed Blanche, 'Claws of the bear', The Middle East, April 2007: 6–10.

246. Eamonn Gearon, 'Red Star in the Morning, Business Warming', The Middle East, July 2006: 26 and Mouawwad, 'China's growth shifts the geopolitics of oil'.

247. Daniel Yergin, 'Energy security will be one of the main challenges of foreign policy', Cambridge Eenergy Research Associates (CERA), 19 July 2006 [accessed 10 February 2010] http://www.cera.com/aspx/cda/public1/news/articles/newsArticleDetails.aspx?CID=8248.

248. Phyllis Bennis, Before and After: US Foreign Policy and the September 11th Crisis (New York: Olive Branch Press, 2003): 197.

249. Klare, Rising Powers, Shrinking Planet: 188; Ghaida Ghantous, 'Saudi Qaeda ideologue sets rules for oil wars,' Yahoo News, 2 March 2006 [accessed 2 March 2006] http://news.yahoo.com/s/nm/20060302/ts_nm/energy_saudi_qaeda_dc.

250. 'Saudis foil air attack plotters', BBC News, 27 April 2007 [accessed 27 April 2007] http://news.bbc.co.uk/1/hi/world/middle_east/6599963.stm.

251. Ian Black, 'Frustration for Bush as pledge to Saudis fails to win oil concession,' Guardian, 17 May 2008: 28; 'Saudi Arabia: The proof of a foreign policy that has failed', Independent, 16 January 2008: 28; Michael Abramowitz, 'Oil efforts are best possible, Saudis say,' Washington Post, 17 May 2008 [accessed 17 May 2008] http://www.washingtonpost.com/wp-dyn/content/article/2008/05/16/AR2008051601111.html?wpisrc=newsletter; 'Oil edges towards $128 a barrel,' BBC News, 16 May 2008 [accessed 17 May 2008] http://news.bbc.co.uk/1/hi/business/7404856.stm; 'Saudis resist Bush oil pressure,' BBC News, 16 May 2008 [accessed 17 May 2008] http://news.bbc.co.uk/1/hi/world/middle_east/7404040.stm; Sheryl Gay Stolberg & Jad Moawwad. 'Saudis rebuff Bush, politely, on pumping more oil,' New York Times, 17 May 2008 [accessed 17 May 2008] http://www.nytimes.com/2008/05/17/world/middleeast/17prexy.html?th&emc=th.

252. Bromley, American Power: 139, 140 and 'Saudi Arabia renews refusal of foreign investment in oil exploration', Al Jazeera, 25 March 2007 [accessed 26 March 2007] http://www.aljazeera.net/NR/exeres/E5074422-8F09-49F1-BF69-E2A3A0336C59.htm.

253. 'US Energy Information Administration/Monthly Energy Review May 2010', Energy Information Administration, May 2010 [accessed 28 May 2010] http://www.eia.gov/emeu/mer/pdf/pages/sec11_2.pdf.

254. Klare Blood and Oil: 79 from the US Department of Energy, Energy Information Administration (DOE/EIA) International Energy Outlook 2001, table D6 page 240. However, in the period between 1999 and 2009, total oil production in the Persian Gulf states still averaged around 20 mbpd ('US Energy Information Administration/Monthly Energy Review May 2010', Energy Information Administration, May 2010 [accessed 28 May 2010] http://www.eia.gov/emeu/mer/pdf/pages/sec11_2.pdf)

255. In July, 2007, Washington announced a $30 billion arms deal with Saudi Arabia and other Gulf States during a visit by Rice and Gates. Obviously, a part of the weapons provided to the Gulf States would be used to protect the oil fields, pipelines and refineries. The US also granted a $30 billion military aid package to Israel and a $13 billion military aid package to Egypt (both spread over ten years). See Paul Adams, 'Iran imperative spurs US aid move,' BBC News, 31 July 2007 [accessed 31 July 2007] http:/news.bbc.co.uk/1/hi/world/middle_east/6923347.stm and Paul Reynolds. 'US launches broad strategy in Middle East,' BBC News, 1 August 2007 [accessed 1 August 2007] http://news.bbc.oc.uk/1/hi/world/middle_east/6925643.stm

256. Michael Elliott, 'Time for an honest talk', Time, 19 November 2001: 44.

257. Tony Emerson, 'The Thirst For Oil', Newsweek, 8 April 2002: 34.

258. Graham E. Fuller & Ian O. Lesser. 'Persian Gulf myths', Foreign Affairs, May/June 1997: 41.

259. 'Saudi Arabia dismisses use of oil weapon in Lebanon war', Al Jazeera, 3 August 2006 [accessed 21 June 2008] http://www.aljazeera.net/news/archive/archive?ArchiveId=334374

260. 'Oil price rises on Gaza conflict', BBC News, 5 January 2009 [accessed 5 January 2009] http://news.bbc.co.uk/1/hi/business/7811043.stm; 'Saudi Arabia rejects Iranian call to an oil cut-off to Israel-supporters', Al Jazeera, 8 January 2009 [accessed 8 January 2009] http://www.aljazeera.net/NR/exeres/DE0EF53A-75BC-4C88–8B23-CFA3A9F65D21.htm

261. Clark, Petrodollar Warfare: 148. Emphasis in original text.

262. Ibid: 148, from David E. Spiro, The Hidden Hand of American Hegemony: Petrodollar Recycling and International Markets (New York: Cornell University Press, 1999): 121. It is not clear when this CIA memo was written, but it was reviewed for declassification in May 1985 (Sprio, The Hidden Hand of American Hegemony: 121)

263. Clark, Petrodollar Warfare: 159. But it can be done as a desperate measure (ibid).

264. Williams, The Tragedy of American Diplomacy: passim.

265. Kolko, Main Currents in Modern American History: 243–244, 349.

266. G. John Ikenberry, 'America's Imperial Ambition', Foreign Affairs, September/October 2002.

267. Pinto, Political Islam and the United States: x.

268. Giacomo Luciani, 'Oil and the Political Economy in the International Relations of the Middle East', in Louise Fawcett (ed.) International Relations of the Middle East (Oxford: Oxford University Press, 2009): 99–100.

269. Mark Mazzetti, 'Spy agencies say Iraq war worsens terrorism threat', New York Times, 24 September 2006 [accessed 24 September 2009] http://www.nytimes.com/2006/09/world/middleeast/24terror.html?_r=1&scp=1&sq=national%20intelligence%20estimate%20september%20202006%20iraq%20terror&st=sce and 'US report says Iraq fuels terror', BBC News, 24 September 2006 [accessed 24 September 2009] http://news.bbc.co.uk/1/hi/world/middle_east/5375064.stm

270. Ibid: 219 and Lucas & Ryan, 'Against Everyone and No-One': 170.

Conclusion

1. Michael T. Klare, Blood and Oil (London: Hamish Hamilton, 2004):10.

2. Andrew Bacevich, American Empire: The Realities and Consequences of US Diplomacy (Massachusetts: Harvard University Press, 2002): 11–31; Andrew J. Bacevich, 'Tragedy renewed: William Appleman Williams', World Affairs, Winter 2009 [accessed 22 April 2009] http://www.worldaffairs.org/2009%20-%20Winter/full-Bacevich.html and Thomas McCormick 'What would William Appleman Williams say now?' History News Network, 24 September 2007 [accessed 22 April 2009] http://hnn.us/articles/42971.html.

3. Vassilis Fouskas & Bülent Gökay, The New American Imperialism: Bush's War on Terror and Blood for Oil (Connecticut: Praeger Security International, 2005): 28, 71, 72, 135, 136, 321; Bacevich, 'Tragedy renewed' and McCormick, 'What would William Appleman Williams say now?'

4. Bacevich, 'Tragedy renewed'.

5. Simon Bromley, American Power and the Prospects for International Order (Cambridge: Polity, 2008): 101–102.

6. Klare, Blood and Oil: 70; Jane Mayer. 'Contract Sport: What did the Vice-President do for Halliburton?', The New Yorker, 16 February 2004 [accessed 17 May 2007] http://web.lexis-nexis.com/executive/form?_index=exec_en.html&_lang=en&ut=3361601930 and Jane Mayer, 'Jane Mayer on her article in The New Yorker about Dick Cheney's relationship with Halliburton', 19 February 2004, web.lexis-nexis.com/executive/form?_index=exec_en.html&_lang=en&ut=3361601930

7. Roger Burbach & Jim Tarbell, *Imperial Overstretch: George W. Bush and the Hubris of Empire* (London: Zed Books, 2004): 15.

8. McCormick, 'What would William Appleman Williams say now?'

9. Burbach & Tarbell, *Imperial Overstretch*: 202; Bacevich, 'Tragedy renewed' and McCormick 'What would William Appleman Williams say now?'.

10. Burbach & Tarbell, *Imperial Overstretch*: 16, 17, 51, 199, 200.

11. Ibid: 173 and 200.

12. Juhasz, Antonia, *The Bush Agenda: Invading the World, One Economy at a Time* (London: Duckworth, 2006): 276.

13. Bromley, *American Power*: 142.

14. Brzezinski, *Second Chance*: 159.

15. Fareed Zakaria, 'We're fighting the wrong war', *Newsweek*, 28 January 2008: 19.

16. A US Geological Survey report published in 2005 estimated that 4.6 billion barrels of oil and 9.8 trillion cubic feet of natural gas could lie within the North Cuba Basin of the Gulf of Mexico. Canadian firm Sherritt, India's ONGC and Norway's Norsk Hydro already have investments in this region. (Laura Smith-Spark, 'Cuba oil prospects cloud US horizon', *BBC News*, 11 September 2006 [accessed 12 September 2006] http://news.bbc.co.uk/2/hi/americas/5321594.stm)

17. Burbach & Tarbell, *Imperial Overstretch*: 201.

18. Ibid: 202.

19. Bacevich, 'Tragedy renewed.'

20. Juhasz, *The Bush Agenda*: 224, 225, 226, 238, 317, 321 and Richard Seymour, 'The Real Cost of the Iraq War', *The Middle East*, May 2009: 50.

21. Stiglitz, Joseph & Linda Blimes, 'The $3 Trillion War', *Vanity Fair*, April 2008 [accessed 10 May 2010] http://www.vanityfair.com/politics/features/2008/04/stiglitz200804.

22. Martin Chulov, 'Shell wins rights over vast Iraqi oilfield as foreign firms get access', *Guardian*, 12 December 2009: 38; Patrick Cockburn, 'Rush for Iraq's oil in defiance of bombers', *Independent*, 12 December 2009: 28; Carola Hoyos, 'Shell wins 'gold rush' Iraqi oilfields auction', *Financial Times*, 12 December 2009: 8; 'Iraq oil development rights contacts awarded', *BBC News*, 11 December 2009 [accessed 11 December 2009] http://news.bbc.co.uk/1/hi/business/8407274.stm; 'Lukoil wins Iraq oilfield contract', *Al Jazeera English*, 12 December 2009 [accessed 13 December 2009] http://english.aljazeera.net/news/middleeast/2009/12/2009121294447518191.html; 'Oil firms awarded Iraq contracts', *Al Jazeera English*, 11 December 2009 [accessed 11 December 2009] http://english.aljazeera.net/news/middleeast/2009/12/200912117243440687.html.

23. Bill Richardson, 'Foreword', in Jan H. Kalicki and David L. Goldwyn (eds) *Energy and Security: Toward a New Foreign Policy Strategy* (Washington: Woodrow Wilson Centre Press, 2005): xvii.

24. James Schlesinger & John Deutch, 'The petroleum deterrence', *Newsweek Issues 2007*, December 2006–January 2007: 22.

25. George W. Bush, 'State of the Union Address by the President', *White House*, 31 January 2006 [accessed 23 December 2010] http://georgewbush-white-house.archives.gov/stateoftheunion/2006/

26. George W. Bush, 'President Bush delivers State of the Union address', *White House*, 23 January 2007 [accessed 9 January 2008] http://www.whitehouse.gov/news/releases/2007/01/20070123–2.html

27. Scheuer, *Imperial Hubris*: passim; Stiglitz, *The Three Trillion Dollar War*: 128; Mark Mazzetti, 'Spy agencies say Iraq war worsens terrorism threat', *New York Times*, 24 September 2006 [accessed 24 September 2009] http://www.nytimes.com/2006/09/world/middleeast/24terror.html?_r=1&scp=1&sq=national%20intelligence%20estimate%20september%20202006%20iraq%20terror&st=sce and 'US report says Iraq fuels terror', *BBC News*, 24 September 2006 [accessed 24 September 2009] http://news.bbc.co.uk/1/hi/world/middle_east/5375064.stm.

28. Ed Blanche, 'Pipeline politics', *The Middle East*, October 2008: 25.

29. Rashid, 'Central Asia': 118.

30. 'Ex-envoy attacks Afghan strategy', *BBC News*, 9 September 2008 [accessed 9 September 2008] http://news.bbc.co.uk/1/hi/world/south_asia/7605869.stm and Paul Reynolds, 'Countering the Taleban's 20-year war', *BBC News*, 9 September 2008 [accessed 9 September 2008] http://news.bbc.co.uk/1/hi/world/south_asia/7605888.stm

31. 'Giant Caspian oil pipeline opens', *BBC News*, 25 May 2005 [accessed 26 May 2005] http://news.bbc.co.uk/1/hi/business/4577497.stm

32. Klare, *Blood and Oil*: 130–131; Stephen Williams, 'The paradox of plenty', *The Middle East*, April 2007: 44–47 and Kieran Cooke, 'Caspian oil set for fast flow to the West,' *BBC News*, 5 May 2005 [accessed 5 May 2008] http://news.bbc.co.uk/1/hi/business/4508633.stm.

33. Blum, 'America's Caspian policy under the Bush Administration'.

34. Rutledge, *Addicted to Oil*: 119.

35. Klare, *Rising Powers, Shrinking Planet*: 144.

36. Rutledge, *Addicted to Oil*: 119, 197 and Joseph Stanislaw & Daniel Yergin, 'Oil: Reopening the door', *Foreign Affairs*, September/October 1993: 85.

37. Clark, *Petrodollar Warfare*: 60.

38. 'US Senate blocks Uzbek payment', *BBC News*, 6 October 2005 [accessed 6 October 2005]. http://news.bbc.co.uk/1/hi/world/asia-pacific/4314432.stm.

39. 'Evicted', *The Economist*, 6 August 2005: 52, 'Kyrgyz closure of US base "final"', *BBC News*, February 6, 2009 [accessed 6 February 2009] http://news.bbc.co.uk/1/hi/world/asia-pacific/7873866.stm.

40. Clifford J. Levy, 'Strategic issues, not abuses, are US focus in Kyrgyzstan', *New York Times*, 22 July 2009 [accessed 7 April 2010] http://www.nytimes.com/2009/07/23/world/asia/23kyrgyz.html?_r=1 and Michael Schwirtz & Clifford J. Levy, 'In reversal, Kyrgyzstan won't close a US base', *New York Times*, 23 June 2009 [accessed 24 June 2009] http://www.nytimes.com/2009/06/24/world/asia/24base.html

41. 'Suppression, China, oil', *The Economist* 9 July 2005: 52.

42. Dan Eggen & Helen DeYoung, 'After warnings to Moscow, US has few options', *Washington Post*, 14 August 2008 [accessed 14 August 2008] http://www.washingtonpost.com/wp-dyn/content/article/2008/08/13/AR2008081303752.html?wpisrc=newsletter; Clifford J. Levy & C.J. Chivers, 'Kremlin signs truce but resists quick pullout', *New York Times*, 17 August 2008 [accessed 17 August 2008] http://www.nytimes.com/2008/08/17/world/europe/17georgia.html?th&emc=th.

43. Ed Blanche, 'Oil's troubled waters', *The Middle East*, November 2008: 44; Richard Seymour, 'A tangled web', *The Middle East*, November 2008: 26; 'The dangers of the safe route', *The Economist*, 16 August 2008: 25; Claire Soares, 'Battle for oil: EU's hope to bypass Russian energy may be a pipe dream', *Independent*, 12 August 2008 [accessed 12 September 2009] http://www.independent.co.uk/news/world/europe/the-battle-for-oil-eursquos-hope-to-bypass-russian-energy-may-be-a-pipe-dream-891499.html; 'Oil prices rise after Turkish pipeline is bombed', Al Jazeera, 8 August 2008 [accessed 13 August 2008] http://www.aljazeera.net/NR/exeres/F4347784–182F-4C18-AC50–3BFD3250AA9B.htm, and Jad Mouawad, 'Conflict narrows oil options for West', *New York Times*, 13 August 2008 [accessed 14 August 2008] http://www.nytimes.com/2008/08/14/world/europe/14oil.html?ref=europe

44. Zbigniew Brzezinksi, 'How to avoid a New Cold War', *Time*, 18 June 2007: 28–29; Zbigniew Brzezinski, 'Staring down the Russians', *Time*, August 25, 2008: 18–19; 'The Big Chill', *The Economist*, 19 May 2007: 41–42; Ellen Barry, 'Putin calls shots to salve old wounds', *New York Times*, 11 August 2008 [accessed 11 August 2008] http://www.nytimes.com/2008/08/12/world/europe/12putin.html?ref=europe; 'Is This A New Cold War?' *Radio Free Europe, Radio Liberty*, 27 August 27, 2008 [accessed 11 September 11, 2009] http://www.rferl.org/content/Interview_New_Cold_War/1194282.html; Steven Lee Myers, 'No Cold War, but Big Chill over Georgia', *New York Times*, 15 August 2008 [accessed 15 August 2008] http://www.nytimes.

com/2008/08/16/washington/16assess.html?adxnnl=1&ref=europe&adxnnl
x=1218911202-M6zoCDO5+irDcLKYJjl1YAQ.

45. 'The dangers of the safe route': 25; Mouawad, 'Conflict narrows oil options
for West'; Soares, 'Battle for oil', and 'Oil prices rise after Turkish pipeline
is bombed'.

46. 'Balkan boost for Russian gas plan,' *BBC News*, 18 January 2008 [accessed
19 January 2008] http://news.bbc.co.uk/1/hi/world/europe/7195522.stm; Vafa
Fakhri, 'Europe's pipeline politics', *BBC News*, 7 June 2009 [accessed 7 June
2009] http://news.bbc.co.uk/1/hi/world/europe/8083511.stm and 'Nabucco gas
pipeline is approved,' *BBC News*, 27 June 2006 [accessed 28 June 2008] http://
news.bbc.co.uk/1/hi/business/5121394.stm. In 2002, the group of oil executives
from Austria, Turkey, Hungary, Bulgaria and Romania were in Vienna discussing the pipeline project. They went out to watch *Nabucco*, the Verdi opera about
the plight of the Jews expelled from Mesopotamia by King Nebuchadnezzar,
and decided to name the pipeline after the opera (Daniel Freifeld, 'The great
pipeline opera', *Foreign Policy*, September/October 2009: 123–124).

47. Freifeld, 'The great pipeline opera': 123–124; 'Dead souls', *The Economist*,
17 May 2008: 52; 'Balkan boost for Russian gas plan'; Fakhri, 'Europe's
pipeline politics'; Tristana Moore 'Gas pipeline stirs up Baltic fears', *BBC
News*, 31 December 2007 [accessed 5 May 2008] http://news.bbc.co.uk/1/
hi/business/7153924.stm; 'Nabucco gas pipeline is approved', 'Nord Stream
gas pipeline underwater construction starts', *BBC News*, 9 April 2010
[accessed 10 April 2010] http://news.bbc.co.uk/2/hi/business/8607214.stm,
and 'Russian gas pipeline to Germany faces Baltic opposition', Al Jazeera,
29 October 2007 [accessed 21 June 2008] http://www.aljazeera.net/news/
archive/archive?ArchiveId=1073179.

48. Luke Harding, 'Russia votes to allow private armies for energy giants',
Guardian, 5 July 2007: 20 and Emma Simpson, 'The changing nature of
Russia's Gazprom,' *BBC News*, 25 March 2007 [accessed 25 March 2007]
http://news.bbc.co.uk/1/hi/business/6485065.stm.

49. Luke, 'Russia votes to allow private armies for energy giants': 20 and
'Ukrainian gas row deadline looms', *BBC News*, 31 December 2008
[accessed 31 December 2008] http://news.bbc.co.uk/1/hi/business/7805770.
stm. Imports account for 61% of EU gas consumption – and 42% of those
imports come from Russia ('EU seeks to expand energy grids', *BBC News*,
13 November 2008 [accessed 13 November 2008] http://news.bbc.co.uk/1/
hi/world/europe/7727028.stm). 42% of the EU's gas imports come from
Russia, 24% from Norway, 18% from Algeria and 16% from other countries
('Russia alarmed over new EU pact', *BBC News*, 22 May 2009 [accessed 22
May 2009] http://news.bbc.co.uk/1/hi/world/europe/8061042.stm).

50. Clark, *Petrodollar Warfare*: 137; Nikolas Kozloff, *Hugo Chavez: Oil, politics and the challenge to the US* (Hampshire: Palgrave Macmillan, 2006): 25–27; John Perkins, *Confessions of an Economic Hit Man: The shocking inside story of how America really took over the world* (London: Ebury Press, 2005): 201;Rutledge, *Addicted to Oil*: 91–96; Juan Forero, 'Documents show CIA knew of a Coup Plot in Venezuela', *New York Times*, 3 December 2004 [accessed 2 March 2008] http://www.nytimes.com/2004/12/03/international/americas/03venezuela.ht ml?ex=1259816400&en=93049610b0d32146&ei=5090&partner=rssuserlan d; Greg Palast 'OPEC chief warned Chavez about coup', *Guardian*, 13 May 2002, http://www.guardian.co.uk/oil/story/0,11319,714504,00.html.

51. 'Energy focus for Chavez in China', *BBC News*, 22 August 2006 [accessed 22 August 2006] http://news.bbc.co.uk/1/hi/business/5276260.stm; 'Venezuela seeks to strengthen economic ties with China', *Al Jazeera*, 22 August 2006 [accessed 21 June 2008] http://www.aljazeera.net/news/archive/ archive?ArchiveId=336036; 'Venezuela signs agreements to supply China with oil', Al Jazeera, 27 March 2007 [accessed 21 June 2008] http://www. aljazeera.net/news/archive/archive?ArchiveId=1036886

52. William Branigin, 'Rice says Russia has taken a "dark turn"', *Washington Post*, 19 September 2008 [accessed 20 September 2008] http://www.wash-ingtonpost.com/wp-dyn/content/article/2008/09/18/AR2008091801654. html?wpisrc=newsletter; 'Chavez aims to deepen Russian ties,' *BBC News*, 29 June 2007 [accessed 29 June 2007] http://news.bbc.co.uk/1/hi/world/ Americas/6252370.stm; 'Russia and Venezuela boost ties', *BBC News*, 26 September 2008 [accessed 26 September 2008] http://news.bbc.co.uk/1/hi/ world/europe/7636989.stm; 'Russian navy sails to Venezuela', *BBC News*, 22 September 2008 [accessed 22 September 2008] http://news.bbc.co.uk/1/ hi/world/americas/7628899.stm; 'Venezuela to get Russian missiles', *BBC News*, 12 September 2009 [accessed 12 September 2009] http://news.bbc. co.uk/1/hi/world/americas/8251969.stm.

53. 'Chavez in talks with Iranian ally', *BBC News*, 1 July 2007 [accessed 1 July 2007] http://news.bbc.co.uk/1/hi/world/Americas/6258924.stm; 'Chavez invites Iranian investment', *BBC News*, 30 July 2006 [accessed 30 July 2006] http://news.bbc.co.uk/1/hi/world/middle_east/5226290.stm; 'Iran, Venezuela swap $760 million energy investments', *Reuters*, 7 September 2009 [accessed 23 December 2009] http://www.reuters.com/article/ idUSTRE5862820090907.

54. 'Chavez proposes oil barter scheme', *BBC News*, 22 December 2007 [accessed 22 December 2007] http://news.bbc.co.uk/1/hi/world/Americas/7155706. stm; 'Pipeline to deepen Andean links', *BBC News*, 13 October 2007 [accessed 14 October 2007] http://news.bbc.co.uk/1/hi/world/Americas/7042878.stm;

Daniel Schweimler, 'Investment and insults mark Chavez tour,' *BBC News*, 10 August 2007 [accessed 10 August 2007] http://news.bbc.co.uk/1/hi/world/Americas/6940274.stm; Michael Shifter, 'In search of Hugo Chavez', *Foreign Affairs*, May/June 2006 [accessed 30 July 2006] http://www.foreignaffairs.com/20605011faessay85303/michael-shifter/in-search-of-hugo-ch-vez.html; Michael Voss, 'Cuba and Venezuela: oil and politics,' *BBC News*, 22 December 2007 [accessed 22 December 2007] http://news.bbc.co.uk/1/hi/world/americas/7157113.stm.

55. 'Chavez demands foreign energy companies to abide by nationalization law', Al Jazeera, 2 February 2007 [accessed 21 June 2008] http://www.aljazeera.net/news/archive/archive?ArchiveId=1031487 and 'Two American firms withdraw from Orinoco', Al Jazeera, 27 June 2007 [accessed 21 June 2008] http://www.aljazeera.net/news/archive/archive?ArchiveId=1061853.

56. 'Nervous energy', *The Economist*, 7 January 2006: 65 and Shifter, 'In search of Hugo Chavez'.

57. Shifter, 'In search of Hugo Chavez' and 'Chavez: Washington is free to stop purchasing oil from Caracas', Al Jazeera, 15 February 2007 [accessed 21 June 2008] http://www.aljazeera.net/news/archive/archive?ArchiveId=1032722

58. Simon Romero, 'US says it will oust Venezuela envoy, and names 2 officials as rebel backers', *New York Times*, 12 September 2008 [accessed 13 September 2008] http://www.nytimes.com/2008/09/13/world/americas/13venez.html?th&emc=th and 'Venezuela renews threat to stop oil shipments to the United States', Al Jazeera, 12 September 2008 [accessed 12 September 2008] http://www.aljazeera.net/NR/exeres/0E86CD52–7871-4CF9-A47F-DD5EB8EA0D06.htm.

59. Clark, *Petrodollar Warfare*: 79, and Tim Parsons, 'What is Peak Oil', Johns Hopkins University, 29 December 2008 [accessed 8 September 2009] http://www.jhsph.edu/publichealthnews/articles/2008/schwartz_peak_oil_background.html. The article can also be found on *Energy Bulletin*, http://www.energybulletin.net/node/47670. In fact, the Ghawar oilfield in Saudi Arabia, the largest oilfield in the world, has an EROEI of 100, which is 'extraordinarily high' (James Buckley, 'Energy in, value out', *Arabian Business.com*, 2 June 2007 [accessed 8 September 2009] http://www.arabianbusiness.com/13766)

60. Nathanial Gronewold, 'Tar sands' climate threat, security promise both exaggerated – report', *New York Times*, 22 May 2009 [accessed 8 September 2009] http://www.nytimes.com/gwire/2009/05/22/22greenwire-tar-sands-climate-threat-security-promise-both-12208.html and Michael A. Levy, 'The Canadian Oil Sands: Energy Security vs. Climate Change', *Council on Foreign Relations*, May 2009 [accessed 8 September 2009] http://www.cfr.org/content/publications/attachments/Oil_Sands_CSR47.pdf: 20.

61. Abraham Lustgarten, 'The dark magic of oil sands', *Fortune*, 10 October, 2005: 38; 'Upgraded', *The Economist*, 5 September 2009: 72; Joel Brinkley, 'Canada's smiles for camera mask chill in ties with US', *New York Times*, 25 October 2005 [accessed 1 September 2009] http://www.nytimes.com/2005/10/25/international/americas/25diplo.html?ex=1287892800&en=079309bf1711bf39&ei=5088&partner=rssnyt&emc=rss; 'China invests in Canada oil sands', *BBC News*, 1 September 2009 [accessed 1 September 2009] http://news.bbc.co.uk/1/hi/business/8231006.stm; 'Enbridge and PetroChina sign Gateway Pipeline cooperation project', 14 April 2005 [accessed 4 September 2009] http://findarticles.com/p/articles/mi_m0EIN/is_2005_April_14/ai_n13609601/; Scott Haggett, 'UPDATE 4-PetroChina takes C$1.9 bln stake in Canada oil sands', *Reuters*, 31 August 2009 [accessed 5 September 2009] http://www.reuters.com/article/companyNews/idUKN3123351120090831?sp=true.

62. Clark, *Petrodollar Warfare*: 82 and Klare, *Rising Powers, Shrinking Planet*: 14, 41.

63. Stephen Williams, 'Coming in from the Cold', *The Middle East*, July 2009: 38; Glenn Kessler, 'Rice and Gaddafi hammer at wall built by decades of animosity', *Washington Post*, 6 September 2008 [accessed 7 September 2008] http://www.washingtonpost.com/wp-dyn/content/article/2008/09/05/AR2008090501149.html?wpisrc=newsletter; Glenn Kessler, 'Rice makes historic visit to Libya', *Washington Post*, 5 September 2008 [accessed 5 September 2008] http://www.washingtonpost.com/wp-dyn/content/article/2008/09/05/AR2008090501149.html?wpisrc=newsletter; Tom Pfeiffer, 'Libya daunting for US firms despite better ties', *Reuters*, 3 September 2008 [accessed 4 September 2008] http://www.reuters.com/article/reutersEdge/idUSLT13607620080903?sp=true; 'Gaddafi to hand out oil money', *BBC News*, 1 September 2008 [accessed 1 September 2008] http://news.bbc.co.uk/1/hi/world/africa/7591458.stm, and 'Rice excited ahead of historic Libya visit', *Independent*, 5 September 2008 [accessed 5 September 2008] http://www.independent.co.uk/news/world/africa/rice-set-for-historic-libya-visit-919929.html.

64. Neil Ford, 'US firms back in Libya,' *The Middle East*, February 2006: 36–37; Josh Martin, 'Uncle Sam goes to Libya,' *The Middle East*, April 2006: 30–34, and 'Major oil firms compete for Libya concessions', Al Jazeera, 5 March 2007 [accessed 21 June 2008] http://www.aljazeera.net/news/archive/archive?ArchiveId=1034634. The application of the Iran Libya Sanctions Act (ILSA) to Libya was terminated on 23 April 2004, when Bush decided that Libya had fulfilled the requirements of all UN resolutions on Pan Am 103. Therefore, the act is now known as the Iran Sanctions Act (ISA). See

Kenneth Katzman, 'The Iran Sanctions Act (ISA)', *Foundation of American Scientists*, 12 October 2007 [accessed 3 July 2008] http://www.fas.org/sgp/crs/row/RS20871.pdf, and Kenneth Katzman, 'The Iran-Libya Sanctions Act (ILSA)', *The US Department of State, Foreign Press Centres*, 26 April 2006 [accessed 3 July 2008] http://fpc.state.gov/documents/organization/66441.pdf).

65. G. John Ikenberry, *Liberal Order and Imperial Ambition* (Cambridge: Polity, 2006): 237.

66. Tisdall, Simon, 'Africa united in rejected US request for military HQ', *Guardian*, 26 June 2007: 24; 'Americans go a-wooing', *The Economist*, 12 April 2008: 62; Adam Mynott, 'US Africa command battles sceptics', *BBC News*, 1 October 2008 [accessed 1 October 2008] http://news.bbc.co.uk/1/hi/world/africa/7644994.stm and 'US Africom has "no hidden agenda" ', *BBC News*, 1 October 2008 [accessed 1 October 2008] http://news.bbc.co.uk/1/hi/world/africa/7645714.stm; Rob Watson, 'Pentagon launches Africa command,' *BBC News*, 30 September 2007 [accessed 30 September 2007] http://news.bbc.co.uk/1/hi/world/Americas/7021379.stm.

67. 'US to get Africa command centre,' *BBC News*, 6 February 2007 [accessed 6 February 2007] http://news.bbc.co.uk/1/hi/world/Americas/6336063.stm and Karen DeYoung, 'US Africa command trims its aspirations', *Washington Post*, 1 June 2008 [accessed 2 June 2008] http://www.washingtonpost.com/wp-dyn/content/article/2008/05/31/AR2008053102055.html?wpisrc=newsletter

68. 'Americans go a-wooing': 62 and 'Policing the under-governed spaces', *The Economist*, 16 June 2007: 61.

69. Watson, 'Pentagon launches Africa command'.

70. Tisdall, Simon, 'Africa united in rejected US request for military HQ', *Guardian*, 26 June 2007: 24, and 'Americans go a-wooing': 62.

71. DeYoung, 'US Africa command trims its aspirations', and Watson, 'Pentagon launches Africa command.'

72. Nicholas Shaxson, *Poisoned Wells: The Dirty Politics of African Oil* (New York: Palgrave Macmillan, 2007): 2.

73. US Deputy Assistant Energy Secretary John R. Brodman testified before Congress in July 2004 that 'it is unlikely that Africa or West Africa will ever take the place of the Middle East in its importance to the world's oil and gas market, but it will nevertheless continue to be an important source of additional supplies to the United States and the world market' (Klare, *Rising Powers, Shrinking Planet*: 148).

74. Rutledge, *Addicted to Oil*: 4–6.

75. Klare, *Blood and Oil*: 126.

76. Ibid: 132.

77. Ibid: 145 and Rutledge, *Addicted to Oil*: 197.

78. Dick Cheney, 'Full text of Dick Cheney's speech at the IP Autumn lunch', *London Institute of Petroleum*, Autumn 1999 [accessed 15 October 2009] http://web.archive.org/web/20000414054656/http://www.petroleum.co.uk/speeches.htm

79. Rutledge, *Addicted to Oil*: 102, 119. She said that as an argument to invade Iraq, adding that 'one of the best things for our supply security would be to liberate Iraq' (ibid).

80. Megan K. Stack & Borzou Daragahi, 'Nations with vast oil wealth gaining clout', *Los Angeles Times*, 17 July 2009 [accessed 17 July 2009] http://www.latimes.com/news/printedition/front/la-fg-oil17-2008jul17,0,6710073.story

81. Carl Hulse, 'House to rethink drilling, Pelosi says', *New York Times*, 16 August 2008 [accessed 17 August 2008] http://www.nytimes.com/2008/08/17/washington/17pelosi.html?th&emc=th

82. Stephen Mihm, 'Dr. Doom', *New York Times*, 15 August 2008 [accessed 17 August 2008] http://www.nytimes.com/2008/08/17/magazine/17pessimist-t.html?ei=5124&en=562d8083f648f9eb&ex=1376539200&partner=facebook&exprod=facebook&pagewanted=all.

83. Johnson, *The Sorrows of Empire*: 257.

84. Jonathan Clarke, 'Viewpoint: The end of the neocons?', *BBC News*, 13 January 2009 [accessed 13 January 2009] http://news.bbc.co.uk/1/hi/world/americas/7825039.stm

85. John Gray, 'A shattering moment in America's fall from power', *The Observer*, 28 September 2008 [accessed 7 October 2008] http://www.guardian.co.uk/commentisfree/2008/sep/28/usforeignpolicy.useconomicgrowth

86. Gray, 'A shattering moment in America's fall from power'.

87. Joseph Stiglitz & Linda Blimes, *The Three Trillion Dollar War: The True Cost of the Iraq Conflict*. (London: Allen Lane, 2008): 114–131 and passim; Blimes & Stiglitz, 'Is this any way to rebuild Iraq?', *Los Angeles Times*, 15 August 2008 [accessed 10 May 2010] http://www.latimes.com/news/opinion/la-oe-bilmes15-2008aug15,0,4432303.story; Blimes & Stiglitz, 'The Iraq war will cost us $3 trillion, and much more', *Washington Post*, 9 March 2008 [accessed 10 May 2010] http://www.washingtonpost.com/wp-dyn/content/article/2008/03/07/AR2008030702846.html; Blimes & Stiglitz, 'The US in Iraq: An economics lesson', *Los Angeles Times*, 2 July 2009 [accessed 10 May 2010] http://articles.latimes.com/2009/jul/02/opinion/oe-bilmes2; Blimes & Stiglitz, 'War's price tag', *Los Angeles Times*, 16 March 2008 [accessed 10 May 2010] http://articles.latimes.com/2008/mar/16/opinion/op-bilmes16; Stiglitz & Blimes, '$3 trillion may be too low', *Guardian*, 6 April 2008 [accessed 10 May 2010] http://www.guardian.co.uk/commentisfree/2008/

apr/06/3trillionmaybetoolow; Stiglitz & Blimes, 'Ask a question: The Three Trillion Dollar War', *McClatchy*, 15 April 2008 [accessed 10 May 2010] http://www.mcclatchydc.com/qna/forum/three_trillion_dollar_war/index.html; Joseph Stiglitz, 'It's not the economy, stupid', *Financial Times*, 28 September 2004 [accessed 10 May 2010] http://www2.gsb.columbia.edu/faculty/jstiglitz/download/opeds/Economy_Stupid.pdf; Joseph Stiglitz, 'The Three Trillion Dollar War', *Project Syndicate*, 7 March 2008 [accessed 10 May 2010] http://www.project-syndicate.org/commentary/stiglitz97/English; Stiglitz & Blimes, 'The Three Trillion Dollar War', *Times Online*, 23 February 2008 [accessed 10 May 2010] http://www.timesonline.co.uk/tol/comment/columnists/guest_contributors/article3419840.ece; Stiglitz & Blimes, 'The $3 Trillion War', *Vanity Fair*, April 2008 [accessed 10 May 2010] http://www.vanityfair.com/politics/features/2008/04/stiglitz200804; Joseph Stiglitz, 'War at any cost? The total economic costs of the war beyond the federal budget: Testimony before the Joint Economic Committee', *Columbia University*, 28 February 2008 [accessed 10 May 2010] http://www2.gsb.columbia.edu/faculty/jstiglitz/download/papers/Stiglitz_testimony.pdf; Peter Wilson, 'Stiglitz: Iraq war "caused slowdown in the US"', *The Australian*, 28 February 2008 [accessed 10 May 2010] http://www.theaustralian.com.au/news/iraq-war-caused-slowdown-in-the-us/story-e6frg6tf-1111115661208.

88. 'Annual US imports of crude oil and petroleum products', Energy Information Administration, 29 July 2010 [accessed 24 December 2010] http://tonto.eia.doe.gov/dnav/pet/hist/LeafHandler.ashx?n=pet&s=mttimus2&f=a

89. Joseph Stiglitz, *Freefall: Free Markets and the Sinking of the Global Economy* (London: Allen Lane, 2010): 4.

90. Mamdouh Salameh, 'Oil prices forecasted to rise again – 08 September 2008', *Al Jazeera English*, 9 September 2008 [accessed 10 September 2008] http://www.youtube.com/watch?v=G4B0YwGBqoc.

91. Loretta Napoleoni, *Terrorism and the Economy: How the War on Terror is bankrupting the world* (New York: Seven Stories Press, 2010): passim.

92. James D. Hamilton, 'Oil prices and the economic recession of 2007–08', *Vox*, 16 June 2009 [accessed 20 May 2010] http://voxeu.org/index.php?q=node/3664.

93. Nouriel Roubini & Stephen Mihm, *Crisis Economics: A Crash Course in the Future of Finance* (London: Allen Lane, 2010): 123.

94. *Newsweek*, 6 March 2006: 5. The American economy is no longer the strong engine of global economic growth that it used to be, as India and China are now able to sustain their high growth rates based on their strong domestic demand for goods and services, even if the US economy is going through difficult times (Daniel Gross, 'Why it's worse than you think', *Newsweek*,

16 June 2008: 25). Jeffrey Garten of the Yale School of Management agrees that economic recovery in the US cannot be strong enough to be the locomotive for global recovery, and that if any country can accomplish that, it would have to be China where growth is more than 7% (Jeffrey E. Garten, 'America still rules: Why the United States will come out of the crisis on top', *Newsweek*, 3 August 2009: 28.) Indeed, the crisis has reinforced the shift of economic power from east to west, as Asia's emerging economies lead the global economic recovery since they grew at an average of 10% in 2009 while the US GDP fell by 1% ('On the rebound', *The Economist*, 15 August 2009: 57, 59, and Jonathan Lynn & Kazunori Takada, 'World trade to shrink 10 percent, Asia leads recovery: WTO', *Reuters*, 22 July 2009 [accessed 17 August 2009] http://www.reuters.com/article/businessNews/idUSSP481137 20090722?feedType=RSS&feedName=businessNews&sp=true)

95. Paul Kennedy, 'Georgia is important. But what it tells us about global politics is far more so', *Guardian*, 16 August 2008: 34.

96. Nouriel Roubini, 'The decline of the American Empire', *RGE Monitor*, 13 August 2008 [accessed 17 August 2009] http://www.rgemonitor.com/roubini-monitor/253323/the_decline_of_the_american_empire. The article could also be found on *Fabius Maximus* http://fabiusmaximus.wordpress.com/2008/08/18/roubini/. Also see Stephen Mihm, 'Dr. Doom'.

97. Roubini, 'The decline of the American Empire'. Actually, it was Larry Summers who first used the term 'balance of financial terror' to describe China's holdings of US dollar reserves (Lawrence Summers, 'The United States and the global adjustment process', *Peterson Institute for International Economics*, 23 March 2004 [accessed 28 October 2008] http://www.iie.com/publications/papers/paper.cfm?ReserachID=200). Also see James Miles, 'A wary respect: A special report on China and America', *The Economist*, 24 October 2009: 3, 5; 'The odd couple', *The Economist*, 24 October 2009: 13; Jeffrey E. Garten, 'America still rules': 28–29; Rana Foroohar, 'A new age of global capitalism starts now', *Newsweek*, 13 October 2008: 25; 'America seeks Chinese cooperation to solve financial crisis', Al Jazeera, 1 June 2009 [accessed 1 June 2009] http://www.aljazeera.net/NR/exeres/7ECEBED1–8E5D-4DED-8505–3BC3A74452F4.htm; Michael Wines & Keith Bradsher, 'China's leader says he is 'worried' over US treasuries', *New York Times*, 13 March 2009 [accessed 14 March 2009] http://www.nytimes.com/2009/03/14/world/asia/14china.html?_r=1&hp.

98. George Soros, *The Crash of 2008 and What it Means: The New Paradigm for Financial Markets* (Chatham: PublicAffairs, 2009): 126.

99. Paul Kennedy, 'Georgia is important. But what it tells us about global politics is far more so', *Guardian*, 16 August 2008: 34.

100. Richard Haass, 'The world that awaits', *Newsweek*, 3 November 2008: 36–38.

101. Charles Kupchan, *The End of the American Era: US Foreign Policy and the Geopolitics of the Twenty-First Century* (New York: Vintage Books, 2003): xx.

102. Joby Warrick & Walter Pincus, 'Reduced dominance is predicted for US', *Washington Post*, 10 September 2008 [accessed 10 September 2008] http://www.washingtonpost.com/wp-dyn/content/article/2008/09/09/AR2008090903302.html?hpid=sec-nation

103. 'Global Trends 2025: A Transformed World', *National Intelligence Council*, *BBC News*, 21 November 2008 [accessed 21 November 2008] http://news.bbc.co.uk/1/shared/bsp/hi/pdfs/21_11_08_2025_Global_Trends_Final_Report.pdf: 93.

104. 'Global Trends 2025': 2.

105. Ibid: iv.

106. Ibid: viii. Emphasis in original text.

107. Warrick and Pincus, 'Reduced dominance is predicted for US'.

108. 'Global Trends 2025': 66.

109. Rana Foroohar, 'The coming energy wars', *Newsweek*, 9 June 2008: 28.

110. 'Global Trends 2025': 41.

111. Bryce, *Gusher of Lies*: 86, 105; Clarke, *Empires of Oil*: 50, 51, 56, 309 and passim, and Klare, *Rising Powers, Shrinking Planet*: 17.

112. Klare, *Rising Powers, Shrinking Planet*: 37; '2020 vision',*The Economist*, 12 December 2009: 81; Matt Ford, 'Scaling the peak', *The Big Issue*, 15–21September 2008: 10; Salameh, 'Oil prices forecasted to rise again', and 'Warning: Oil supplies are running out fast', *Independent*, 3 August 2009 [accessed 4 August 2009] http://www.independent.co.uk/news/science/warning-oil-supplies-are-running-out-fast-1766585.html.

113. Klare, *Rising Powers, Shrinking Planet*: 13, 34–35, 37, 39, 40 and passim, and Michael Klare, 'Entering the era of tough oil', *Znet*, 17 August 2007 [accessed 26 December 2010] http://www.zcommunications.org/entering-the-tough-oil-era-by-michael-t-klare

114. Bryce, *Gusher of Lies*: 37, 109–110 and passim.

115. Ivo Bozon, 'McKinsey conversations with global leaders: Jeroen van der Veer of Shell', McKinsey Quarterly, July 2009 [accessed 26 December 2010] http://www.mckinseyquarterly.com/McKinsey_conversations_with_global_leaders_Jeroen_van_der_Veer_of_Shell_2410 and Lee Hudson Teslik, 'Royal Dutch Shell CEO on the end of "easy oil" ', Council on Foreign Relations, 7 April 2008 [accessed 26 December 2010] http://www.cfr.org/publication/15923/royal_dutch_shell_ceo_on_the_end_of_easy_oil.html.

116. Klare, 'Energy Security': 488–496.

117. Klare, *Rising Powers, Shrinking Plane.*.

118. Roubini, 'The decline of the American Empire'. Jeffrey Garten of the Yale School of Management argues that even if the US does not lead the global economic recovery by being the global engine of growth, Washington can still lead the global recovery using its political power, since 'Washington acted quickly and decisively to save the global economy from collapse ... without having to do all the heavy lifting economically' (Garten, 'America still rules': 28–31).

119. America's obvious successor is China, which surpassed Japan as the world's second largest economy in the second quarter of 2010 (Pei, 'Asia's rise': 35; Minxin Pei, 'Why China won't rule the world', *Newsweek Issues 2010*: 17 and 'China second global economy, but ... ', Al Jazeera, 17 August 2010 [accessed 18 August 2010] http://www.aljazeera.net/NR/exeres/883AB789–328A-4F23-AAA7-E19DEA47C02B.htm). China has the world's largest foreign currency reserve, estimated at $2 trillion (Wines & Bradsher, 'China's leader says he is 'worried' over US treasuries'). But Beijing is not ready yet to assume the position of global leadership. Furthermore, China still has no option but to continue buying US treasury bonds in large amounts, or else the dollar would fall and the value of its existing treasury bonds would decline (Martin Jacques, 'No one rules the world', *New Statesman*, 30 March 2009: 23).

120. Juhasz, *The Bush Agenda*: 310.

121. 'Empire – Running on empty?' *Al Jazeera English*, 1 July 2010 [accessed 20 December 2010] http://www.youtube.com/watch?v=JlE-i4QzT04

122. Catherine Lutz, 'Obama's Empire', *New Statesman*, 3 August 2009: 22–27 and Simon Reid-Henry, 'Coming soon to Obama's backyard', *New Statesman*, 3 August 2009: 31.

123. Christopher Dickey, 'The Empire Burden', *Newsweek*, 22 June 2009: 48.

124. Richard Haass, 'No Exit', *Newsweek*, 14 December 2009: 57.

125. For more information on America's military and energy interests in Africa during the Obama Administration, see Scott Baldauf, 'Clinton manoeuvres testy times for US, Africa', *Christian Science Monitor*, 5 August 2009 [accessed 26 August 2009] http://www.csmonitor.com/2009/0805/p06s01-woaf.html; Richard Dowden, 'Fine words, but will Africa listen?', *Independent*, 13 July 2009 [accessed 13 July 2009] http://www.independent.co.uk/opinion/commentators/richard-dowden-fine-words-but-will-africa-listen-1743371.html; Richard Dowden, 'We want him to rule all African countries', *Independent*, 11 July 2009 [accessed 11 July 2009] http://www.independent.co.uk/news/world/africa/we-want-him-to-rule-all-african-countries-1741907.html;

Jeffrey Gettleman, Howard LaFranchi, 'What Clinton seeks to achieve in Africa', *Christian Science Monitor*, 4 August 2009 [accessed 26 August 2009] http://www.csmonitor.com/2009/0805/p02s01-usfp.html;   Gerlad LeMelle, 'Straight talk: revealing the real US-Africa policy', *Foreign Policy in Focus*, http://www.fpif.org/fpiftxt/6236; Jeff Mason & Matt Spetalnick, 'Obama meets Ghana's leader to stress governance', *Reuters*, 11 July 2009 [accessed 11 July 2009] http://www.reuters.com/article/homepageCrisis/idUSLB215447._CH_.2400); 'Preaching reform, Clinton wins African hearts', *New York Times*, 13 August 2009 [accessed 26 August 2009] http://www.nytimes.com/aponline/2009/08/13/world/AP-AF-Clinton-in-Africa.html?scp=10&sq=hillary%20clinton%20africa%20oil&st=cse;   Tracey D. Samuelson, 'A stop-by-stop account of Clinton's Africa trip', *Christian Science Monitor*, 15 August 2009 [accessed 26 August 2009] http://www.csmonitor.com/2009/0815/p02s10-usfp.html; Emira Woods, 'Obama visits Africa's "oil gulf"', *Foreign Policy in Focus*, 13 July 2009 [accessed 14 August 2009] http://www.fpif.org/fpiftxt/6252.

126. Gerlad LeMelle, 'Straight talk: revealing the real US–Africa policy', *Foreign Policy in Focus*, http://www.fpif.org/fpiftxt/6236

127. 'Colombia – US base accord reached', *BBC News*, 15 August 2009 [accessed 15 August 2009] http://news.bbc.co.uk/1/hi/world/americas/8202724.stm and 'Chavez threatens to cut ties with Colombia', Al Jazeera, 26 August 2009 [accessed 26 August 2009] http://www.aljazeera.net/NR/exeres/B037A2F4-C583-4AA3-B34E-FA7970D233FB.htm.

128. 'The Obama-Biden Plan', *The Obama-Biden Transition Project* [accessed 27 December 2010] http://change.gov/agenda/energy_and_environment_agenda/; Perry Bacon, Jr. & Michael D. Shear, 'Obama urges opening up oil reserves', *Washington Post*, 5 August 2008 [accessed 5 August 2008] http://www.washingtonpost.com/wp-dyn/content/article/2008/08/04/AR2008080400477.html?wpisrc=newsletter.

129. Michael Abramowitz & Juliet Eilperin, 'Bush calls for offshore oil drilling', *Washington Post*, 19 June 2008 [accessed 19 June 2008] http://www.washingtonpost.com/wp-dyn/content/article/2008/06/18/AR2008061800312.html?wpisrc=newsletter; Bacon & Shear, 'Obama urges opening up oil reserves'; 'Bush lifts offshore drilling ban', *BBC News*, 14 July 2008 [accessed 29 August 2009] http://news.bbc.co.uk/1/hi/business/7506346.stm; 'House backs fresh US oil drilling', *BBC News*, 17 September 2008 [accessed 29 August 2009] http://news.bbc.co.uk/1/hi/business/7620282.stm; 'Obama urges opening oil reserves', *BBC News*, 5 August 2008 [accessed 29 August 2009] http://news.bbc.co.uk/1/hi/world/americas/7541291.stm; Sheryl Gay Stolberg, 'Will $4 gasoline trump a 27-year-old ban?' *New York Times*, 19

June 2008 [accessed 19 June 2008] http://www.nytimes.com/2008/06/19/washington/19energy.html?th&emc=th; Jonathan Weisman, 'Obama says energy compromise is necessary', *Washington Post*, 3 August 2008 [accessed 4 August 2008] http://www.washingtonpost.com/wp-dyn/content/article/2008/08/02/AR2008080201538.html?wpisrc=newsletter.

130. Barack Obama, 'Remarks by the President on jobs, energy independence, and climate change,' *White House*, 26 January 2009 [accessed 24 August 2009] http://www.whitehouse.gov/blog_post/Fromperiltoprogress/

131. Obama, 'Remarks by the President on jobs, energy independence, and climate change'.

132. Matthew E. Kahn, 'The green economy', *Foreign Policy*, May June 2009: 34 and Andrew C. Revkin, 'Obama: Climate plan firm amid economic woes', *New York Times*, 18 November 2008 [accessed 26 August 2009] http://dotearth.blogs.nytimes.com/2008/11/18/obama-climate-message-amid-economic-woes/?scp=1&sq=it%20will%20also%20help%20us%20transform%20our%20industries%20generating%205%20million%20new%20green%20jobs&st=cse.

133. These include former Senate majority leader and co-chairman of the Obama campaign Tom Daschle who serves on the boards of three ethanol companies and works at a Washington law firm where, according to his online job description, 'he spends a substantial amount of time providing strategic and policy advice to clients in renewable energy' (Larry Rohter, 'Obama camp closely linked with ethanol', *New York Times*, 23 June 2008 [accessed 23 June 2008] http://www.nytimes.com/2008/06/23/us/politics/23ethanol.html?th&emc=th)

134. Obama, 'Remarks by the President on jobs, energy independence and climate change'; Brian Merchant, '$60 billion for green in the stimulus plan: Where the money will go', *Treehugger*, 16 February 2009 [accessed 26 August 2009] http://www.treehugger.com/files/2009/02/green-stimulus-bill-60-billion.php and Nancy Pelosi, 'American Recovery and Reinvestment Act', *US Congress* [accessed 26 August 2009] http://www.speaker.gov/newsroom/legislation?id=0273

135. Barack Obama, 'A historic energy bill', *White House*, 26 June 2009 [accessed 18 September 2009] http://www.whitehouse.gov/blog/A-Historic-Energy-Bill/; John M. Broder, 'House passes bill to address threat of climate change', *New York Times*, 26 June 2009 [accessed 26 June 2009] http://www.nytimes.com/2009/06/27/us/politics/27climate.html?th&emc=th; John M. Broder, 'Obama opposes trade sanctions in climate bill', *New York Times*, 28 June 2009 [accessed 28 June 2009] http://www.nytimes.com/2009/06/29/us/politics/29climate.html?th&emc=th;

Steven Mufson, 'Obama praises climate bill's progress but opposes its tariffs', *Washington Post*, 29 June 2009 [accessed 29 June 2009] http://www.washingtonpost.com/wp-dyn/content/article/2009/06/28/AR2009062801229.html?wpisrc=newsletter and Steven Mufson, David A. Fahrenthold & Paul Kane, 'In close vote, House passes climate bill', *Washington Post*, 27 June 2009 [accessed 27 June 2009] http://www.washingtonpost.com/wp-dyn/content/article/2009/06/26/AR2009062600444.html?wpisrc=newsletter.

136. George W. Bush, 'President delivers State of the Union', *White House*, 28 January 2003 [accessed 26 June 2006] www.whitehouse.gov/news/releases/2003/01/20030128-19.html.

137. Barack Obama, 'The American Promise: Acceptance Speech at the Democratic Convention', *Obama Speeches*, 28 August 008 [accessed 2 July 2010] http://obamaspeeches.com/E10-Barack-Obama-The-American-Promise-Acceptance-Speech-at-the-Democratic-Convention-Mile-High-Stadium--Denver-Colorado-August-28-2008.htm.

138. Matthew E, Kahn, 'The green economy', *Foreign Policy*, May/June 2009: 34 and Gross, 'The recession is over': 35, 37. Myron Ebell of the Competitive Enterprise Institute, a Washington group that opposes state mandates requiring that a certain percentage of power come from renewable sources, argues that creating green jobs often does not create jobs on a net basis. 'If you create jobs in wind power or ethanol,' he said, 'that will take away jobs in other industries' like building and operating conventional gas turbine power plants (Steven Greenhouse, 'Millions of jobs of a different collar', *New York Times*, 26 March 2008 [accessed 26 August 2009] http://www.nytimes.com/2008/03/26/business/businessspecial2/26collar.html?_r=1)

139. Bryce, *Gusher of Lies*: 10–11, 60–65 and passim. Crude oil contains about 18,400 British Thermal Units (BTUs) per pound, coal contains 10,400 BTUs per pound, corn contains about 7,000 BTUs per pound and ethanol about 6,400 BTUs per pound (Bryce, *Gusher of Lies*: 128).

140. Bryce, *Gusher of Lies*: 128, 147–167, 230 and passim and Michael Grunwald, 'Seven myths about alternative energy', *Foreign Policy*, September/October 2009: 130–133.

141. 'National Security Consequences of US Oil Dependency', *Council on Foreign Relations*, October 2006 [accessed 28 December 2010] http://www.cfr.org/publication/11683/national_security_consequences_of_us_oil_dependency.html?cid=rss-taskforcereports-national_security_consequences-101206

142. 'National Security Consequences of US Oil Dependency.'

143. ' "Hard truths" about Global Energy Detailed in New Study by the National Petroleum Council', *National Petroleum Council*, 18 July 2007 [accessed 28 December 2010] http://www.npc.org/7–18_Press_rls-post.pdf

144. Marc Labonte & Gail Makinen, 'Energy Independence: Would It Free The United States From Oil Price Shocks?', *CRS Report For Congress*, 17 November 2000 [accessed 27 December 2010] http://www.globalsecurity.org/military/library/report/crs/RS20727.pdf

145. Bryce, *Gusher of Lies*: 230, 303 and passim, and Grunwald, 'Seven myths about alternative energy': 130–133.

BIBLIOGRAPHY

Author's note: The list below is not the complete list of sources which I have used for this book. It is just a selection of the sources used in my research. I have used hundreds of newspaper, magazine and internet articles for my research mainly from *The New York Times*, *The Washington Post*, BBC, *Al Jazeera*, *The Independent*, *The Guardian* and others, and, due to the limitations of space, I could not include them all here. I only include the ones which I found most important.

Primary sources

Bush, George H.W., 'National Security Directive 26: US Policy Toward the Persian Gulf', *Foundation of American Scientists*, 2 October 1989, http://www.fas.org/irp/offdocs/nsd/nsd26.pdf [accessed 13 October 2006]
— 'National Security Directive 45: US Policy in Response to the Iraqi Invasion of Kuwait', *Foundation of American Scientists*, 20 August 1990, http://www.fas.org/irp/offdocs/nsd/nsd_45.htm [accessed 13 October 2006]
— 'National Security Directive 54: Responding to Iraqi Aggression in the Gulf', *Washington Post*, 15 January 1991, http://www.washingtonpost.com/wp-srv/inatl/longterm/fogofwar/docdirective.htm [accessed 13 October 2006]
— 'National Security Strategy of the United States', Foundation of American Scientists, August 1991, http://www.fas.org/man/docs/918015-nss.htm [accessed 19 October 2006]
Bush, George W., 'Governor George W. Bush, "A Distinctly American Internationalism"', Ronald Reagan Presidential Library, Simi Valley, California, 19 November 1999, http://www.mtholyoke.edu/acad/intrel/bush/wspeech.htm [accessed 10 October 2008].
— 'President Bush Delivers Graduation Speech at West Point, United States Military Academy, West Point, New York', White House, 1 June 2002,

http://www.whitehouse.gov/news/releases/2002/06/20020601-3.html [accessed 30 June 2007]

— 'President Bush Delivers State of the Union Address', White House, 23 January 2007, http://www.whitehouse.gov/news/releases/2007/01/20070123-2.html [accessed 9 January 2008]

— 'President Bush Discusses Freedom in Iraq and Middle East', White House, 6 November 2003, http://www.whitehouse.gov/news/releases/2003/11/20031106-2.html [accessed 22 February 2008]

— 'President Bush Discusses Iraq,' White House, 10 April 2008, http://www.whitehouse.gov/news/releases/2008/04/20080410–2.html [accessed 11 April 2008]

— 'President Bush Participates in Press Availability with President Kufuor of Ghana', White House, 20 February 2008, http://www.whitehouse.gov/news/releases/2008/02/20080220-2.html [accessed 21 February 2008]

— 'President Delivers State of the Union Address', White House, 29 January 2002, http://www.whitehouse.gov/news/releases/2002/01/20020129-11.html [accessed 30 June 2004]

— 'President Delivers "State of the Union"', White House, 28 January 2003, www.whitehouse.gov/news/releases/2003/01/20030128–19.html [accessed 26 June 2006]

— 'President Promotes Energy Efficiency through Technology', White House, 25 February 2002, http://www.whitehouse.gov/news/releases/2002/02/20020225-5.html [accessed 2 March 2008]

— 'President Sworn-in to Second Term', White House, 20 January 2005, http://www.whitehouse.gov/news/releases/2005/01/20050120–1.html [accessed 30 June 2007]

— 'Remarks by the President in Photo Opportunity after Meeting with National Energy Policy Development Group', White House, 19 March 2001, http://www.whitehouse.gov/news/releases/2001/03/20010320-1.html [accessed 6 February 2008]

— 'Remarks by the President to Capital City Partnership', River Centre Convention Centre St. Paul Minnesota, White House, 17 May 2001, http://www.whitehouse.gov/news/releases/2001/05/20010517-2.html [accessed 29 January 2008]

— 'Remarks by the President While Touring Youth Entertainment Academy', White House, 14 March 2001, http://www.whitehouse.gov/news/releases/2001/03/20010314-2.html [accessed 6 February 2008]

— 'State of the Union Address', White House, 2 February 2005, http://www.whitehouse.gov/news/releases/2005/02/20050202–11.html [accessed 30 June 2007]

— 'State of the Union Address by the President', White House, 31 January 2006, http://georgewbush-whitehouse.archives.gov/stateoftheunion/2006/ [accessed 23 December 2010]

— 'The National Security Strategy of the United States of America', White House, September 2002, http://www.whitehouse.gov/nsc/nss.html [accessed 3 April 2006]

Clinton, William J., 'Executive Order 12957: Prohibiting certain transactions with respect to the development of Iranian petroleum resources', White House, Office of the Press Secretary, Iranian Trade, 15 March 1995, [accessed 12 June 2005] http://www.iraniantrade.org/12957.htm, and http://frwebgate.access.gpo.gov/cgi-bin/getdoc.cgi?dbname=1995_register&docid=fr17mr95-136.pdf [accessed 4 July 2008]

— 'Executive Order 12959: Prohibiting Certain Transactions with Respect to Iran', White House, Office of the Press Secretary, 6 May 1995, http://frwebgate.access.gpo.gov/cgi-bin/getdoc.cgi?dbname=1995_register&docid=fr09my95-132.pdf [accessed 4 July 2008]

— 'Executive Order 13059: Prohibiting certain transactions with respect to Iran', *Iranian Trade*, 19 August 1997, http://www.iraniantrade.org/13059.htm [accessed 30 June 2005]

— *'Iran and Libya Sanctions Act Of 1996', Foundation of American Scientists, 18 June 1996, http://www.fas.org/irp/congress/1996_cr/h960618b.htm {accessed 30 June 2005}*

— 'Iraq Liberation Act of 1998', *Endnet*, 31 October 1998 http://ednet.rvc.cc.il.us/~PeterR/IR/docs/IraqLib.htm [accessed 4 July 2008]

'Foreign suitors for Iraqi oilfield contracts as of 5 March 2001 part 1', *Judicial Watch*, http://www.judicialwatch.org/IraqOilFrgnSuitors.pdf [accessed 11 November 2007]

'Foreign suitors for Iraqi oilfield contracts as of 5 March 2001 part 2', *Judicial Watch*, http://www.judicialwatch.org/IraqOilGasProj.pdf [accessed 11 November 2007]

'Future of Iraq: Oil and Energy Working Group: Considerations relevant to an oil policy for a liberated Iraq', 27 January 2003, http://www.thememory-hole.org/state/future_of_iraq/future_oil.pdf [accessed 11 January 2008]

'Future of Iraq Project, Oil and Energy Working Group, Subcommittee on Oil Policy, Summary Paper', 20 April 2003, http://www.thememoryhole.org/state/future_of_iraq/future_oil.pdf [accessed 11 January 2008]

'Iraqi oilfields and exploration blocks', *Judicial Watch*, http://www.judicialwatch.org/IraqOilMap.pdf [accessed 11 November 2007]

'National Energy Policy: Report of the National Energy Policy Development Group', White House, May 2001, http://www.whitehouse.gov/energy/National-Energy-Policy.pdf [accessed 20 December 2005]

'Nuclear Posture Review Report', US Department of Defence, November 2001, http://www.defenselink.mil/news/Jan2002/d20020109npr.pdf [accessed 28 January 2008]

Obama, Barack, 'A historic energy bill', White House, 26 June 2009, http://www.whitehouse.gov/blog/A-Historic-Energy-Bill/ [accessed 18 September 2009]

— 'Remarks by the President on jobs, energy independence, and climate change', White House, 26 January 2009, http://www.whitehouse.gov/blog_post/Fromperiltoprogress/ [accessed 24 August 2009]

— 'The American Promise: Acceptance Speech at the Democratic Convention', Obama Speeches, 28 August 2008, http://obamaspeeches.com/E10-Barack-Obama-The-American-Promise-Acceptance-Speech-at-the-Democratic-Convention-Mile-High-Stadium--Denver-Colorado-August-28-2008.htm [accessed 2 July 2010]

'Quadrennial Defence Review Report', 30 September 2001, US Department of Defence, http://www.defenselink.mil/pubs/

qdr2001.pdf#search='quadrennial%20defence%20review%202001' [accessed 4 February 2006]

'Saudi Arabia major oil and natural gas development projects', *Judicial Watch*, http://www.judicialwatch.org/SAOilProj.pdf [accessed 11 November 2007]

'Selected oil facilities in Saudi Arabia', *Judicial Watch*, http://www.judicialwatch.org/SAOilMap.pdf [accessed 11 November 2007]

'Selected oil facilities of the United Arab Emirates', *Judicial Watch*, [http://www.judicialwatch.org/UAEOilMap.pdf [accessed 11 November 2007]

'United Arab Emirates major oil and gas development projects', *Judicial Watch*, http://www.judicialwatch.org/UAEOilProj.pdf [accessed 11 November 2007]

Academic papers

Boys, James, 'An Evaluation of Engagement and Enlargement: The Clinton Doctrine (1993–1997)', PhD thesis submitted to the University of Birmingham, UK, 2006.

Cormier, Pierre Raymond Joseph, 'Understanding Bush's New World Order: Three Perspectives', MA thesis submitted to the University of Manitoba, May 1995.

Hollo, Reuven, 'Oil and American Foreign policy in the Persian Gulf 1947–1991', PhD thesis submitted to the University of Texas, May 1995.

Books

Atwan, Abdel Bari, *The Secret History of Al-Qaida*. London: Abacus, 2007.

Bacevich, Andrew, *American Empire: The Realities and Consequences of US Diplomacy*. Massachusetts: Harvard University Press, 2002.

Bennis, Phyllis, *Before and After: US Foreign Policy and the September 11th Crisis*. New York: Olive Branch Press, 2003.

— *Calling the Shots: How Washington dominates today's UN*. New York: Olive Branch Press, 2000.

Blum, William, *Rogue State: A Guide to the World's Only Superpower*. London: Zed Books, 2001.

Bremer, Paul, *My Year in Iraq: The struggle to build a future of hope*. New York: Simon and Schuster, 2006.

Briody, Dan, *The Halliburton Agenda: The politics of oil and money*. New Jersey: Wiley, 2004.

Brisard, Jean-Charles and Guillaume Dasquié, *Forbidden Truth: US–Taliban secret oil diplomacy and the failed hunt for bin-Laden*. New York: Nation Books, 2002.

Bromley, Simon, *American Hegemony and World Oil*. Pennsylvania State University Press, 1991.

— *American Power and the Prospects for International Order*. Cambridge: Polity, 2008.

Bronson, Rachel, *Thicker Than Oil: America's uneasy partnership with Saudi Arabia*. New York: Oxford University Press, 2006.

Brown, Anthony, *Oil, God and Gold: The Story of Aramco and the Saudi Kings*. New York: Houghton Mifflin Company, 1999.

Bryce, Robert, *Cronies: Oil, the Bushes, and the Rise of Texas, America's Superstate*. New York: Public Affairs, 2004.

— *Gusher of Lies: The Dangerous Delusions of Energy Independence*. New York: PublicAffairs: 2008.

Brzezinski, Zbigniew, *The Choice: Global Domination or Global Leadership*. New York: Basic Books, 2005.

— *The Grand Chessboard: American Primacy and its Geostrategic Imperatives*. New York: Basic Books, 1997.

— *Second Chance: Three Presidents and the Crisis of American Superpower*, New York: Basic Books, 2008.

Burbach, Roger and Jim Tarbell, *Imperial Overstretch: George W. Bush and the Hubris of Empire*, London: Zed Books, 2004.

Chandrasekaran, Rajiv, *Imperial Life in the Emerald City: Inside Baghdad's Green Zone*. London: Bloomsbury Publishing, 2008.

Chaterjee, Pratap, *Iraq, Inc.: A profitable occupation*. New York: Seven Stories Press, 2004.

Chomsky, Noam, *Hegemony or Survival: America's Quest for Global Dominance*. London: Penguin Books, 2004.

— *Imperial Ambitions: Conversations with Noam Chomsky on the Post-9/11 World*. London: Penguin Books, 2006.

— *Interventions*. London: Penguin Books, 2007.

— *Middle East Illusions*. Maryland: Rowman & Littlefield Publishing Group, 2003.

— *The Essential Chomsky*. London: The Bodley Head, 2008.

— *Understanding Power: The Indispensable Chomsky*. London: Vintage Books, 2003.

— *What We Say Goes*. London: Penguin Books, 2009.

Chomsky, Noam and Gilbert Achcar, *Perilous Power: The Middle East and US Foreign Policy*. London: Penguin Books, 2008.

Clark, Wesley K., *A Time to Lead: For Duty, Honor and Country*. Hampshire: Palgrave Macmillan, 2007

Clark, William, *Petrodollar Warfare: Oil, Iraq, and the Future of the Dollar*. Gabriola Island: New Society Publishers, 2005.

Clarke, Duncan, *Empires of Oil: Corporate Oil in Barbarian Worlds*. London: Profile Books, 2007: 292.

Clarke, Richard, *Against All Enemies: Inside America's War on Terror*. New York: Free Press, 2004.

Clinton, Bill, *My Life*. London: Arrow Books, 2005.

Coll, Steve, *Ghost Wars: The Secret History of the CIA, Afghanistan and Bin Laden, From the Soviet Invasion to September 10, 2001*, London: Penguin Books, 2005.

Daalder, Ivo and James Lindsay, *America Unbound: The Bush Revolution in Foreign Policy* (1st edn). Washington, DC: Brookings Institution Press, 2003.

— *America Unbound: The Bush Revolution in Foreign Policy* (2nd edn). Washington, DC: Brookings Institution Press, 2005.

Dekmejian, R. Hrair and Hovann H. Simonian, *Troubled Waters: The Geopolitics of the Caspian Region*. London: I.B.Tauris, 2003.

Draper, Robert, *Dead Certain: The Presidency of George W. Bush*. New York: Free Press, 2008.

Fouskas, Vassilis and Bülent Gökay, *The New American Imperialism: Bush's War on Terror and Blood for Oil*. Connecticut: Praeger Security International, 2005.

Freedman, Lawrence and Efraim Karsh, *The Gulf Conflict 1990–1991: Diplomacy and War in the New World Order*. London: Faber and Faber, 1994.

Friedman, Alan, *Spider's Web: The Secret History of How the White House Illegally Armed Iraq*. New York: Bantam Books, 1993.

Frum, David, *The Right Man: An Inside Account of the Surprise Presidency of George W. Bush*. London: Weidenfeld and Nicolson, 2003.

Gellman, Barton, *Angler: The shadow presidency of Dick Cheney*. London: Penguin Books, 2009.

Gordon, Michael and Bernard Trainor, *Cobra II: The Inside Story of the Invasion and Occupation of Iraq*. London: Atlantic Books, 2006.

Greenspan, Alan, *The Age of Turbulence*. London: Penguin Books, 2008.

Griffin, David Ray, *The 9/11 Commission Report: Omissions and distortions*. Gloucestershire: Arris Books, 2005.

Halper, Stefan and Jonathan Clarke, *America Alone: The Neo-conservatives and the Global Order*. Cambridge: Cambridge University Press, 2004.

Heinberg, Richard, *The Party's Over: Oil, War and the Fate of Industrial Societies*. Gabriola Island: New Society Publishers, 2005.

Hiro, Dilip, *Blood of the Earth: The global battle for vanishing oil resources*. London: Politico's, 2007.

— *Neighbours, Not Friends: Iraq and Iran after the Gulf War*. London: Routledge, 2001.

— *Secrets and Lies: Operation Iraqi Freedom and After*. New York: Nation Books, 2004.

— *Secrets and Lies: The True Story of the Iraq War*. London: Politico's, 2005.

Hyland, William G., *Clinton's World: Remaking American Foreign Policy.* London: Praeger, 1999.

Ikenberry, G. John, *Liberal Order and Imperial Ambition.* Cambridge: Polity, 2006: 207.

Johnson, Chalmers, *The Sorrows of Empire: Militarism, Secrecy and the End of the Republic.* London: Verso, 2006.

Juhasz, Antonia, *The Bush Agenda: Invading the World, One Economy at a Time.* London: Duckworth, 2006.

— *The Tyranny of Oil: The World's Most Powerful Industry – And What We Must Do To Stop It.* New York: Harper Collins, 2008.

Kaplan, Lawrence and William Kristol, *The War Over Iraq: Saddam's Tyranny and America's Mission.* California: Encounter Books, 2003.

Karlsson, Svante, *Oil and the World Order: American Foreign Oil Policy.* Warwickshire: Berg Publishers Limited, 1986.

Kean, Thomas et al. *The 9/11 Commission Report: Final Report of the National Commission on Terrorist Attacks Upon the United States.* New York: Norton, 2004.

Kennedy, Paul, *The Rise and Fall of the Great Powers: Economic Change and Military Conflict From 1500 to 2000.* London: Fontana Press, 1989.

Kinzer, Stephen, *Overthrow: America's Century of Regime Change from Hawaii to Iraq.* New York: Times Books, 2006.

Kissinger, Henry, *Does America Need A Foreign Policy? Toward a Diplomacy for the 21st Century.* New York: Simon & Schuster, 2001.

Klare, Michael T., *Blood and Oil: How America's Thirst for Petrol is Killing Us.* London: Hamish Hamilton, 2004.

— *Resource Wars: The New Landscape of Global Conflict.* New York: Owl Books, 2002.

— *Rising Powers, Shrinking Planet: How Scarce Energy is Creating a New World Order.* Oxford: Oneworld, 2008.

Klein, Naomi, *The Shock Doctrine: The Rise of Disaster Capitalism.* London: Penguin, 2008.

Kleveman, Lutz, *The New Great Game: Blood and Oil in Central Asia.* London: Atlantic Books, 2004.

Kolko, Gabriel, *Main Currents in Modern American History.* New York: Harper and Row Publishers, 1976.

— *The Age of War: The United States Confronts the World.* London: Lynne Rienner Publishers, 2006.

— *The Politics of War: Allied Diplomacy and the World Crisis of 1943–1945.* London: Weidenfeld and Nicolson, 1969.

— *The Roots of American Foreign Policy.* Boston: Beacon Press, 1971.

— *The Triumph of Conservatism: A Reinterpretation of American History, 1900–1916.* London: The Free Press of Glencoe, 1963.

Kolko, Joyce and Gabriel Kolko, *The Limits of Power: The World and United States Foreign Policy, 1945–1954.* New York: Harper and Row Publishers, 1972.

Krugman, Paul, *The Great Unravelling.* London: Penguin Books, 2004.

Lawrence, Robert Z., *A US–Middle East Trade Agreement: A Circle of Opportunity?* Washington, DC: Peterson Institute for International Economics, 2006.

Lieber, Robert J. *The American Era: Power and Strategy for the 21st Century.* Cambridge: Cambridge University Press, 2005.

Little, Douglas, *American Orientalism: The United States and the Middle East since 1945.* London: I.B.Tauris, 2005.

Lusane, Clarence, *Colin Powell and Condoleezza Rice: Foreign Policy, Race and the New American Century.* Connecticut: Praeger, 2006.

Mann, James, *Rise of the Vulcans: The history of Bush's war cabinet.* London: Penguin Books, 2004.

McCormick, Thomas J., *America's Half-Century: United States Foreign Policy in the Cold War and After* (2nd edition). Baltimore: Johns Hopkins University Press: 1995.

McCrisken, Trevor B. *American Exceptionalism and the Legacy of Vietnam: US Foreign Policy Since 1974.* Hampshire: Palgrave Macmillan, 2003.

Miller, T. Christian, *Blood Money: Wasted Billions, Lost Lives and Corporate Greed in Iraq.* New York: Little, Brown and Company, 2006.

Moore, Michael, *The Official Fahrenheit 9/11 Reader: The Must-Read Book of the Must-See Box Office Smash.* London: Penguin Books, 2004.

Montgomery, Bruce P., *The Bush-Cheney Administration's Assault on Open Government.* Connecticut: Praeger, 2008.

Napoleoni, Loretta, *Terrorism and the Economy: How the War on Terror is bankrupting the world.* New York: Seven Stories Press, 2010.

Nixon, Richard, *Beyond Peace.* New York: Random House, 1994.

Packer, George, *The Assassins' Gate: America in Iraq.* London: Faber and Faber Limited, 2007.

Painter, David, *Oil and the American Century: The Political Economy of US Foreign Oil Policy, 1941–1954.* London: The Johns Hopkins Press Ltd, 1986.

Palast, Greg, *The Best Democracy Money Can Buy.* London: Robinson, 2002.

Pelletière, Stephen, *America's Oil Wars.* Connecticut: Praeger, 2004.

Phillips, David L. *Losing Iraq: Inside the postwar reconstruction fiasco.* New York: Basic Books, 2005.

Pilger, John, *Hidden Agendas.* London: Vintage, 1998.

— *The New Rulers of the World.* London: Verso, 2002.

Pinto, Maria Do Ceau, *Political Islam and the United States: A Study of U.S. Policy towards Islamist Movements in the Middle East.* Ithaca Press, 1999.

Pollack, Kenneth, *The Persian Puzzle: The Conflict Between Iran and America.* New York: Random House, 2005.

— *The Threatening Storm: The Case for Invading Iraq.* New York: Random House, 2002.

Quandt, William, *Peace Process: American Diplomacy and the Arab–Israeli Conflict since 1967.* Washington, DC: Brookings Institution Press, 2005.

Rashid, Ahmed, *Jihad: The rise of militant Islam in Central Asia.* London: Yale University Press, 2002.

— *Taliban: The Story of the Afghan Warlords*. London: Pan MacMillan, 2001.

Rutledge, Ian, *Addicted to Oil: America's Relentless Drive For Energy Security*. London: I.B.Tauris, 2005.

Sampson, Anthony, *The Seven Sisters: The Great Oil Companies and The World They Shaped*. Kent: Hodder and Stoughton Limited, 1980.

Scheuer, Michael, *Imperial Hubris: Why the West is Losing the War on Terror*. Washington, DC: Potomac Books, 2005.

Schwarzkopf, Norman H., *It Doesn't Take a Hero*. London: Bantam Books, 1992.

Shaxson, Nicholas, *Poisoned Wells: The Dirty Politics of African Oil*. New York: Palgrave Macmillan, 2007.

Simpson, Christopher, *National Security Directives of the Reagan and Bush Administrations: The Declassified History of US Political and Military Policy, 1981–1991*. Oxford: Westview Press, 1995.

Sniegoski, Stephen J., *The Transparent Cabal: The Neoconservative Agenda, War in the Middle East, and the National Interest of Israel*. Virginia: Enigma Editions, 2008.

Solomon, Lewis D., *Paul D. Wolfowitz: Visionary Intellectual, Policymaker, and Strategist*. London: Praeger Security International, 2007.

Stiglitz, Joseph, *Freefall: Free Markets and the Sinking of the Global Economy*. London: Allen Lane, 2010.

Stiglitz, Joseph and Linda Blimes, *The Three Trillion Dollar War: The True Cost of the Iraq Conflict*. London: Allen Lane, 2008.

Strange, Susan, *States and Markets*. London: Pinter, 1988.

— *States and Markets* (2nd edn). London: Pinter, 1997.

Suskind, Ron, *The Price of Loyalty*. London: Free Press, 2004.

Tenet, George and Bill Harlow, *At the Centre of the Storm: My Years at the CIA*. London: Harper Press, 2007.

Tertzakian, Peter, *A Thousand Barrels a Second: The Coming Oil Break Point and the Challenges Facing an Energy Dependent World*. New York: McGraw Hill, 2006: 134.

Unger, Craig, *House of Bush, House of Saud*. London: Gibson Square, 2007.

Warde, Ibrahim, *The Price of Fear: Al-Qaeda and the Truth Behind the Financial War on Terror*. London: I.B.Tauris, 2007.

Weisberg, Jacob, *The Bush Tragedy: The Unmaking of a President*. London: Bloomsbury, 2008.

Wheeler, Nicholas J. *Saving Strangers: Humanitarian intervention in international society*. Oxford: Oxford University Press, 2000.

Williams, William Appleman, *Empire as a Way of Life*. New York: Ig Publishing, 2007.

— *The Contours of American History*. Chicago: Quadrangle Books, 1966.

— *The Roots of the Modern American Empire: A Study in the Growth and Shaping of Social Consciousness in a Marketplace society*. New York: Vintage Books, 1970.

— *The Tragedy of American Diplomacy*. New York: Norton, 1984.

Woodward, Bob, *Bush at War*. London: Pocket Books, 2002.

— *Plan of Attack*. New York: Simon and Schuster, 2004.

— *State of Denial: Bush at War, Part III*. London: Pocket Books, 2006.

Yergin, Daniel, *The Prize: The Epic Quest for Oil, Money and Power*. New York: Free Press, 1992.

Book chapters

Fraser, Cary, 'The Middle East and the Persian Gulf as the gateway to imperial crisis', in David Ryan and Patrick Kiely (eds), *America and Iraq: Policy-Making, Intervention and Regional Politics* (Oxford: Routledge, 2009): 200–216.

Klare, Michael T. 'Energy Security', in Paul D. Williams (ed.), *Security Studies: An Introduction* (Oxford: Routledge, 2008): 483–496.

— 'Resources', in John Feffer (ed.), *Power Trip: US Unilateralism and Global Strategy After September 11* (New York: Seven Stories Press, 2003): 50–60.

Litwak, Robert S. 'Iraq and Iran: From Dual to Differentiated Containment', in Robert J. Lieber (ed.) *Eagle Rules? Foreign policy and American primacy in the twenty-first century* (New Jersey: Prentice Hall, 2002): 173–193.

Lucas, Scott and Maria Ryan, 'Against Everyone and No-One: The Failure of the Unipolar in Iraq and Beyond', in David Ryan and Patrick Kiely (eds), *America and Iraq: Policy-Making, Intervention and Regional Politics* (Oxford: Routledge, 2009): 154–180.

Morrissey, John, 'The Geoeconomic Pivot of the Global War on Terror: US Central Command and the War in Iraq', in David Ryan and Patrick Kiely (eds), *America and Iraq: Policy-Making, Intervention and Regional Politics* (Oxford: Routledge, 2009): 103–122.

Nacht, Michael, 'Weapons Proliferation and Missile Defence: New Patterns, Tough Choices', in Robert J. Lieber (ed.) *Eagle Rules? Foreign Policy and American Primacy in the Twenty-First Century* (New Jersey: Prentice Hall, 2002): 282–298.

Perkins, Bradford, 'The Tragedy of American Diplomacy: Twenty Five Years After', in William Appleman Williams, *The Tragedy of American Diplomacy* (New York: Norton, 1984): 313–334.

Pinto, Maria Do Ceau, 'Persian Gulf Instability: A Threat to Western Interests', in L.C. Montanheiro, R.H. Haigh and D.S. Morris (eds) *Essays on International Co-operation and Defence* (Sheffield: Sheffield Hallam University Press, 1998): 123–145.

Rashid, Ahmed, 'Central Asia', in John Feffer (ed.) *Power Trip: US Unilateralism and Global Strategy after September 11* (New York: Seven Stories Press, 2003): 117–128.

Strange, Susan, 'The Future of the American Empire', in Richard Little and Michael Smith (eds) *Perspectives on World Politics* (London: Routledge, 2006): 352–358.

Williams, William Appleman, 'The Large Corporation and American Foreign Policy', in David Horowitz (ed.) *Corporations and the Cold War* (London: Monthly Review Press, 1969): 71–104.

Periodicals, newspapers and magazine articles

Brzezinski, Zbigniew, 'A geostrategy for Eurasia', *Foreign Affairs*, September/October 1997: 50–63.

Dodge, Toby, 'The ideological roots of failure: The application of kinetic neo-liberalism to Iraq', *International Affairs*, Vol. 86, No. 6, November 2010: 1269–1286.

'Does he know where it's leading?' *The Economist*, 30 July 2005: 23–25.

Emerson, Tony, 'The Thirst for Oil', *Newsweek*, 8 April 2002: 32–35.

Fineman, Howard and Michael Isikoff. 'Big energy at the table', *Newsweek*, 14 May 2001: 48–50.

Fisk, Robert, 'A financial revolution with profound political implications', *Independent*, 7 October 2009: 11.

Fisk, Robert, 'Economic balance of power shifts to the east', *Independent*, 6 October 2009: 2.

Fuller, Graham E. and Ian O. Lesser, 'Persian Gulf myths', *Foreign Affairs*, May/June 1997: 42–52.

Grunwald, Michael, 'Seven myths about alternative energy', *Foreign Policy*, September/October 2009: 130–133.

'Hot air rising', *The Economist*, 20 September 2003: 52–53.

Ikenberry, John, 'America's Imperial Ambition: The Lures of Preemption', *Foreign Affairs*, September/October 2002, Vol. 81, No. 5.

Jaffe, Amy Myers and Robert A. Manning, 'Shocks of a world of cheap oil', *Foreign Affairs*, January/February 2000: 16–29.

Krauthammer, Charles, 'Coming ashore', *Time*, 17 February 2003: 27.

— 'The Unipolar Moment', *Foreign Affairs*, Vol. 70, No. 1, 1990: 23–31.

Morse, Edward L. and James Richard, 'The Battle for Energy Dominance', *Foreign Affairs*, March/April 2002: 16–31.

Pilger, John, 'Iraq: The Lying Game', *The Daily Mirror*, 27 August 2002: 8–9.

Pollack, Kenneth. 'Securing the Gulf', *Foreign Affairs*, July/August 2003.

'Revolution Delayed', *The Economist*, 17 September 2002: 30–31.

Rice, Condoleezza, 'Promoting the National Interest', *Foreign Affairs*, January/February 2000: 45–62.

Sachs, Jeffrey, 'America has passed on the baton', *Financial Times*, 30 September 2009: 13.

Schlesinger, James and John Deutch, 'The petroleum deterrence', *Newsweek Issues 2007*, December 2006–January 2007: 22.

Seymour, Richard, 'The real cost of the Iraq war', *The Middle East*, May 2009: 49–50.

Stanislaw, Joseph and Daniel Yergin, 'Oil: Reopening the door', *Foreign Affairs*, September/October 1993: 81–93.

Stewart, Heather, 'Economic crisis will end the global reign of US dollar, warns World Bank president', *Guardian*, 29 September 2009: 24.

— 'Oil pricing rumour turns screw on dollar', *Guardian*, 7 October 2009: 27.

Wolfee, Richard and Gretel C. Kovach, 'Let's make an oil deal: A Bush family friend may be undermining Iraqi peace', *Newsweek*, 1 October 2007: 26.

Zakaria, Fareed, 'Don't blame the Saudis', *Newsweek*, 6 September 2004: 17.

— 'Elections are not democracy', *Newsweek*, 7 February 2005: 13.

— 'We're fighting the wrong war', *Newsweek*, 28 January 2008: 19.

Internet material

'Abraham Sees Nation Threatened by Energy Crisis', *USA Today*, 9 March 2001, http://www.usatoday.com/news/washington/2001–03-19-energy.htm [accessed 27 June 2008]

Al-Amir, Fouad, 'Discussion on the Iraq law', *Al-Ghad*, http://www.al-ghad.org/2007/02/20/discussion-on-the-iraq-oil-law/ [accessed 1 March 2007]

Bacevich, Andrew J., 'Tragedy renewed: William Appleman Williams', *World Affairs*, Winter 2009, http://www.worldaffairs.org/2009%20-%20Winter/full-Bacevich.html [accessed 22 April 2009]

Blimes, Linda and Joseph Stiglitz, 'Is this any way to rebuild Iraq?', *Los Angeles Times*, 15 August 2008 http://www.latimes.com/news/opinion/la-oe-bilmes15–2008aug15,0,4432303.story [accessed 10 May 2010]

—'The Iraq war will cost us $3 trillion, and much more', *Washington Post*, 9 March 2008 http://www.washingtonpost.com/wp-dyn/content/article/2008/03/07/AR2008030702846.html [accessed10 May 2010]

— 'The US in Iraq: An economics lesson', *Los Angeles Times*, 2 July 2009, http://articles.latimes.com/2009/jul/02/opinion/oe-bilmes2 [accessed 10 May 2010]

— 'War's price tag', *Los Angeles Times*, 16 March 2008, http://articles.latimes.com/2008/mar/16/opinion/op-bilmes16 [accessed 10 May 2010]

Chomsky, Noam, 'Iraq is a trial run', *Chomsky-Info*, 21 March 2003, http://chomsky.info/interviews/20030321.htm [accessed 16 August 2009]

Hoagland, James, 'Now an Iraqi war in Washington', *Washington Post*, 9 April 2001, http://www.globalpolicy.org/security/sanction/iraq1/turnpoint/2001/0409us.htm [accessed 21 March 2008]

Klare, Michael, 'Bush-Cheney energy strategy: Procuring the rest of the world's oil', *Foreign Policy in Focus*, January 2004, http://www.fpif.org/papers/03petropol/politics_body.html [accessed 10 June 2004]

— 'Entering the era of tough oil', *Znet*, 17 August 2007, http://www.zcommunications.org/entering-the-tough-oil-era-by-michael-t-klare [accessed 26 December 2010]

— 'Oil fix – Bush will act globally to lock in US supply', *Pacific News*, 15 April 2002,   http://www.pacificnews.org/content/pns/2002/apr/0415oilfix.html [accessed 26 January 2004]

— 'The Bush/Cheney energy strategy: Implications for US foreign and military policy', *Information Clearing House*, 26–27 May 2003, http://www.informationclearinghouse.info/article4458.htm [accessed 25 January 2010]

Krauthammer, Charles, 'The New Unilateralism', *Washington Post*, 8 June 2001,   http://www.washingtonpost.com/ac2/wp-dyn/A38839–2001Jun7?language=printer [accessed 28 December 2009]

Mayer, Jane, 'Contract Sport: What did the Vice-President do for Halliburton?', *The New Yorker*, 16 February 2004, http://web.lexis-nexis.com/executive/form?_index=exec_en.html&_lang=en&ut=3361601930 [accessed 17 May 2007]

— 'Jane Mayer on her article in The New Yorker about Dick Cheney's relationship with Halliburton', 19 February 2004, http://web.lexis-nexis.com/executive/form?_index=exec_en.html&_lang=en&ut=3361601930 [accessed 17 May 2007]

McCormick, Thomas, 'What would William Appleman Williams say now?' History News Network, 24 September 2007, http://hnn.us/articles/42971.html [accessed 22 April 2009]

Mihm, Stephen, 'Dr. Doom', *New York Times*, 15 August 2008, http://www.nytimes.com/2008/08/17/magazine/17pessimist-t.html?ei=5124&en=562d8083f648f9eb&ex=1376539200&partner=facebook&exprod=facebook&pagewanted=all [accessed 17 August 2009]

Pasternak, Judy, 'Going backwards: Bush's energy plan bares industry clout', *Los Angeles Times*, 26 August 2001, http://www.commondreams.org/headlines01/0826–02.htm [accessed 6 February 2008]

Ricks, Thomas E., 'Army Historian Cites Lack of Postwar Plan', *Washington Post*, 25 December 2004, http://www.washingtonpost.com/wp-dyn/articles/A24891–2004Dec24.html [accessed 26 May 2006]

Rohter, Larry, 'Obama camp closely linked with ethanol', *New York Times*, 23 June 2008 http://www.nytimes.com/2008/06/23/us/politics/23ethanol.html?th&emc=th [accessed 23 June 2008]

Rosett, Claudia, 'One purely evil cartel', *The Wall Street Journal*, 30 July 2003, http://www.opinionjournal.com/columnists/cRosett/?id=110003813 [accessed 9 August 2009]

Roubini, Nouriel, 'The decline of the American Empire', *RGE Monitor*, 13 August 2008,   http://www.rgemonitor.com/roubini-monitor/253323/the_decline_of_the_american_empire [accessed 17 August 2009]; also found on *Fabius Maximus*, http://fabiusmaximus.wordpress.com/2008/08/18/roubini/,

Rumsfeld, Donald and Rik Kirkland, 'Don Rumsfeld talks guns and butter', *Fortune*, 18 November 2002, http://money.cnn.com/magazines/fortune/fortune_archive/2002/11/18/332271/ [accessed 4 July 2008] and http://web.ebscohost.com/ehost/detail?vid=5&hid=15&sid=f1a96962–8b4e-4c6e-86ff-5e1ae5fc7590%40sessionmgr2